S0-CGP-221

CALIFORNIA

An
Environmental
Atlas & Guide

Bear Klaw Press Environmental Guide Series

Volume 1 - California: An Environmental Atlas & Guide

CALIFORNIA

An Environmental Atlas & Guide

Bern Kreissman

Assisted by
Barbara Lekisch

Copyright © 1991 Bern Kreissman

Library of Congress Catalog Card Number: 90-083315
International Standard Book Number: 0-9627489-9-4

All rights reserved. No part of this publication may be reproduced or used in any form or by any means, electronic, or mechanical, including photocopying, recording, taping, or information storage and retrieval systems without the prior written permission of the copyright owners.

Designed and illustrated by Giovanni Bacigalupi
Santa Rosa, California

Manufactured in the United States of America

10 9 8 7 6 5 4 3 2 1

Library of Congress Cataloging in Publication Data
Kreissman, Bern,
California: an environmental atlas and guide.

Bibliography: p.
Includes index.

Bear Klaw Press
926 Plum Lane
Davis, California 95616

Phone: 916 753 7788
Fax: 415 594 0411

Dedicated to the Future

Starrett, Jude, Gregory

and

Giovanna, Sonya, Doran

"Whatever befalls the earth, befalls the sons of the earth. If men spit upon the ground, they spit upon themselves. Man did not weave the web of life, he is merely a strand in it. Whatever he does to the web, he does to himself."

Chief Seattle

Contents

Introduction ix
Preface xi
Counties and County Seats 2
Regional Councils of Government 4
Local Agency Formation Commissions 6
State Symbols 7
Land Ownership 8
Air Resources Board 10
Department of Fish and Game 12
Department of Forestry and Fire Protection 14
Department of Parks and Recreation 16
State Park System 20
State Water Resources Control Board 22
Department of Water Resources 24
Landforms 26
Mountain Ranges and Peaks 28
Glaciers and Glaciated Areas 32
Caves 34
Earthquakes 36
Volcanic Areas 40
Hot Springs and Pools 42
Rivers and Streams 44
Wild and Scenic Rivers 48
Waterfalls 50
Lakes and Reservoirs 52
Dams and Reservoirs 54
The Delta 56
San Francisco Bay 58
Coastal Wetlands 60
Interior Wetlands 62
Farallon Islands and Año Nuevo Island 64
Channel Islands 66
Giant Kelp Forests 68
Deserts 70
Natural Vegetation 72
Habitat Sites 74
Important Habitat Areas-U.S. Fish and Wildlife Service 78
Habitat Types-California Interagency Wildlife Task Group 80
Habitat Types-University of California 80
Habitat Losses 81
State Forests-Department of Forestry and Fire Protection 82
National Forests-Forest Service 84
Trails-National Trails System 86
National Wilderness Preservation System 88
National Park System-National Park Service 92
Historic Parks-Department of Parks and Recreation 94
Underwater Parks-Department of Parks and Recreation 96
Reserves-Department of Parks and Recreation 98
Natural Preserves-Department of Parks and Recreation 100
Wilderness Areas-Department of Parks and Recreation 102
Redwood Parks and Preserves 104
Wildlife Management Areas-Department of Fish and Game 106
Wildlands Program Areas-Department of Fish and Game 110
Marine Life Refuges and Reserves-Department of Fish and Game 112
Fish Hatcheries-Department of Fish and Game 116
Wild Trout Waters-Department of Fish and Game 118
Catch and Release Waters-Department of Fish and Game 120
Ecological Reserves-Department of Fish and Game 122
Artificial Reefs-Department of Fish and Game 126
Biosphere Reserves-UNESCO 128
World Heritage Sites-UNESCO 130
National Wildlife Refuges-U.S. Fish and Wildlife Service 132
Research Natural Areas-Forest Service 134
Research Natural Areas-Bureau of Land Management 136
Research Natural Areas-National Park Service 138
Research Natural Areas-U.S. Fish and Wildlife Service 138
National Natural Landmarks-NationalPark Service 140
Experimental Forests-Forest Service 144
Special Interest Areas-Forest Service 146
Outstanding Natural Areas-Bureau of Land Management 148
Protected Natural Areas-National Park Service 148
Experimental Ecological Areas-National Park Service 148
Public Use Natural Areas-U.S. Fish and Wildlife Service 148
National Conservation Areas 150
Riparian Habitat Demonstration Areas-
Bureau of Land Management 152
Areas of Critical Environmental Concern-
Bureau of Land Management 154
Areas of Special Biological Significance-
Water Resources Control Board 158
National Estuary Program-Environmental Protection Agency 160
National Estuarine Reserve Research System-
National Oceanic and Atmospheriic Administration 160
National Marine Sanctuaries-
National Oceanic and Atmospheric Administration 162
State Coastal Conservancy 164
Friends Of The River 165
Natural Reserve System-University of California 166
Sanctuaries-Audubon Society 170
Preserves and Easements-California Nature Conservancy 172
California Native Plant Society 176
National Wildlife Federation 177
Trails-Rails to Trails Conservancy 178
Sierra Club 180
The Trust For Public Land 182
Planning & Conservation League 183
Wildlife Rehabilitation Centers 184
Williamson Act Counties 188
Timberland Production Zones 190
Water Bank Counties 192
Sno-Park Sites-Department of Parks and Recreation 194
Off-Highway Vehicle Areas 196
California Highway Information Network-
Department of Transportation 200
Directory-Federal Agencies 206
Directory-State Agencies 214
Directoty-Regional Agencies 217
Directory-County Agencies 219
Directory-United Nations Agencies 225
Directory-Private Agencies 226
Acronyms and Initialisms 230
Bibliography 232
Index 240

Introduction

When I first plunged into California environmental activities, I was fairly confident that my lifelong experience as a conservationist had prepared me for duty as an environmental activist. Almost immediately, however, I made dual discoveries: California is probably the greatest ecological treasure house of the United States, and its natural grandeur and astonishing diversity are matched by the volume of its ecological names, boundaries, dates, offices, designations, initialisms, practices, addresses, acronyms, legislation, programs, and locations—most of them unknown to me. I also discovered that many of my conservationist colleagues were equally baffled, and that even the most knowledgeable activists were generally unfamiliar with areas outside their particular spheres of expertise. A search of several major libraries convinced me that nothing had been published to assist us in these matters. To paraphrase "Rare Ben Jonson" our efforts to serve in the conservationist crusade had "an enemy called ignorance."

I, therefore, determined to list, locate, and describe all those features of California of most interest to the environmental community. That list of features is here before you, not all of course, but the most significant natural elements.

In the pursuit of information, I visited and spent long hours, even days, in every library of ecological interest, from Palo Alto and Menlo Park, to San Francisco and Marin County, to Davis and Sacramento; every pertinent state and federal office in that region; and many private organizations, such as the Sierra Club and The California Nature Conservancy. Those visits were supplemented by correspondence and innumerable phone calls across the entire western region, New York, and Washington, D.C. At almost every instance I was impressed by the unfailing courtesy, the evident desire to help, and the readiness to go beyond office routines demonstrated by federal and state office personnel. Their assistance can only be acknowledged in superlatives.

In similar fashion, scores of librarians: state and public librarians, special librarians in governmental agencies and environmental organizations, and academic librarians at Stanford and the University of California campuses of Davis and Berkeley shared their expertise and knowledge with zeal and unfailing good will. As representatives of all the librarians who deserve my regard and for their very special assistance, my deepest thanks go to Linda Hoffmann and her staff of the Government Documents and Maps Department of the University of California, Davis.

My thanks go also to several others who assisted me in the development of the project. Barbara Lekisch, the project assistant, provided many hours of work and more important, the knowledge of California environment gained through her years as librarian at the Sierra Club and through her recent publication, *Tahoe Place Names.* Her assistance in the preparation of this work including the index is deeply appreciated. Giovanni Bacigalupi, publications designer, has designed and produced the book. Gail Schlachter of Reference Service Press and Stuart Hauser of Publishers Support Services gave unstintingly of their time and expertise to guide me in the technicalities of publication. Along the way Sarah Marvin, Steve Greco, and Stephanie Mandel made helpful contributions. My wife, Shirley, deserves more than thanks for her assistance and for her willingness to close her eyes to the mess. Finally, I wish to thank my conservation colleagues, the thousands of unnamed crusaders whose lives are devoted to the protection, preservation, and restoration of planet earth. Their devotion and enthusiasm have spurred me to the completion of this work.

Bern Kreissman
Davis, California
December, 1990

Preface

California, "Glorious California" as John Muir called it, runs 824 miles long and about 250 miles wide. It is the third largest state in the union, but ecologically it is the richest and most diverse area of the United States. No other region of the country holds as many native plant species or as many mountain ranges. No other state boasts as many national forests or more wilderness sites. Federal and state agencies have set aside hundreds of special "natural areas" in California for particular forms of protection, research, management, or recreation. The state holds 4,955 lakes over 1 acre in size, and, though they are now being recounted, a probability of more than 40,000 stream miles including 2,000 miles of "wild and scenic rivers." Several dozen sanctuaries providing safekeeping areas for animal and plant species have been established in the state, and even the United Nations has identified a score of California areas for their world-wide significance, extraordinary beauty, and ecological vulnerability.

Practically all of these areas are "terra incognita," not only for the layman, but also for knowledgeable environmentalists, librarians, and research specialists. With the exception of a few guides to the most popular of these areas, such as directories to state parks, very little of this information is available in convenient form. Nowhere else, except for one out-of-print technical inventory of California natural areas, has a compilation as comprehensive or as detailed as this *Atlas and Guide* been attempted.

As a library reference tool or as an environmentalist's "vade mecum," this publication provides the answers to location and description queries on glaciers, volcanoes, caves, kelp forests, sea otters, wild rivers, waterfalls, and a myriad of other ecological features of the state. As a reading text, it will supply fascinating details and pleasure for all Californiaphiles—How many mountain ranges does California hold? Where do monarch butterflies go in the spring? Is Yosemite a National Natural Landmark? Where have all the condors gone?

The body of the *Atlas* section consists of 3 parts:

1) A map on the right-hand page showing the numbered locations of members of a single ecological feature. The location symbols are accurately situated but, by their very nature, they do not pretend to geologic-survey precision.

2) The facing left-hand page carries an accompanying descriptive statement. Federal agency names are taken from the *United States Government Manual* and state names from the California *Roster*.

3) A table, or tables, follow immediately after the descriptive statement: a numeric table identifying the numbered map sites, and in most cases where 25 or more elements appear on a single map, an alphabetic table for convenience in locating a known name. Basic statistical information is given in the tables when appropriate.

On occasion, the descriptive narrative or the tables may run to a second page. The narrative is continued on the verso of the map page and the map is repeated on the facing right-hand page.

The descriptive statements often end with the statement "See: Bibliography" followed by a number. The number refers to that entry in the "Bibliography" section. Thus, "See: Bibliography 5" means, for further information see bibliographic entry number 5.

The "Guide" section contains the listings, addresses, and telephone numbers of the federal, state, and major private agencies of interest to the California conservation community. In addition, the acronyms and initialisms provide a list of substitutes, such as ACEC, BLM, and OCS, which have practically supplanted the original titles, and also the important though lesser-used abbreviations current in environmental jargon.

The "Bibliography" entries are numbered, and the "Journal Articles" numbers follow consecutively after the "Books" listing. The entries are referenced in the text by the lead number as explained above.

The "Index" covers every name of consequence appearing in the tables, and every important informational item in the descriptive statements. Consequently, references to items such as the California Wilderness Act or the Wildlife Habitat Relationships System will be found in the index. Familiar compound names such as "Lake Tahoe" or "Mount Goddard" are indexed under " Lake" or "Mount," with a reference from "Tahoe" or "Goddard." Sites with slight name variations are listed together. Thus, "Bolsa Chica Ecological Reserve Marine Life Refuge" and "Bolsa Chica Ecological Reserve" are both indexed as "Ecological Reserve" and "Marine Life Refuge" under "Bolsa Chica." In order to maintain consistency with the usage by the Department of Fish and Game, the Department of Parks and Recreation, and other state agencies, all titles are alphabetized and indexed under the first word of the title (except for the English word "the"). Thus, "El Segundo Dunes" appears under "El," "Upper Santa Clara River" appears under "Upper," and "Julia Pfeiffer Burns State Park" appears under "Julia."

Since it is impractical to include all the ecological features of California in a single volume, this first volume is devoted, primarily, to natural features such as rivers, faultlines, habitat, and sanctuaries. Volume 2 of the Bear Klaw Press Environmental Guide Series will show the man-made elements of interest to environmentalists such as power transmission lines, energy generating plants, and toxic dump sites. Where that distinction is blurred as in the cases of dams and their associated lakes and fish hatcheries with their attendant sanctuary waters, those features are given in the present volume.

Toward that second volume in this series and to a possible second edition of Volume 1, the author would be pleased to receive additional or corrected information on California environmental features. Your correspondence is invited. Letters should be addressed to the author in care of Bear Klaw Press.

Preface

California Atlas

2 Counties and County Seats

Counties are the legal subdivisions of the state by declaration of the State Constitution. As a result, county government plays a stronger role in California than in many of the other 49 states. When California joined the Union in 1850, 27 counties were established. Over the next 57 years 31 additional counties were created, for a total of 58, including 1 merged city-county—San Francisco. That number has remained fixed since that date, 1907, despite some later attempts to form additional counties.

There are 2 types of county in California: the predominant form is the General Law County (46) formed under provisions of the Constitution and supplementary statutes. However, the Constitution does allow a county to adopt its own charter, subject to basic limitations, and the Charter Counties (12) have greater leeway in structuring their government. For instance, Charter Counties are required to elect only the members of the Board of Supervisors and Sheriff, while General Law Counties must elect those officials plus 11 other county officers: assessor, auditor, controller, coroner, county clerk, district attorney, public administrator, recorder, superintendent of public schools, tax collector, and treasurer.

See: Bibliography 19, 31.

	County		County		County		County		County
1	Del Norte	13	Sierra	25	Alpine	37	Stanislaus	49	San Luis Obispo
2	Siskiyou	14	Lake	26	Solano	38	Mariposa	50	Kern
3	Modoc	15	Colusa	27	Amador	39	Santa Cruz	51	San Bernardino
4	Humboldt	16	Sutter	28	Marin	40	Santa Clara	52	Santa Barbara
5	Trinity	17	Yuba	29	Contra Costa	41	Merced	53	Ventura
6	Shasta	18	Nevada	30	San Joaquin	42	Madera	54	Los Angeles
7	Lassen	19	Placer	31	Calaveras	43	San Benito	55	Orange
8	Tehama	20	Sonoma	32	Tuolumne	44	Fresno	56	Riverside
9	Plumas	21	Napa	33	Mono	45	Monterey	57	San Diego
10	Mendocino	22	Yolo	34	San Francisco	46	Kings	58	Imperial
11	Glenn	23	Sacramento	35	San Mateo	47	Tulare		
12	Butte	24	El Dorado	36	Alameda	48	Inyo		

County	#	County Seat	Type	Cities
Alameda	36	Oakland	Charter	14
Alpine	25	Markleeville	General Law	- -
Amador	27	Jackson	General Law	5
Butte	12	Oroville	Charter	5
Calaveras	31	San Andreas	General Law	1
Colusa	15	Colusa	General Law	2
Contra Costa	29	Martinez	General Law	16
Del Norte	1	Crescent City	General Law	1
El Dorado	24	Placerville	General Law	2
Fresno	44	Fresno	Charter	15
Glenn	11	Willows	General Law	2
Humbolt	4	Eureka	General Law	7
Imperial	58	El Centro	General Law	7
Inyo	48	Independence	General Law	1
Kern	50	Bakersfield	General Law	11
Kings	46	Hanford	General Law	4
Lake	14	Lakeport	General Law	2
Lassen	7	Susanville	General Law	1
Los Angeles	54	Los Angeles	Charter	83
Madera	42	Madera	General Law	2
Marin	28	San Rafael	General Law	11
Mariposa	38	Mariposa	General Law	- -
Mendocino	10	Ukiah	General Law	4
Merced	41	Merced	General Law	6
Modoc	3	Alturas	General Law	1
Mono	33	Bridgeport	General Law	- -
Monterey	45	Salinas	General Law	12
Napa	21	Napa	General Law	4
Nevada	18	Nevada City	General Law	2
Orange	55	Santa Ana	General Law	26
Placer	19	Auburn	Charter	5
Plumas	9	Quincy	General Law	1
Riverside	56	Riverside	General Law	19
Sacramento	23	Sacramento	Charter	4
San Benito	43	Hollister	General Law	2
San Bernardino	51	San Bernardino	Charter	17
San Diego	57	San Diego	Charter	16
San Francisco	34	San Francisco	Charter	1
San Joaquin	30	Stockton	General Law	6
San Luis Obispo	49	San Luis Obispo	General Law	7
San Mateo	35	Redwood City	Charter	19
Santa Barbara	52	Santa Barbara	General Law	5
Santa Clara	40	San Jose	Charter	15
Santa Cruz	39	Santa Cruz	General Law	4
Shasta	6	Redding	General Law	2
Sierra	13	Downieville	General Law	1
Siskiyou	2	Yreka	General Law	9
Solano	26	Fairfield	General Law	7
Sonoma	20	Santa Rosa	General Law	8
Stanislaus	37	Modesto	General Law	9
Sutter	16	Yuba City	General Law	2
Tehama	8	Red Bluff	Charter	3
Trinity	5	Weaverville	General Law	- -
Tulare	47	Visalia	General Law	8
Tuolumne	32	Sonora	General Law	1
Ventura	53	Ventura	General Law	9
Yolo	22	Woodland	General Law	4
Yuba	17	Marysville	General Law	2

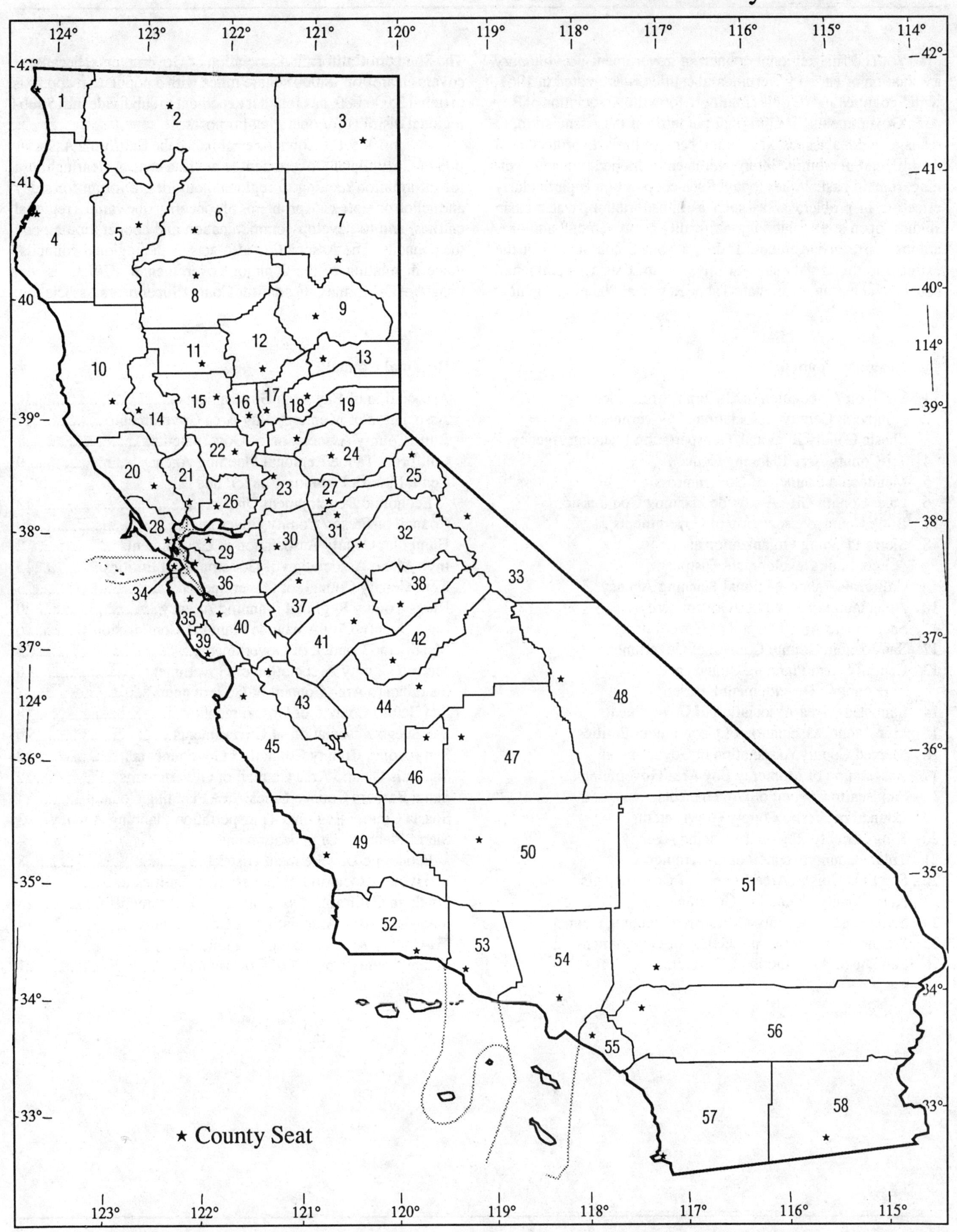

124°
123°
122°
121°
120°
119°
118°
117°
116°
115°
114°
42°
41°
40°
39°
38°
37°
36°
35°
34°
33°
1
2
3
4
5
6
7
8
9
10
11
12
13
14
15
16
17
18
19
20
21
22
23
24
25
26
27
28
29
30
31
32
33
34
35
36
37
38
39
40
41
42
43
44
45
46
47
48
49
50
51
52
53
54
55
56
57
58
★ County Seat

4 Regional Councils of Government

The first regional council of governments, a voluntary association of varied governmental entities, was created in 1961, with 5 counties and 42 cities joining to form the Association of Bay Area Governments. The original purposes of the Association, to manage regional issues on a comprehensive basis, to protect local control, and to promote intergovernmental cooperation and planning, remain basic issues today. Such cooperation is particularly effective in problem areas such as transportation, water basin studies, open space planning, air quality improvement, and hazardous waste management. Today, all but 8 counties (6 in the extreme north end of the state) are involved in 26 such governmental councils including the bi-state Tahoe Regional Planning Agency. The Southern California Association of Governments, because it covers an area of 38,000 square miles with a population approximately 13,000,000, has found it expedient to subdivide into 5 subregional districts for operating purposes.

In 1983, the Councils established the California Association of Councils of Government to act as a central clearinghouse for information relating to regional activities, a forum for study and action on statewide problems of concern to the various regional entities, and to develop communication and cooperation among the councils. The Association also acts as the regional councils' voice in relating to the 2 major governmental affiliations, the League of California Cities and the County Supervisors Association.

	Regional Council
1	Siskiyou Association of Government Entities
2	Humboldt County Association of Governments
3	Shasta County Regional Transportation Planning Agency
4	Tri-County Area Planning Council
5	Mendocino Council of Governments
6	Lake County-City Areawide Planning Commission
7	Butte County Association of Governments
8	Sierra Planning Organization and Economic Development District
9	California Tahoe Regional Planning Agency
10	Association of Bay Area Governments
11	Sacramento Area Council of Governments
12	San Joaquin County Council of Governments
13	Central Sierra Planning Council and Economic Development District
14	Stanislaus Area Association of Governments
15	Inyo-Mono Association of Government Entities
16	Merced County Association of Governments
17	Association of Monterey Bay Area Governments
18	San Benito Council of Governments
19	Council of Fresno County Governments
20	Kings County Regional Planning Agency
21	Tulare County Council of Governments
22	San Luis Obispo Area Council of Governments
23	Kern County Council of Governments
24	Santa Barbara County-Cities Area Planning Council
25	Southern California Association of Governments
26	San Diego Association of Governments

Regional Council	**#**
Association of Bay Area Governments	10
Association of Monterey Bay Area Governments	17
Butte County Association of Governments	7
California Tahoe Regional Planning Agency	9
Central Sierra Planning Council and Economic Development District	13
Council of Fresno County Governments	19
Humboldt County Association of Governments	2
Inyo-Mono Association of Governmental Entities	15
Kern County Council of Governments	23
Kings County Regional Planning Agency	20
Lake County-City Areawide Planning Commission	6
Mendocino Council of Governments	5
Merced County Association of Governments	16
Sacramento Area Council of Governments	11
San Benito Council of Governments	18
San Diego Association of Governments	26
San Joaquin County Council of Governments	12
San Luis Obispo Area Council of Governments	22
Santa Barbara County-Cities Area Planning Council	24
Shasta County Regional Transportation Planning Agency	3
Sierra Planning Organization and Economic Development District	8
Siskiyou Association of Government Entities	1
Southern California Association of Governments	25
Stanislaus Area Association of Governments	14
Tri-County Area Planning Council	4
Tulare County Council of Governments	21

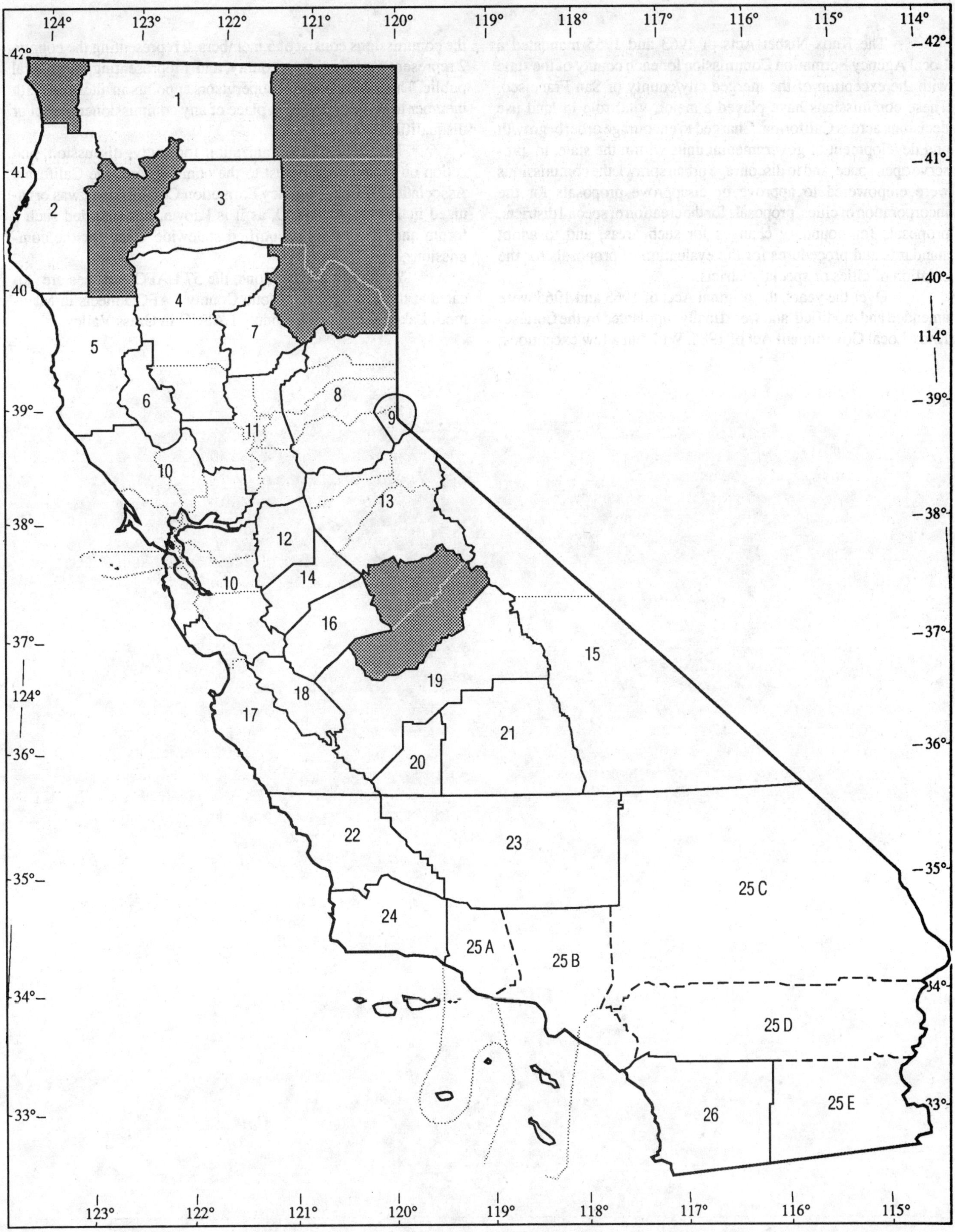
124°
123°
122°
121°
120°
119°
118°
117°
116°
115°
114°
42°
41°
40°
39°
38°
37°
36°
35°
34°
33°
1
2
3
4
5
6
7
8
9
10
11
12
13
14
15
16
17
18
19
20
21
22
23
24
25 A
25 B
25 C
25 D
25 E
26

6 Local Agency Formation Commissions

The Knox-Nisbet Acts of 1963 and 1965 mandated a Local Agency Formation Commission for each county of the state with the exception of the merged city/county of San Francisco. These commissions have played a major, vital role in land use decisions across California. Charged to encourage orderly growth and development of governmental units within the state, to preserve open space, and to discourage urban sprawl, the commissions were empowered to approve or disapprove proposals for the incorporation of cities, proposals for the creation of special districts, proposals for boundary changes for such areas, and to adopt standards and procedures for the evaluation of proposals for the creation of cities or special districts.

Over the years, the original Acts of 1963 and 1965 were amended and modified, and were finally supplanted by the Cortese-Knox Local Government Act of 1985. With but a few exceptions, the commissions consist of 5 members, 2 representing the county, 2 representing cities in the county, and 1 representing the general public. The county board of supervisors appoints an alternate sixth member to serve and vote in place of any commissioner absent or disqualified.

To provide an organization for study, discussion, and action on matters of interest to the commissions, the California Association of Local Agency Formation Commissions was organized in 1972. CALAFCO, as it is known, has supplied such a forum and, in addition, a unified statewide voice for the commissions.

With only 2 exceptions, the 57 LAFCO offices are located at the county seats: Mono County LAFCO meets in Mammoth Lakes and Nevada County LAFCO in Grass Valley.

An official state symbol is a designation by the legislature of a particularly unique, beautiful, or culturally or historically important element of California. These symbols represent the state when employed in appropriate situations. Non-environmental symbols include the state colors, blue and gold, adopted in 1951; state song, "I Love You, California," by Silverwood and Frankenstein, first sung publicly by Mary Garden in 1913; and the state flag, designed by William Todd, raised at Sonora on June 14, 1846 by settlers rebelling against Mexican rule. After it was replaced by the stars and stripes, it was chosen as the state flag in 1911.

Environmental State Symbols

Animal

Grizzly Bear, designated in 1953

The Grizzly Bear existed in greater numbers in California than anywhere else in the United States before the encroachment of the settlers. The bear was tracked and killed throughout the state, and in August, 1922, the last California Grizzly Bear was killed in Tulare County.

Bird

California Quail (Valley Quail), designated in1931

Widely distibuted throughout the state, the California Quail is noted for its hardiness and adaptability. It is a highly prized game bird.

Fish

Golden Trout, designated in 1947

Native to California alone, the Golden Trout is the most beautiful of all the trouts. Originally, it occurred only in several mountain streams of the Kern River drainage in Tulare County, but hatchery plantings have now extended its range to many other high elevations in the Sierra Nevada and even to other states.

Flower

California Poppy, designated in 1903

The poppy was highly prized by California Indians as a source for food and oil. Its botanical name was given by Albert Van Chamisso, a naturalist member of the Royal Prussian Academy of Science who was struck by the beauty of vast fields of the poppy when his ship dropped anchor in San Francisco Bay in 1816. The poppy grows wild throughout the state.

Fossil

Saber-Toothed Tiger, designated in 1968

A tiger sized carnivorous cat, the Saber-Tooth was common in California 40,000,000 years ago. Remains of the Saber-Toothed Cat may be seen at the Rancho La Brea Tar Pits.

Insect

Dog-Face Butterfly, designated in 1973

Found only in California, the Dog-Face Butterfly ranges from the foothills of the Sierra Nevada to the Coast Ranges, and from Sonoma to San Diego. The male butterfly sports a dog's head silhouette on its wings.

Marine Mammal

Gray Whale, designated in 1975

The California Gray Whale is the best known of the great whales that occur off the west coast of North America. It measures from 35 to 50 feet in length and weighs from 20 to 40 tons. Its black body is dappled with gray and the gray effect is heightened by encrusted barnacles and whitish scars.

Mineral

Native Gold, designated in 1965

California has produced more gold than any other state in the union. The gold rush of 1849 led to California statehood and gave rise to the nickname, " The Golden State."

Reptile

Desert Tortoise, designated in 1973

Native to California, the Tortoise was to be found in abundance in the Mojave Desert and the uplands east of the Salton Sea. However, specimen collecting, predation, illness, and, particularly habitat destruction have severely reduced the numbers of these reptiles. Today, the Desert Tortoise is an endangered species.

Rock

Serpentine, designated in 1965

Serpentine, the green and blue stone with shiny outcroppings is found in many areas of the state. It is the host rock for most of the state's deposits of chromite, magnesite, and cinnabar.

Tree

California Redwood, designated in 1953

The tallest and largest living organisms in the world (Coast Redwood and Giant Sequoia), the redwood is to be found, naturally, only in a remote area of western China, a small patch in southern Oregon, and in California.

Land Ownership

California, the third largest state of the union with 101,563,500 acres (158,693 square miles), of which 100,206,720 acres (156,573 square miles) are land surface and 1,356,780 acres (2,120 square miles) are water surface, runs about 640 miles straight north to south, 824 miles northwest to southeast. The last reliable source for statistics regarding public land ownership in California was a publication with that exact title, *Public Land Ownership in California,* produced by the Public Lands Commission in 1977. At present, the Department of General Services is attempting to compile another such list, but cannot reconcile individual agency reports of land holdings with recognized sum totals of federal, state, or local holdings. Part of the difficulty may be attributed to overlapping jurisdictions and part to questions regarding the submerged lands of the state which are owned, technically, by the State Lands Commission. However, the contradictions between agency reports and total holdings are evident as the statistics shown here indicate.

Reportedly, 48.87 percent of California, 48,968,690 acres (76,514 square miles), is public land, of which the federal government owns the lion's share, 45,076,382 acres. The state owned lands account for 1,953,456 acres; county lands, 617,132 acres; and local and regional agencies, 1,321,720 acres.

The Forest Service reports the largest federal holdings with 24,323,818 acres, followed by the Bureau of Land Management with 17,149,997 acres, and the National Park Service with 4,571,316 acres. The Bureau of Reclamation manages 831,449 acres; the Bureau of Indian Affairs has jurisdiction over 446,902 acres; and U.S. Fish and Wildlife Service, 128,845 acres. The Department of Energy reports 74,174 acres and the remainder of the federal lands are held by various other non-environmental agencies with the Department of Defense as the major owner with 3,479,051 acres.

State agencies presumably control 1,953,456 acres with the Department of Parks and Recreation reportedly the largest state landholder with 1,280,785 acres. The State Lands Commission reports 587,000 acres, Fish and Game 479,201 acres, and the Department of Forestry and Fire Protection 69,384 acres. Other agencies of the state hold additional lands.

Privately held farming and grazing lands account for 31,351,000 acres. Private timber and wildlands cover 10,111,200 acres, residential ownerships hold 5,000,200 acres. The remaining 6,132,410 privately held acres are in other miscellaneous uses.

124° 123° 122° 121° 120° 119° 118° 117° 116° 115° 114°

42° 41° 40° 39° 38° 37° 36° 35° 34° 33°

U.S. Forest Service

NationalPark Service

Department of Defense

Bureau of Indian Affairs

Bureau of Land Management

Private or State of California

123° 122° 121° 120° 119° 118° 117° 116° 115°

The Air Resources Board was created in 1968 by combining the Bureau of Air Sanitation (established 1955) and the Motor Vehicle Pollution Control Board (established 1960). The Air Resources Board is responsible for achieving and maintaining satisfactory air quality in California. This responsibility requires the board to establish ambient air quality standards for certain pollutants, regulate vehicle emissions, identify and control toxic air pollutants, administer air pollution research studies, provide technical expertise to help county and regional control officials set emission limits for industrial causes of air pollution, develop and oversee implementation plans for the attainment and maintenance of both state and federal air quality standards, and oversee the regulation of sources of pollution by air pollution control districts.

The board consists of a full-time chairperson and 8 part-time members, all of whom are appointed by the Governor and serve at his pleasure. The chairperson of the board also serves as the Governor's Secretary of Environmental Affairs and, as such, has an advisory and coordinating role in the environmental area.

Air quality is monitored by stations located in the 14 air basins. An air basin is a region characterized by somewhat homogeneous climatological and geographical characteristics.

Air Resources Board
1102 Q Street
Sacramento, CA 95814

1 North Coast Air Basin
North Coast Unified Air Quality Monitoring District
5630 South Broadway
Eureka, CA 95501

2 San Francisco Bay Air Basin
Bay Area Air Quality Monitoring District
939 Ellis Street
San Francisco, CA 94109

3 North Central Coast Air Basin
Monterey Bay Unified Air Pollution Control District
1164 Monroe Street, Suite D
Salinas, CA 93906-3596

4 South Central Coast Air Basin
San Luis Obispo County Air Pollution Control District
2156 Sierra Way, Suite B
San Luis Obispo, CA 93401

5 South Coast Air Basin
South Coast Air Quality Monitoring District
9150 Flair Drive
El Monte, CA 91731

6 San Diego Air Basin
San Diego Air Pollution Control District
9150 Chesapeake Drive
San Diego, CA 92123-1095

7 Northeast Plateau
Lassen County Air Pollution Control Board
175 Russell Avenue
Susanville, CA 96130

8 Sacramento Valley Air Basin
Sacramento Metropolitan Air Quality Monitoring District
8475 Jackson Road, Suite 215
Sacramento, CA 95826

9 San Joaquin Valley Air Basin
Fresno County Air Pollution Control District
P.O. Box 11867
Fresno, CA 93775

10 Great Basin Valleys Air Basin
Great Basin Unified Air Pollution Control District
157 Short Street, Suite 6
Bishop, CA 93514

11 Southeast Desert Air Basin
San Bernardino County Air Pollution Control District
15428 Civic Drive, Suite 200
Victorville, CA 92392

12 Mountain Counties Air Basin
Northern Sierra Air Quality Monitoring District
540 Searls Avenue
Nevada City, CA 95959

13 Lake County
Lake County Air Quality Monitoring District
883 Lakeport Boulevard
Lakeport, CA 95453

14 Lake Tahoe Air Basin
El Dorado County Air Pollution Control District
7536 Green Valley Road
Placerville, CA 95667-4197

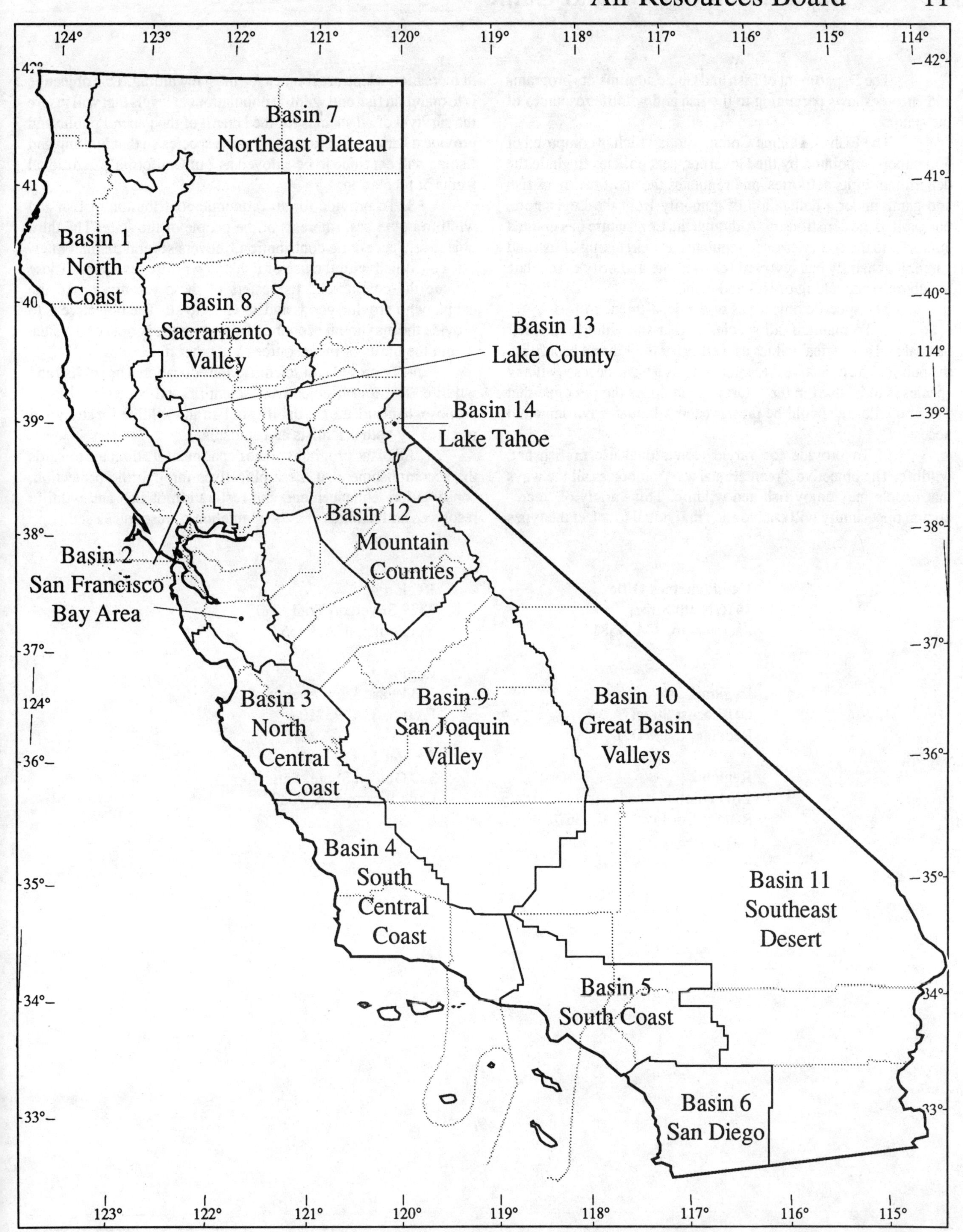
124°
123°
122°
121°
120°
119°
118°
117°
116°
115°
114°
42°
41°
40°
39°
38°
37°
36°
35°
34°
33°
Basin 1
North
Coast
Basin 2
San Francisco
Bay Area
Basin 3
North
Central
Coast
Basin 4
South
Central
Coast
Basin 5
South Coast
Basin 6
San Diego
Basin 7
Northeast Plateau
Basin 8
Sacramento
Valley
Basin 9
San Joaquin
Valley
Basin 10
Great Basin
Valleys
Basin 11
Southeast
Desert
Basin 12
Mountain
Counties
Basin 13
Lake County
Basin 14
Lake Tahoe

The Department of Fish and Game administers programs and enforces laws pertaining to the fish and wildlife resources of the state.

The Fish and Game Commission, which is composed of 5 members appointed by the Governor, sets policies to guide the department in its activities, and regulates the sport taking of fish and game under a delegation of authority from the Legislature, pursuant to the constitution. Although the Legislature has granted authority to the commission to regulate the sport taking of fish and game, it generally has reserved for itself the authority to regulate the commercial taking of fish and game.

The specific objectives of the department are:

1. To maintain all species of fish and wildlife for their natural and ecological values as well as for their direct benefits to the public. The objective, "to maintain," is a basic necessity if any species is to be used in the future; this includes the principle that fish and wildlife should be preserved as a human environmental necessity.

2. To provide for varied recreational use of fish and wildlife. The objective, "recreational use," embraces all the ways that people may enjoy fish and wildlife. This variety of recreational opportunity will enable each individual to select the types of recreation which are found to be most rewarding. This objective is to maintain fish and wildlife populations at levels that will insure the survival of all species for the benefit of the general public and provide a harvestable surplus of game species so that hunting and fishing will continue to be enjoyed as 2 of California's traditional forms of recreation.

3. To provide for an economic contribution of fish and wildlife in the best interests of the people of the state. The third objective, "economic contribution," covers several distinct interests concerned with the use of fish and wildlife resources. These include the commercial harvesters of these resources, and the people who provide goods and services to all. The objective is to provide the maximum economic benefits to the people of the state within the limits of the resources and other objectives.

4. To provide for scientific and educational use of fish and wildlife. The fourth objective, "scientific and educational use," proposes to insure the availability of fish and wildlife for study and research by both scientists and students.

All of the programs of the department are directed towards the accomplishment of these objectives through the protection, conservation, enhancement, and restoration of fish and wildlife resources and habitats and the regulation of resources used.

Headquarters Office
1416 Ninth Street
Sacramento, CA 95814

Region 1
601 Locust Street
Redding, CA 96001

Region 2
1701 Nimbus Road
Rancho Cordova, CA 95670

Region 3
7329 Silverado Trail
Yountville, CA 94558

Region 4
1234 East Shaw Avenue
Fresno, CA 93710

Region 5
330 Golden Shore, Suite 50
Long Beach, CA 90802

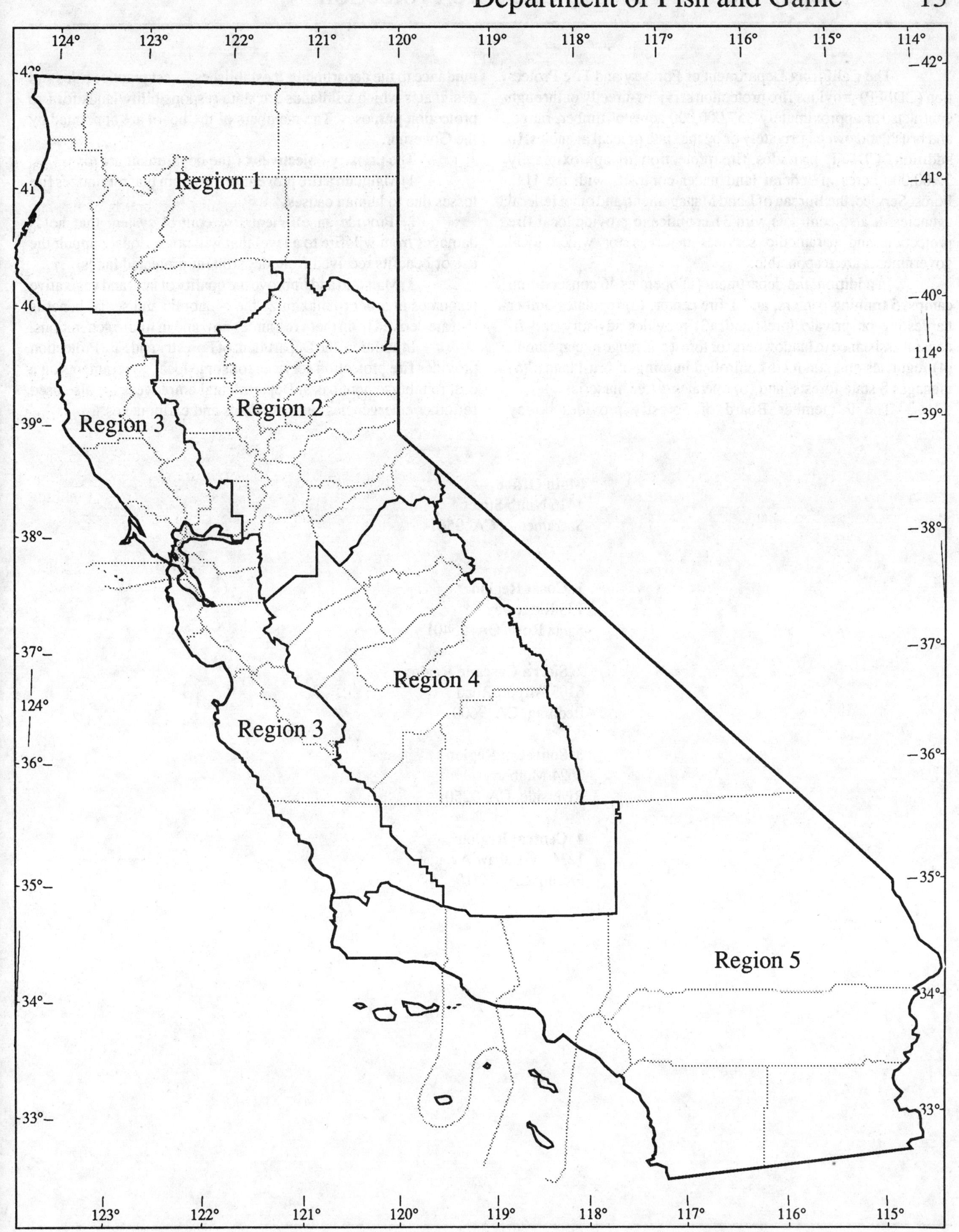
Region 1
Region 2
Region 3
Region 3
Region 4
Region 5
124°
123°
122°
121°
120°
119°
118°
117°
116°
115°
114°
42°
41°
40°
39°
38°
37°
36°
35°
34°
33°

The California Department of Forestry and Fire Protection (CDFFP) provides fire protection services directly or through contracts for approximately 357,000,000 acres of timber, range, and brushland owned privately or by the state or local agencies. In addition, CDFFP provides fire protection to approximately 3,700,000 acres of federal land under contracts with the U.S. Forest Service, the Bureau of Land Management, and other federal agencies. It also contracts with 31 counties to provide local fire protection and paramedic services in areas for which local governments are responsible.

In addition, the department (1) operates 46 conservation camps, 5 training centers, and 1 fire center, (2) regulates timber harvesting on private forestland, (3) provides advisory and financial assistance to landowners for forest and range management, (4) regulates and conducts controlled burning of brushlands, (5) manages 8 state forests, and (6) operates 3 tree nurseries.

The 9 member Board of Forestry provides policy guidance to the department. It establishes forest practice rules and designates which wildlands are state responsibility lands for fire protection purposes. The members of the board are appointed by the Governor.

The primary objectives of the department are to:

1) Maintain a fire prevention program that minimizes fire losses due to human causes.

2) Provide an efficient fire control system that holds damages from wildfire to a level that will not seriously impair the use or benefits received from department protected lands.

3) Maintain and improve the quality of land and vegetative resources in order to maximize the economic and social benefits that are derived from these resources now and in future generations.

In addition, the Department of Forestry and Fire Protection provides fire protection services for some local governments on a cost reimbursement basis. Departmental employees are also used for other emergencies such as floods and earthquakes.

Main Office
1416 Ninth Street
Sacramento, CA 95814

1 Coast Region
135 Ridgeway
Santa Rosa, CA 95401

2 Sierra Cascade Region
6105 Airport Road
Redding, CA 9600

3 Southern Region
2524 Mulberry
Riverside, CA 92501

4 Central Region
1234 East Shaw Avenue
Fresno, CA 93710

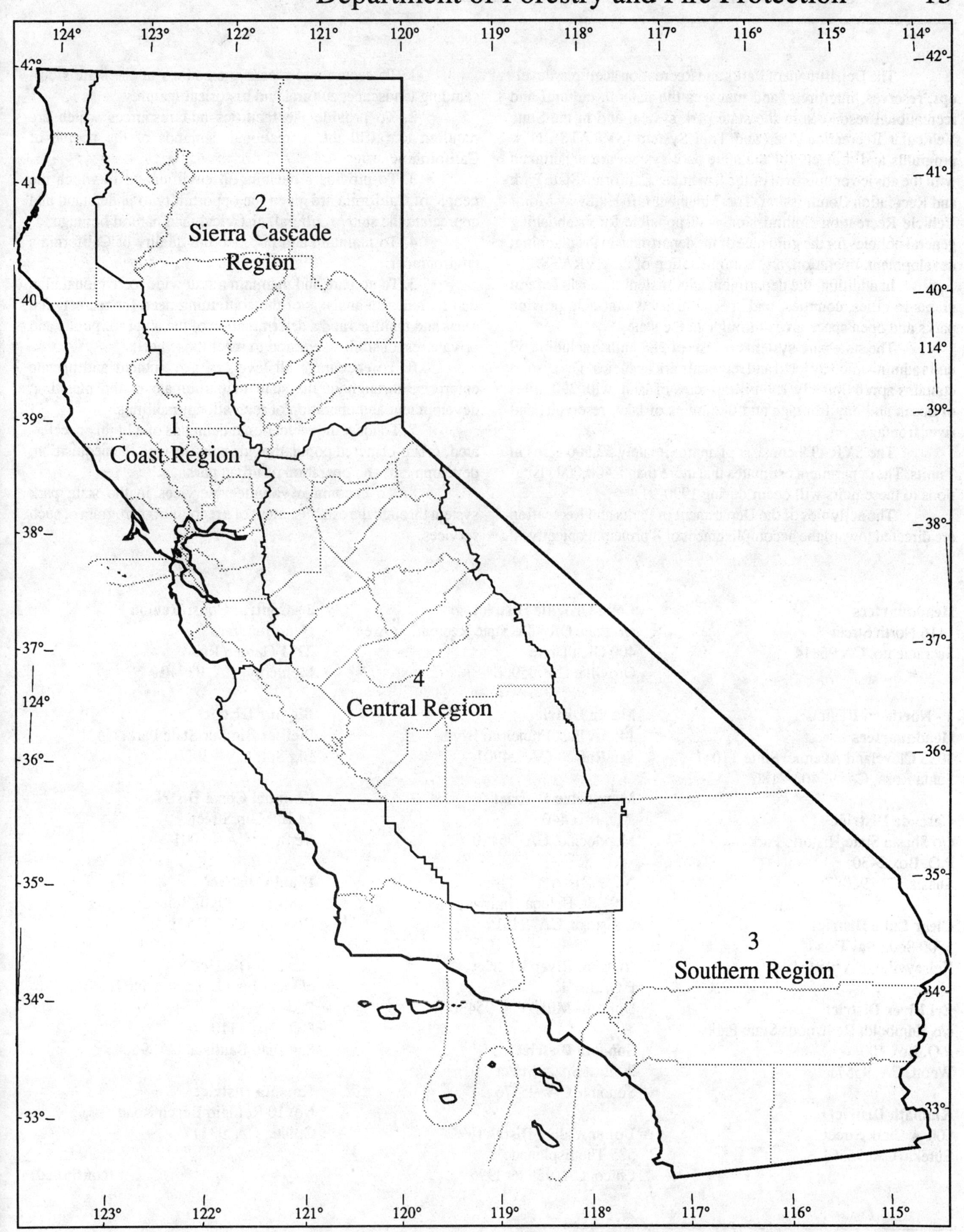
2
Sierra Cascade
Region
1
Coast Region
4
Central Region
3
Southern Region
124°
123°
122°
121°
120°
119°
118°
117°
116°
115°
114°
42°
41°
40°
39°
38°
37°
36°
35°
34°
33°

The Department of Parks and Recreation acquires, develops, reserves, interprets, and manages the natural, cultural and recreational resources in the state park system and in the State Vehicular Recreation Area and Trail System (SVRATS). New programs and projects for the state park system are undertaken with the advice or approval of the 8 member California State Park and Recreation Commission. The 7 member Off-Highway Motor Vehicle Recreation Commission is responsible for establishing general policies for the guidance of the department in the planning, development, operation, and administration of the SVRATS.

In addition, the department administers state and federal grants to cities, counties, and special districts that help provide parks and open space areas throughout the state.

The state park system consists of 287 units, including 39 units administered by local and regional park agencies. The system contains approximately 1.4 million acres of land, with 290 miles of ocean and bay frontage and 686 miles of lake, reservoir, and river frontage.

The SVRATS consists of approximately 52,500 acres in 7 units. The department estimates that more than 1,400,000 visitations to these units will occur during 1990-91.

The activities of the Department of Parks and Recreation are directed toward the accomplishment of 8 principal objectives:

1. To secure and preserve elements of the state's outstanding landscape, cultural and historical features.

2. To provide the facilities and resources which are required to fulfill the recreational demands of the people of California.

3. To provide a meaningful environment in which the people of California are given the opportunity to understand and appreciate the state's cultural, historical, and natural heritage.

4. To maintain and improve the quality of California's environment.

5. To prepare and maintain a statewide recreational plan that includes an analysis of the continuing need for recreational areas and facilities and a determination of the levels of public and private responsibility required to meet those needs.

6. To encourage all levels of government and private enterprise throughout the state to participate in the planning, development and operation of recreational facilities.

7. To meet the recreational demands of a highly accelerated, urban-centered population growth, through the acquisition, development, and operation of urban parks.

8. To encourage volunteer services in the state park system through the establishment of a recognition program of such services.

Headquarters
1416 North Street
Sacramento, CA 95814

1 - Northern Region Headquarters
3033 Cleveland Avenue, Suite 110
Santa Rosa, CA 95403-2186

Cascade District
c/o Shasta State Historic Park
P.O. Box 2430
Shasta, CA 96087

Clear Lake District
5300 Soda Bay Road
Kelseyville, CA 95451

Eel River District
c/o Humboldt Redwoods State Park
P.O. Box 100
Weott, CA 95571

Klamath District
600-A Clark Street
Eureka, CA 95501

Lake Oroville District
c/o Lake Oroville State Recreation Area
400 Glen Drive
Oroville, CA 95966

Marin District
1455A East Francisco Blvd.
San Rafael, CA 94901

Mendocino District
P.O. Box 440
Mendocino, CA 95460

Napa District
3801 St. Helena Highway N.
Calistoga, CA 94515

Russian River District
P.O. Box 123
Duncans Mills, CA 95430

Sonoma District
20 East Spain Street
Sonoma, CA 95476

Upper Valley District
525 The Esplanade
Chico, CA 95926-3996

2 - Central Coast Region Headquarters
2211 Garden Road
Monterey, CA 93940

Big Sur District
Pfeiffer Big Sur State Park, No. 1
Big Sur, CA 93920

Channel Coast District
24 E. Main Street
Ventura, CA 93001

Diablo District
4180 Treat Blvd. Suite D
Concord, CA 94521

Gavilan District
c/o San Juan Bautista State Historic Park
P.O. Box 1110
San Juan Bautista, CA 95045

Gaviota District
No. 10 Refugio Beach Road
Goleta, CA 93117

(continued)

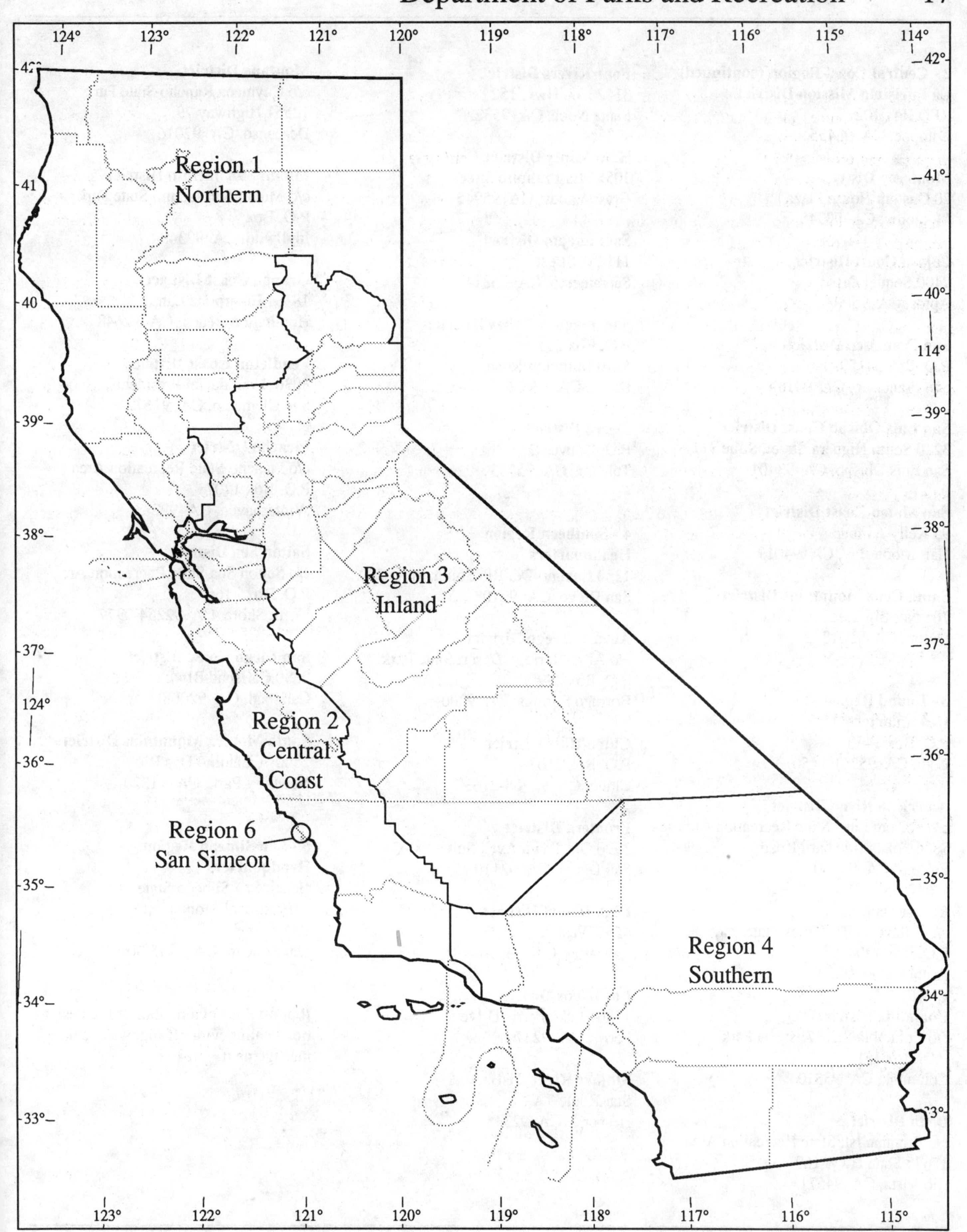
Region 1
Northern
Region 2
Central
Coast
Region 3
Inland
Region 4
Southern
Region 6
San Simeon
124°
123°
122°
121°
120°
119°
118°
117°
116°
115°
114°
42°
41
40
39°
38°
37°
36°
35°
34°
33°

2 - Central Coast Region (continued)
La Purisima Mission District
RFD Box 102
Lompoc, CA 93436

Monterey District
20 Custom House Plaza
Monterey, CA 93940

Pajaro Coast District
7400 Soquel Drive
Aptos, CA 95003

San Francisco District
P.O. Box 34159
San Francisco, CA 94134

San Luis Obispo Coast District
3220 South Higuera Street, Suite 311
San Luis Obispo, CA 93401

San Mateo Coast District
95 Kelly Avenue
Half Moon Bay, CA 94019

Santa Cruz Mountains District
101 No. Big Trees Park Rd.
Felton, CA 95018

3 - Inland Region
Headquarters
P.O. Box 1450
Lodi, CA 95241-1450

American River District
c/o Folsom Lake State Recreation Area
7806 Folsom-Auburn Road
Folsom, CA 95630

Calaveras District
c/o Calaveras Big Trees State Park
P.O. Box 120
Arnold, CA 95223

Columbia District
c/o Columbia State Historia Park
P.O. Box 151
Columbia, CA 95310

Delta District
c/o Brannan Isl. State Recreation Area
17645 State Hwy. 160
Rio Vista, CA 94571

Four Rivers District
31426 W. Hwy. 152
Santa Nella, CA 95322

Gold Mines District
10556 East Empire Street
Grass Valley, CA 95945

Sacramento District
111 "I" Street
Sacramento, CA 95814

San Joaquin Valley District
P.O. Box 205
5290 Millerton Road
Friant, CA 93626

Sierra District
P.O. Drawer D
Tahoma, CA 95733

4 - Southern Region
Headquarters
1333 Camino Del Rio South, Suite 200
San Diego, CA 92108

Anza-Borrego District
c/o Anza-Borrego Desert State Park
P.O. Box 299
Borrego Springs, CA 92004

Chino Hills District
P.O. Box 2163
Chino, CA 91708-2163

Frontera District
3990 Old Town Ave., Suite 300-C
San Diego, CA 92110

High Desert District
4555 West Avenue G
Lancaster, CA 93536

Los Lagos District
17801 Lake Perris Drive
Perris, CA 92370

Mojave River District
Star Route 7A
Hesperia, CA 92345

Montane District
c/o Cuymaca Rancho State Park
12551 Highway 79
Descanso, CA 92016

Mount San Jacinto District
c/o Mount San Jacinto State Park
P.O. Box 308
Idyllwild, CA 92349

Orange Coast District
18331 Enterprise Lane
Huntington Beach, CA 92648

Pendleton Coast District
3030 Avenida del Presidente
San Clemente, CA 92672

Picacho District
c/o Picacho State Recreation Area
P.O. Box 1207
Winterhaven, CA 92283

Salton Sea District
c/o Salton Sea State Recreation Area
P.O. Box 3166
North Shore, CA 92254-0977

San Diego Coast District
2680 Carlsbad Blvd.
Carlsbad, CA 92008

Santa Monica Mountains District
2860-A Camino Dos Rios
Newbury Park, CA 91320

6 - San Simeon Region
Headquarters
Hearst San Simeon State
Historical Monument
P.O. Box 8
San Simeon, CA 93452-0040

Region 5 is not a region. It is a designation dealing with off-highway vehicle management.

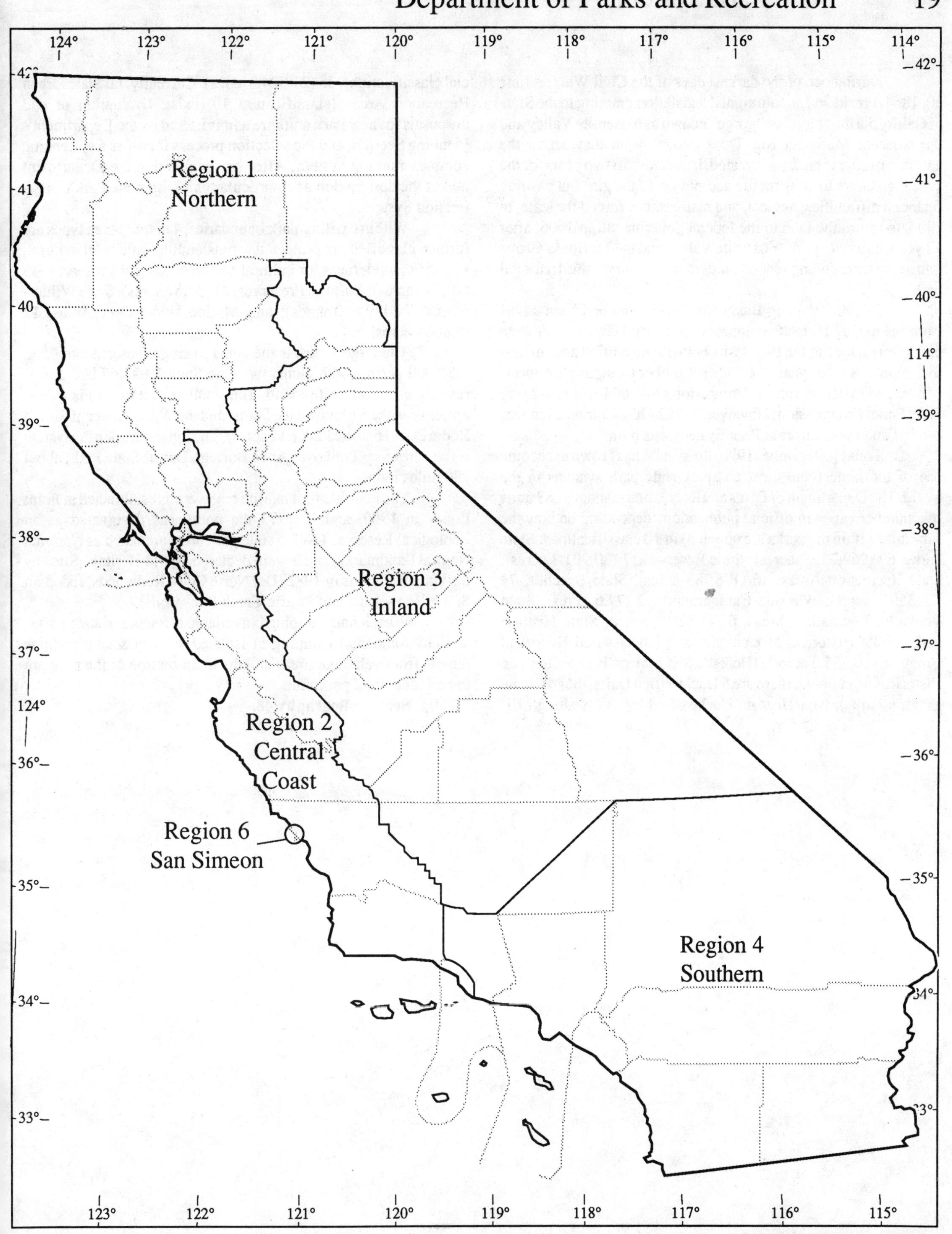
Region 1
Northern
Region 3
Inland
Region 2
Central
Coast
Region 6
San Simeon
Region 4
Southern

During one of the darkest days of the Civil War, on June 30, 1864, President Lincoln signed legislation granting to the State of California the "cleft" or "gorge" known as Yosemite Valley and the adjacent Mariposa Big Trees Grove. With that action the nation's first state park was created. It took almost two years for the state legislature to confirm the acceptance of the gift, but ensuing financial difficulties, neglect, and malfeasance forced the state, in 1905, to return the land to the federal government. In 1906, after 42 years as a state park, Yosemite Valley and the Mariposa Grove joined the surrounding federal lands as the country's third national park.

In 1902, the Big Basin Redwoods area had been saved from the axe by legislative appropriation. By 1925, 4 other state parks were added to the Big Basin, but with no unified administrative control. In that year, the Pacific Lumber Company's opposition helped to defeat long standing efforts to establish a governing State Parks Commission. However, in 1927, the Commission was created and the California Park System was born.

Today, (December 1990) the system has grown to become one of the largest and most complex state park systems in the world. The Department of Parks and Recreation manages 287 units (the number varies in official publications depending on how the sub-units within the parks are counted) in 10 classifications: State Parks, 69 (962,735.2 acres); State Reserves,17 (30,901.3 acres); State Recreation Areas, 36 (178,785 acres); State Beaches, 73 (24,359.5 acres); Wayside Campgrounds, 2 (77.6 acres); State Vehicular Recreation Areas, 6 (41,825.9 acres); State Historic Parks, State Historical Monuments, and Unclassified Historical Units, 48 (10,935.2 acres), (Hearst's San Simeon is the sole State Historical Monument, there are 5 Unclassified Units, thus 42 State Historic Parks); Non-Historic Unclassified Units (awaiting official classification), 36 (21,524.3 acres). Currently, no State Urban Recreation Areas (classification 10) exist. Evaluation of and proposals for new park units are administered by the Department's Planning Section, and the selection process involves a systematic complex ranking system. After site acquisition the Department makes the designation as a particular unit within the Park Classification System.

Within existing park boundaries, 4 distinct area types are further classified to protect the outstanding values of natural, scientific, wilderness, or cultural features: Natural Preserves, 35 (10,790 acres); Cultural Preserves, 11 (3,552 acres); State Wildernesses, 7 (439,610 acres); and Marine Underwater Areas, 12 (9,800+ acres).

The largest park is the Anza Borrego Desert State Park, (550,000 acres), with Henry W. Coe State Park (67,000 acres) running a very distant second. The smallest unit owned is a toss-up between the Bufano Peace Statue and the Watts Tower of Simon Rodia State Historic Park (.11 acres). The longest trail in thesystem is the Anza Sky Trail over Anza Borrego Desert State Park, about 170 miles long.

The nation's first marine reserve was established at Point Lobos in 1960, and in 1973 the point was designated as an Ecological Reserve. Twelve state park units are listed as National Natural Landmarks, and 3 were declared World Heritage Sites by the United Nations in 1982 (Del Norte Coast Redwoods, Jedediah Smith Redwoods, and Prairie Creek Redwoods).

Wheelchair camping is available at a score of state parks, and environmental camping at isolated sites in scenic settings remote from vehicular areas and the main section of the park are to be found at 22 park units.

See: Bibliography 48.

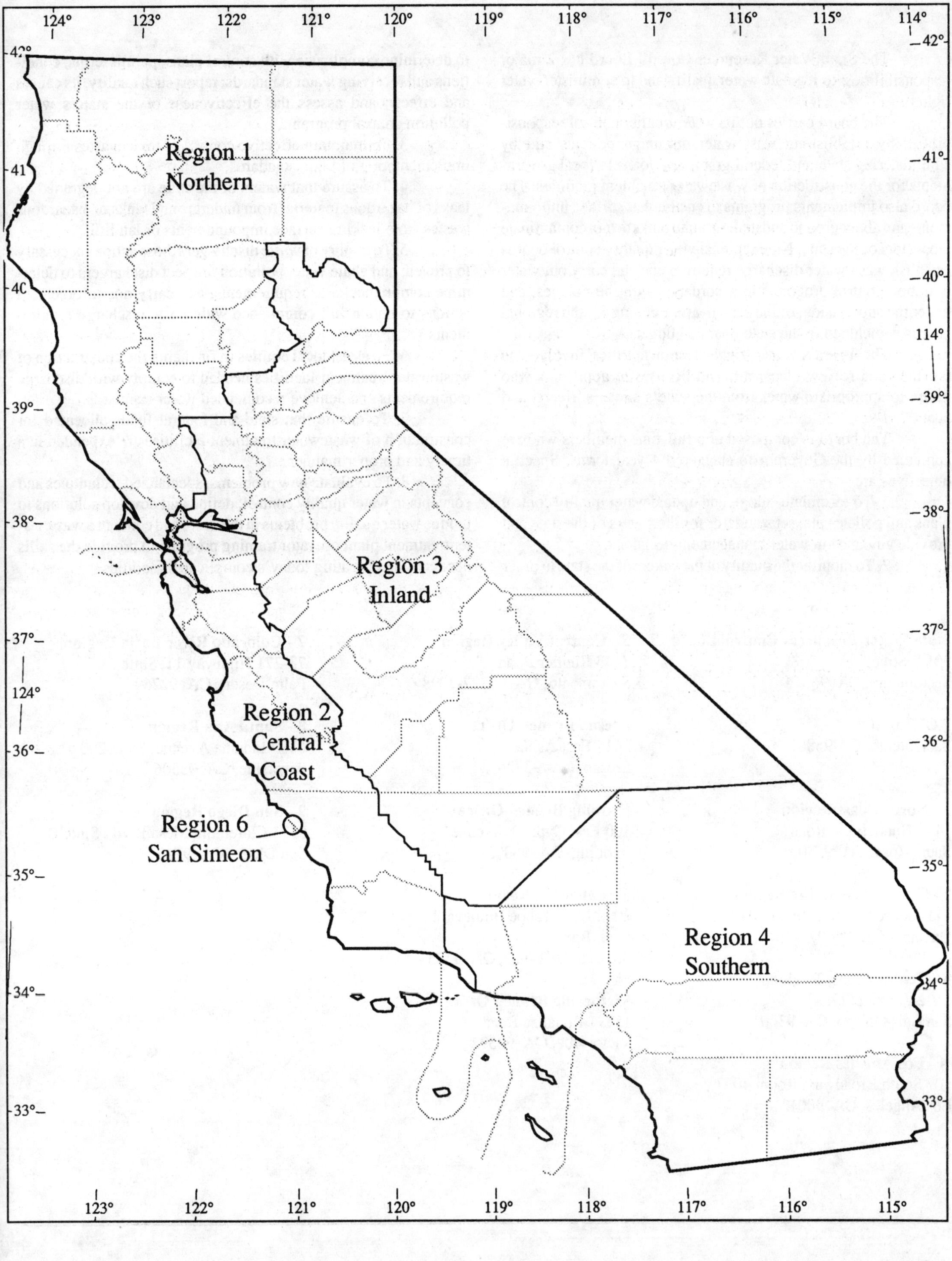

Region 1
Northern

Region 3
Inland

Region 2
Central
Coast

Region 6
San Simeon

Region 4
Southern

The State Water Resources Control Board has 2 major responsibilities: to regulate water quality and to administer water rights.

The board carries out its water quality control responsibilities by establishing wastewater discharge policies and by administering state and federal grants and loans to local governments for the construction of wastewater treatment facilities. The board also implements programs to ensure that surface impoundments and aboveground and underground tanks do not contaminate the waters of the state. Nine regional water quality control boards establish wastewater discharge requirements and carry out water pollution control programs in accordance with the policies, and under the supervision, of the state board. Funding for the regional boards is included in the state board's budget.

The board's water rights responsibilities involve the issuance and review of permits and licenses to applicants who desire to appropriate water from the state's streams, rivers, and lakes.

The board is composed of 5 full-time members who are appointed by the Governor to staggered 4 year terms. Specific objectives are:

1. To formulate, adopt, and update water quality control plans and policies that set standards for the waters of the state and provide guidance in water management decisions.

2. To monitor the quality of the waters of the state in order to determine compliance with control plans, permit terms, conditions and receiving water standards; report such quality, its causes and effects; and assess the effectiveness of the state's water pollution control program.

3. To maintain effective control of toxic wastes through implementation of toxic standards.

4. To assure that waters of the state are not degraded by leaks of hazardous material from underground tanks or hazardous wastes from leaking surface impoundments or landfills.

5. To require of waste dischargers those actions necessary to prevent and abate water pollution, inspect dischargers to determine compliance with requirements, and carry out enforcement actions to obtain full compliance with waste discharge requirements.

6. To assist local entities in financing the construction of wastewater treatment facilities needed to comply with discharge requirements and achieve recomended water standards.

7. To ensure that state and federal funds allocated for construction of wastewater treatment facilities are expended in a timely and proper manner.

8. To evaluate new problems, specialized techniques and concepts in water quality control; define and develop solutions to unique water quality problems in the state; and conduct a wastewater treatment plant operator training program to provide the skills necessary in operating today's complicated facilities.

State Water Resources Control Board
901 P Street
Sacramento, CA 95814

P.O. Box 100
Sacramento, CA 95801

1 - North Coast Region
1440 Guerneville Road
Santa Rosa, CA 95403

2 - San Francisco Bay Region
111 Jackson Street, Room 6040
Oakland, CA 94607

3 - Central Coast Region
1102-A Laurel Lane
San Luis Obispo, CA 93401

4 - Los Angeles Region
107 South Broadway, Room 4027
Los Angeles, CA 90012

5 - Central Valley Region
3443 Routier Road
Sacramento, CA 95827-3098

Fresno Branch Office
3614 East Ashlan Ave.
Fresno, CA 93726

Redding Branch Office
100 East Cypress Avenue
Redding, CA 96002

6 - Lahontan Region
2092 Lake Tahoe Boulevard
P. O. Box 9428
South Lake Tahoe, CA 95731

Victorville Branch Office
15371 Bonanza Road
Victorville, CA 92392

7 - Colorado River Basin Region
73-271 Highway 11, Suite 21
Palm Desert, CA 92260

8 - Santa Ana Region
6809 Indiana Avenue, Suite 200
Riverside, CA 92506

9 - San Diego Region
9771 Claremont Mesa Blvd., Suite B
San Diego, CA 92124

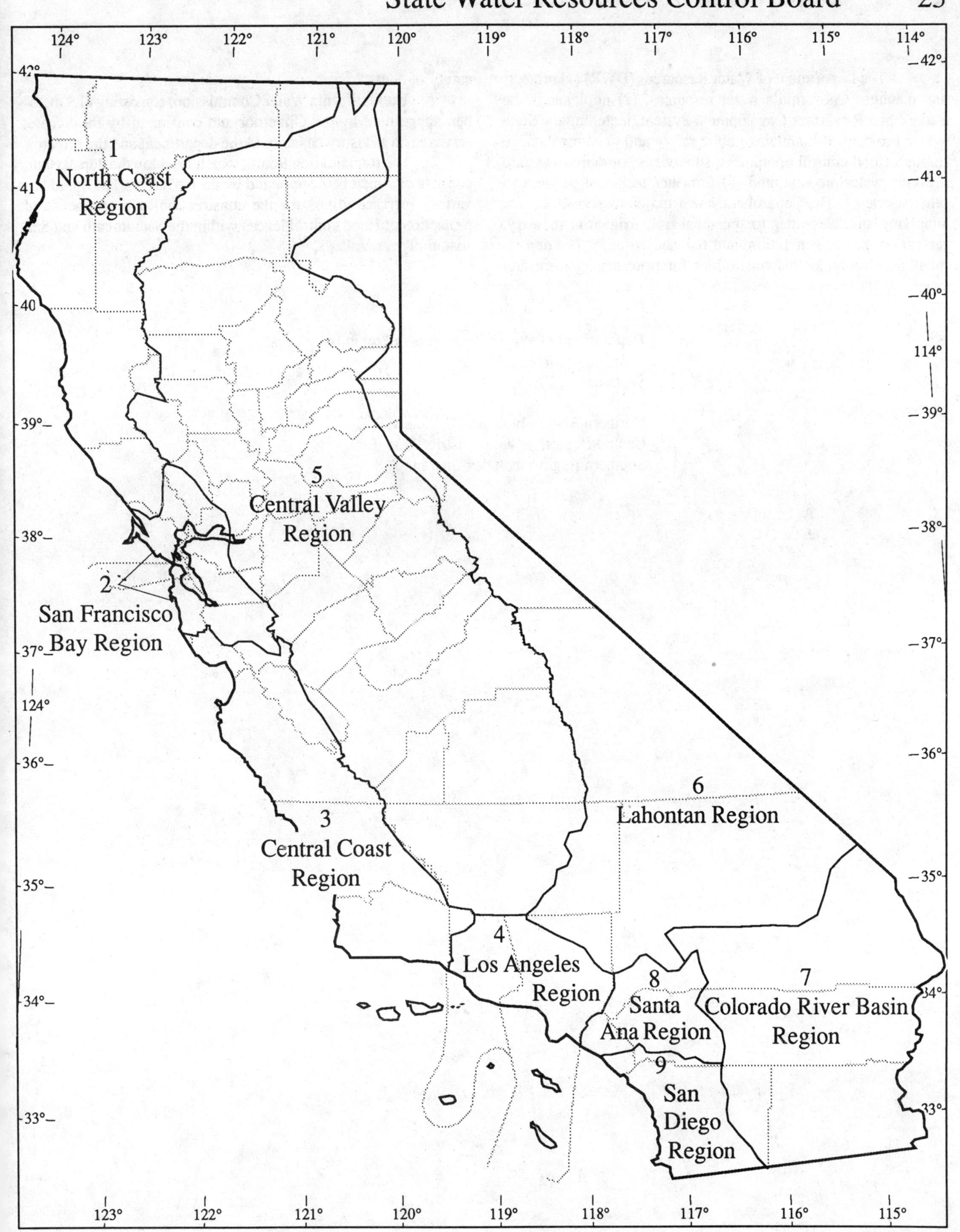
124°
123°
122°
121°
120°
119°
118°
117°
116°
115°
114°
42°
41°
40°
39°
38°
37°
36°
35°
34°
33°
1
North Coast
Region
2
San Francisco
Bay Region
3
Central Coast
Region
4
Los Angeles
Region
5
Central Valley
Region
6
Lahontan Region
7
Colorado River Basin
Region
8
Santa
Ana Region
9
San
Diego
Region

The Department of Water Resources (DWR) (1) protects and manages California's water resources, (2) implements the State Water Resources Development System, including the State Water Project, (3) maintains public safety and prevents damage through flood control operations, supervision of dams, and safe drinking water projects, and (4) furnishes technical services to other agencies. The department has a major responsibility for supplying suitable water for personal use, irrigation, industry, recreation, power generation, and fish and wildlife. The department also has major responsibilities for flow management and safety of dams.

The California Water Commission, consisting of 9 members, appointed by the Governor and confirmed by the Senate, serves in an advisory capacity to the department and the director.

The Reclamation Board, which is within the department, consists of 7 members appointed by the Governor. The board has various responsibilities for the construction, maintenance, and protection of flood control levees within the Sacramento and San Joaquin River Valleys.

Department of Water Resources (all regions)
1416 Ninth Street
Sacramento, CA 95814

Northern Region includes Districts 1, 2, 3
Central Region includes Districts 4, 5, 6
Southern Region includes Districts 7, 8

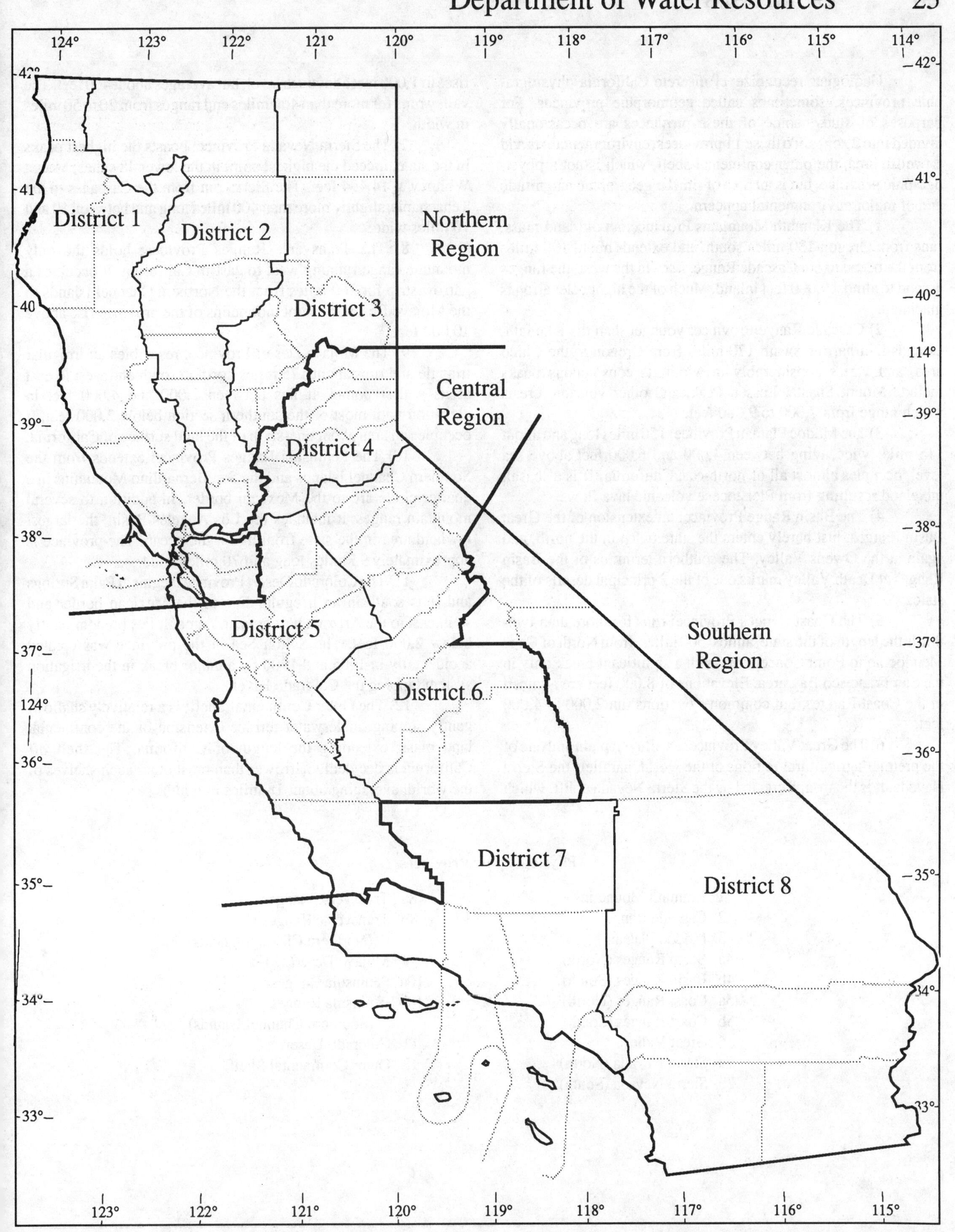

District 1
District 2
Northern Region
District 3
Central Region
District 4
District 5
Southern Region
District 6
District 7
District 8
124°
123°
122°
121°
120°
119°
118°
117°
116°
115°
114°
42°
41°
40°
39°
38°
37°
36°
35°
34°
33°

Geologists recognize 11 discrete California physiographic provinces, sometimes called geomorphic provinces. For purposes of study, some of these provinces are occasionally divided into 2 zones. To these 11 provinces, environmentalists add a twelfth area, the outer continental shelf, which is not a physiographic province, but is an area of similar geographic magnitude and of major environmental concern.

1) The Klamath Mountains Province: an old land mass, runs from Oregon 130 miles south, and extends about 100 miles from the ocean to the Cascade Range. Low in the west, the ranges ascend to almost 9,000 feet inland. Much of the higher elevation is glaciated.

2) Cascade Range Province: younger than the Klamath, extends in a narrow swath 170 miles from Oregon to the Chico area, and varies considerably in width Its conspicuous peaks include Mount Shasta, Lassen Peak, and other volcanic crests which range from 7,000 to 9,000 feet.

3) The Modoc Plateau Province: 150 miles long and about 115 miles wide, lying between 4,000 and 6,000 feet above sea level, occupies almost all of north-east California. It is a basalt tableland resulting from Pleistocene volcanic lava flows.

4) The Basin Range Province: an extension of the Great Basin Ranges, just barely enters the state twice in the north, and again at the Owens Valley. The southern terminus of the Basin Ranges at Death Valley marks one of the 3 principal deserts of the state.

5) The Coast Ranges Province: runs for more than two-thirds the length of the state, almost 500 miles, from North of Cape Mendocino to Point Conception, with a significant break only in the San Francisco Bay area. Elevations of 8,000 feet are attained in the Coast Ranges, but common elevations run 2,000 to 4,000 feet.

6) The Great Valley Province: an alluvial plain and one of the premier agricultural regions of the world, parallels the Sierra Nevada. It is the trough created by the Sierra Nevada uplift, which rises to 1,000 feet above sea level, but averages about 400 feet. The valley runs for more than 400 miles and ranges from 20 to 50 miles in width.

7) The Sierra Nevada Province: boasts the highest peaks in the state, indeed the highest point in the lower 48 states, Mount Whitney at 14,494 feet. The Sierra run from the Cascades to the Tehachapis, slightly more than 400 miles long and between 50 and 70 miles wide.

8) The Transverse Ranges Province: holds the only mountain range running west to east in California. It occupies a narrow strip for 260 miles from the Northern Channel Islands to the Mojave Desert. Several mountains of the province rise above 10,000 feet.

9) The Mojave Desert Province: resembles an irregular triangle and runs about 175 miles north to south and west to east at its widest points. It lies between 2,000 and 4,000 feet in elevation with most of the southern section below 2,000 feet. It occupies an area about one-sixth of the land surface of California.

10) The Peninsula Ranges Province: extends from the Southern Channel Islands and the San Bernardino Mountains in a quarter-circle arc to the Mexican border. In addition to several mountain ranges, it includes the Los Angeles Basin, the largest lowland area in the state fronting on the ocean. The province is approximately 95 miles long and 70 miles wide.

11) The Colorado Desert Province: begins at Palm Springs and runs south in an irregular line to the Mexican border and southeast to the Arizona border near Yuma. It lies predominantly below 2,000 feet. The Salton Sea in the province was created accidentally in 1905 and 1906 by a major break in the irrigation system tapping the Colorado River.

12) The Outer Continental Shelf: is a relatively shallow, gently sloping, underwater terrace extension of the continental land mass, extending the length of California. The shelf off California is decidedly narrower than most other such shelves of the world, averaging about 18 miles in width.

Physiographic Provinces

1 Klamath Mountains
2 Cascade Range
3 Modoc Plateau
4a Basin Ranges (North)
4b Basin Ranges (South)
5a Coast Ranges (North)
5b Coast Ranges (South)
6 Great Valley
7a Sierra Nevada (North)
7b Sierra Nevada (South)
8a Transverse Ranges
8b Transverse Ranges (Northern Channel Islands)
9 Mojave Desert
10a Peninsula Ranges
10b Peninsula Ranges (Southern Channel Islands)
11 Colorado Desert
12 Outer Continental Shelf

124° 123° 122° 121° 120° 119° 118° 117° 116° 115° 114°

42° 41° 40° 39° 38° 37° 36° 35° 34° 33°

1
2
3
4a
4a
4b
5a
5b
6
7a
7b
8a
8b
9
10a
10b
11
12
12
12

Mountain Ranges and Peaks

California holds the most diverse array of mountains in the nation, more mountains than any other state except Alaska, and the highest point in the contiguous 48 states, Mount Whitney at 14,494 feet (only 80 miles distant,is the lowest point in the United States, Bad Water, at 282 feet below sea level). Parenthetically, one might note that the oldest living natural entity, the Bristlecone Pine, exists on California's White Mountain peaks, and in the Sierra Nevada stands the world's largest natural organism, the magnificent Giant Sequoia. The state even boasts the smallest true mountain range in the world, Sutter Buttes.

Most geologists theorize that the mountain range landscape of California is the result of the interaction of the Pacific (tectonic) Plate sliding under (subducting) the North American Plate, with the Pacific Plate moving in a north-westerly direction. As the Pacific Plate pushed under the North American Plate, the enormous masses of rubble created at the juncture point eventuated in the Coast Ranges. The actual contact points moved steadily westward thus resulting in the series of ranges we now know.

To the east, a fault block, a great mass of newly cooled solid granite, the largest unitary block of granite on the planet, was thrust up, creating the Sierra Nevada. The Sierra showed a sharp eastern slope and a more gradual western incline. The trough created by the uplift of the Sierra Nevada was filled in over the millennia with eroded material to form the Great Central Valley.

As the Pacific Plate was forced deeper into the earth's depths it began to remelt and the newly formed magma, gases, and oxygen erupted into what is now the Cascade Range, breaking off and separating the northwest end of the Sierra Nevada to create a separate mountain range area we now call the Klamath region.

The action of the two land masses also resulted in great cracks in the North American Plate at the contact points. One of those cracks, the San Andreas Fault, split off a long western section of California from the rest of the North American plate; that section is being pushed steadily northwest by the underlying plate at the rate of 2 inches or so per year. The San Andreas Fault also takes an eastward bend just to the east of Los Angeles; along that crack the underlying plate wrenched the Coast Range away from its northwest to southeast trend to an west-east direction, creating the Transverse Ranges.

As stated above, geologists also theorize that the Klamath Ranges were once a part of the Sierra Nevada, as the core foundation of the Trinity Alps, a part of the Klamath, seems to be identical to that of the Sierra Nevada. Also the Marble Mountains, another unit of the Klamath, shows a characteristic type of marble found elsewhere only in the Sierra Nevada.

These theories help to explain why the rocks of the Coast Ranges are progressively younger as they move west, and why the rocks of these ranges seem to be more jumbled in their setting than other mountain rock formations. It also explains why the Coast Ranges are among the most erosive in the world. For instance, it has been calculated that the Eel River is eroding the Coast Range at a rate up to 80 times faster than the rivers eroding the Sierra Nevada — 8 inches to .1 inch per century. In a similar manner, it also explains why the southern Coast Ranges are so frequently subject to mud slides. Once created, the mountains were subjected to the reshaping forces of glacial grinding, and the erosive forces of water and wind.

California geologists are not in agreement as to the demarcation of mountain regions or mountain range borders, but a practical arrangement, acceptable for most purposes, divides the state into 8 mountain regions: Klamath, Cascade, Basin, North Coast, Sierra Nevada, South Coast, Transverse, and Peninsular, which together include 112 identifiable mountain ranges, not including some possible range designations which have not been officially sanctioned by the U.S. Geological Survey. Unique unto itself is the Sutter Buttes, rising out of the central valley, which is recognized as the world's smallest mountain range, for a total of 113 distinct mountain ranges.

Eleven of California's ranges attain elevations over 10,000 feet: Benton (highest peak, Glass Mountain, 11,123 feet); Cascade (Mount Shasta, 14,162 feet); Cathedral (Mount Lyell, 13,114 feet); Inyo (Waucoba Mountain, 11,123 feet); Ritter (Mount Ritter, 13,157 feet); San Bernardino (San Gorgonio Peak, 11,502 feet); San Gabriel (Mount San Antonio, 10,064 feet); Sierra Nevada (Mount Whitney, 14,494 feet); Sweetwater (Mount Patterson, 11,673 feet); and the White Mountains (White Mountain, 14,246 feet). The Warner Range with Eagle Peak at 9,892 feet just misses.

Thirteen peaks in California climb higher than 14,000 feet, with 11 of those mountains in the Sierra Nevada. Mount Whitney, the tallest mountain, is surrounded by five other peaks over 14,000 feet, including Mount Williamson, California's second highest point. Another five Sierra mountains over 14,000 feet are centered just west of Big Pine. A somewhat suspect early inventory of California's mountains lists 76 peaks between 13,000 and 14,000 feet, and there are more than 500 crests over 12,000 feet.

Some of the popular mountains of the state are noted more for their scenic and recreational values than their height: Mount Lassen, 10,459 feet; Mount Tamalpais, 2,571 feet; Mount Diablo, 3,849 feet; Sutter Buttes, 2,132 feet; Half Dome, 8,842 feet; and El Capitan, 7,569 feet.

See: Bibliography 43, 44.

(continued)

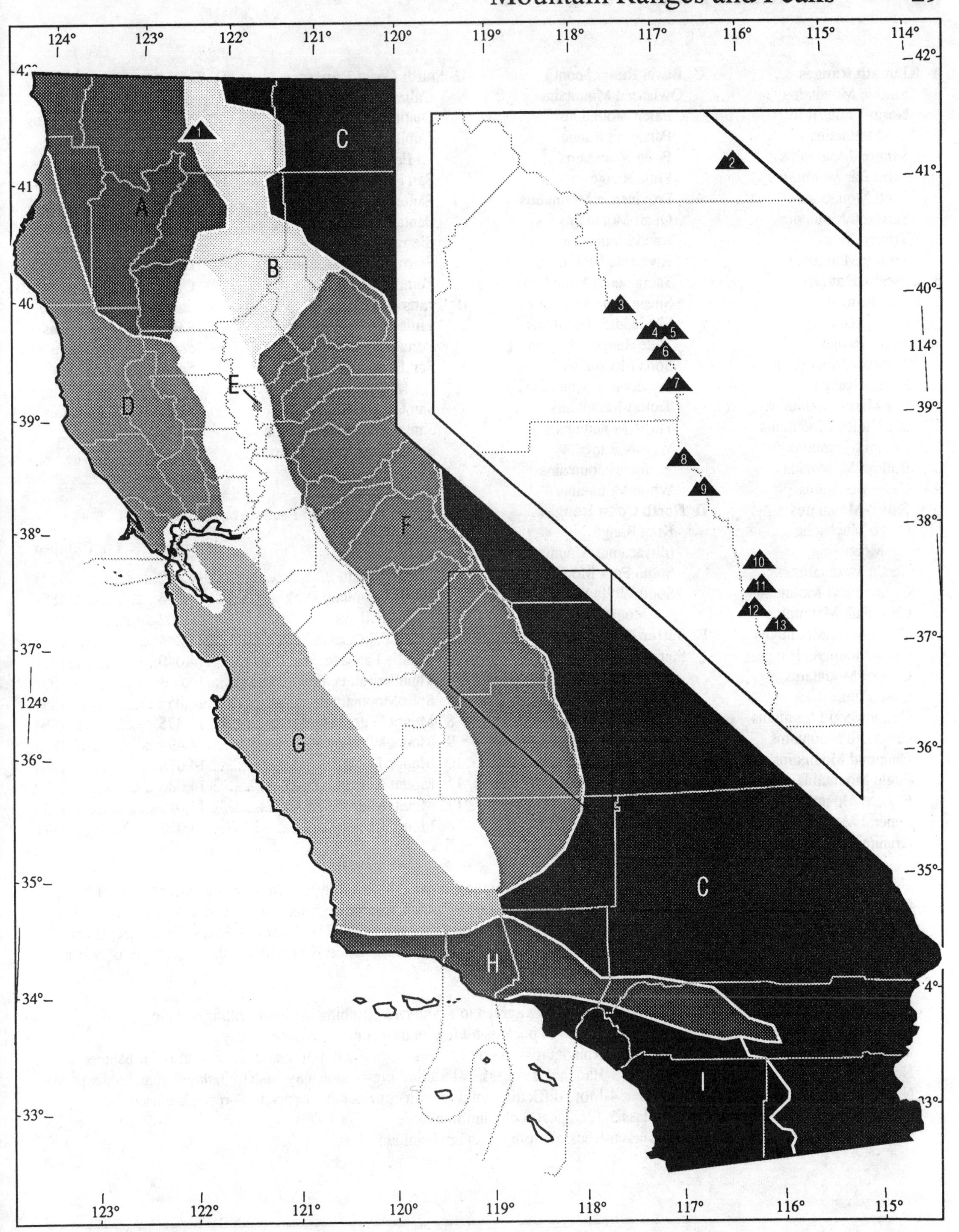
A
B
C
D
E
F
G
H
I
1
2
3
4
5
6
7
8
9
10
11
12
13
124°
123°
122°
121°
120°
119°
118°
117°
116°
115°
114°
42°
41°
40°
39°
38°
37°
36°
35°
34°
33°

A Klamath Ranges
- Marble Mountains
- North Yolla Bolly Mountains
- Salmon Mountains
- Scott Bar Mountains
- Scott Mountains
- Siskiyou Mountains
- Trinity Alps
- Trinity Mountains

B Cascade Range

C Basin Ranges
- Amargosa Range
- Argus Range
- Avawatz Mountains
- Benton Range
- Big Maria Mountains
- Big Valley Mountains
- Bristol Mountains
- Bullion Mountains
- Cady Mountains
- Calico Mountains
- Cargo Muchacho Mountains
- Castle Mountains
- Chemehuevi Mountains
- Chocolate Mountains
- Chuckwalla Mountains
- Clark Mountain Range
- Clipper Mountains
- Coso Range
- Cottonwood Mountains
- Coxcomb Mountains
- Diamond Mountains
- Eagle Mountains
- El Paso Mountains
- Funeral Mountains
- Granite Mountains (4 separate ranges)
- Grapevine Mountains
- Greenwater Range
- Inyo Mountains
- Iron Mountains
- Ivanpah Mountains
- Kingston Range
- Last Chance Range
- Little Maria Mountains
- Marble Mountains
- New York Mountains
- Nopah Range
- Old Woman Mountains
- Orocopia Mountains

C Basin Range(cont.)
- Owlshead Mountains
- Palen Mountains
- Panamint Range
- Pinto Mountains
- Piute Range
- Providence Mountains
- Quail Mountains
- Rand Mountains
- Riverside Mountains
- Sacramento Mountains
- Sheep Hole Mountains
- Skedaddle Mountains
- Slate Range
- Soda Mountains
- Sweetwater Mountains
- Tiefort Mountains
- Turtle Mountains
- Warner Range
- Whipple Mountains
- White Mountains

D North Coast Ranges
- King Range
- Mayacamas Mountains
- South Fork Mountains
- South Yolla Bolly Mountains

E Sutter Buttes

F Sierra Nevada
- Cathedral Range
- Greenhorn Mountains
- Grizzly Mountains
- Piute Mountains
- Ritter Range
- Scodie Mountains
- Sierra Nevada

G South Coast Ranges
- Caliente Range
- Diablo Range
- Gabilan Range
- La Panza Range
- San Rafael Mountains
- Santa Cruz Mountains
- Santa Lucia Range
- Sierra de Salinas
- Sierra Madre Mountains
- Temblor Range

H Transverse Range
- Little San Bernadinos
- Portal Ridge Mountains
- San Bernadino Mountains
- San Gabriel Mountains
- Santa Monica Mountains
- Santa Susana Mountains

H Transverse Range (cont.)
- Santa Ynez Mountains
- Sierra Pelona Mountains
- Tehachapi Mountains
- Topatopa Mountains

I Peninsular Ranges
- Agua Tibia Mountains
- Coyote Mountains
- Jacumba Mountains
- Laguna Mountains
- Oak Ridge Mountains
- San Jacinto Mountains
- San Ysidro Mountains
- Santa Ana Mountains
- Santa Margarita Mountains
- Santa Rosa Mountains
- Vallecito Mountains

Mountain Peaks Above 14,000 Feet

	Peak	Elevation	1st. Climbed
1	Mount Shasta	14,162	1873
2	White Mountain Peak	14,246	1884
3	Mount Sill	14,162	1903
4	North Palisade	14,244	1903
5	Middle Palisade	14,040	1921
6	Thunderbolt Peak	14,003	1931
7	Split Mountain	14,058	1887
8	Mount Williamson	14,375	1884
9	Mount Whitney	14,494	1873
10	Mount Tyndall	14,018	1864
11	Mount Russell	14,086	1926
12	Mount Muir	14,015	1935
13	Mount Langley	14,027	1871

Note: Three sections of the Sierra Club claim the peaks as their special province of interest. The Sierra Peaks Section lists 246 pinnacles, with 35 special interest mountains and 15 high interest summits as their area of major attention. The Hundred Peaks Section registers 270 mountains, and the Desert Peaks Section lists 96 desert summits with 7 mountains of particular interest.

Mountaineers have agreed to a standard climbing difficulty rating system:
Class 1-Hands in pockets hiking on trails or easy cross-country
Class 2-Rough cross-country travel—boulder hopping and use of hands for balance
Class 3-Handholds necessary for climbing—some may wish for belays because of exposure
Class 4-More difficult climbing with considerable exposure—ropes are used
Class 5-Technical rock climbing
Class 6-Rock climbing with artificial aid

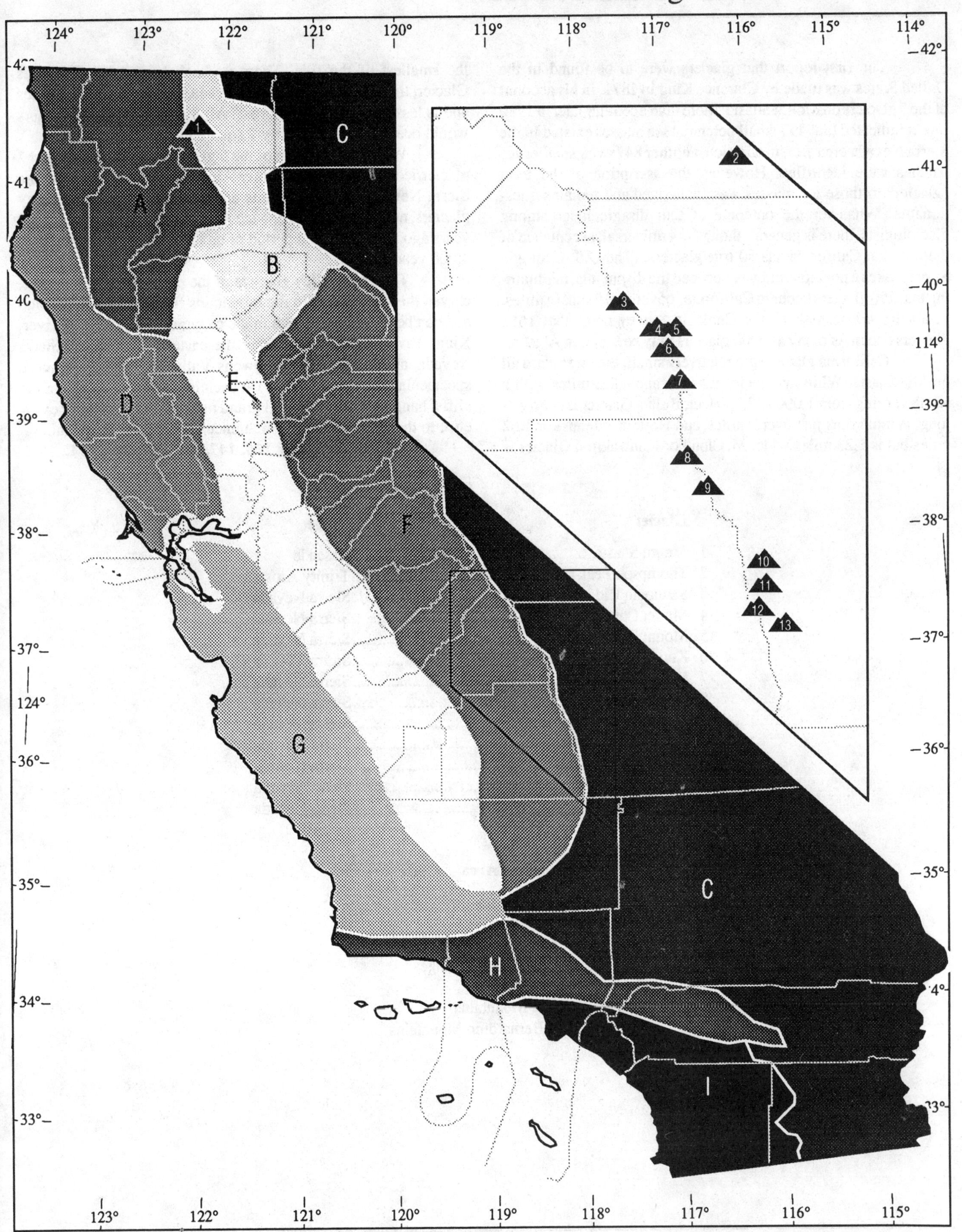
A
B
C
D
E
F
G
H
I
1
2
3
4
5
6
7
8
9
10
11
12
13

The first report that glaciers were to be found in the United States was made by Clarence King in 1871, in his account of the 5 glaciers on Mount Shasta. More than a century later, a 1980 report indicated that 497 small perennial ice masses existed in the Sierra Nevada area and, in addition, another 847 even smaller ice patches were identified. However, the ascription of the term "glacier" to those ice masses was challenged and remains questionable. Whatever the outcome of that disagreement among glaciologists, there is general, though not universal, acceptance of the fact that California has 80 true glaciers. (The U.S. Geologic Survey takes a position midway between the disputants, maintaining that 290 glaciers occur in California, covering 19 square miles. In all the westrern states the Geologic Survey notes that 1651 glaciers occur.As many as 5000 glaciers may be found in Alaska.)

California glaciers are relatively small; the largest are all on Mt. Shasta. Whitney Glacier runs for about 2.2 miles with a width varying from 1,000 to 2,000 feet, Bulam Glacier is nearly as long, Wintun runs just over 2 miles, and Hotlum extends for 1.62 miles but is 1.23 miles wide. McCloud or Konwakiton Glacier is the smallest of the five. They are each larger than Palisades Glacier, the largest in the Sierra Nevada, which runs for 1.5 miles and is less than 1,800 feet across. All 80 of California's glaciers would occupy an area of only 7 square miles.

With the exception of the 5 glaciers on Mt. Shasta, and 2 glacier/ice patches in the Trinity Alps, all are to be found in the Sierra Nevada. These remnants are the vestiges of great valley glaciers, not of the Pleistocene Ice Age, which ended about 11,000 years ago, but rather of a period of an ice advance approximately 4,000 years past.

The great valley glaciers of the earlier period, however, carved their signatures in the mountains from Oregon to Mexico and can be most clearly seen in the canyons of the Merced River, Kings River, Kern River, along the eastern flank of the Sierra Nevada, and in the northern Owens Valley. Nowhere is it more spectacular, however, than in Yosemite, where carved granite cliffs, hanging waterfalls, and grand meadows all owe their existence to the great ice mass which once covered the entire valley.

See: Bibliography 126, 147, 148, 168, 185, 194.

	Glacier	**Range**
1	Mount Shasta	Cascade
2	Thompson Peak and Sawtooth Mountain	Trinity Alps
3	Sawtooth Ridge	Sierra Nevada
4	Mount Conness	Sierra Nevada
5	Mount Dana	Sierra Nevada
6	Kuna Peak and Koip Peak	Sierra Nevada
7	Mount Lyell	Sierra Nevada
8	Ritter Range	Sierra Nevada
9	Mount Abbot	Sierra Nevada
10	Mount Humphreys	Sierra Nevada
11	Glacier Divide	Sierra Nevada
12	The Palisades	Sierra Nevada
13	Mount Goddard	Sierra Nevada

Glaciated Area

A Mount Shasta
B Trinity Alps
C Mount Lassen
D Northern Coast Ranges
E Sierra Nevada
F White Mountains
G San Bernardino Mountains

Existing Glaciers

Ancient Glaciated Areas

A cave is a natural cavity in rock large enough for human entrance. Some speleologists insist that it must contain an area of total darkness to be listed as a true cave. Two basic types of cave exist, those created as the rock itself was being laid down, such as caves in congealed lava flows, and caves created by the removal of materials after the enveloping rock was made, such as limestone caves in which the limestone is dissolved under influences of acid flow. To this second group also belong the shoreline caves near San Diego, in which exposed coasts are subjected to the mechanical erosion of vigorous wave assaults. A rock shelter is a cave whose opening is the largest cross section of the cavity, with a floor area protected by a rock vault roof; sinkhole is a cave without a roof, or a large open ground depression. Fissure caves are formed by the partial roofing over of a large fissure, and talus caves, similar to fissure caves, occur when a narrow gorge is filled with talus such as the caves of the Pinnacles National Monument. Travertine caves, created in manner similar to limestone caves, number less than 5 in California. Cavern is a term synonymous with cave, but it is often used to indicate a large cave or a system of connected caves, such as the California Caverns of Shasta County.

An admittedly incomplete survey in 1962 listed 648 limestone caves in California. Limestone caves, because of their scientific value and aesthetic features, are regarded as most important by speleologists and spelunkers alike, but these caves are outnumbered by lava tubes, sea caves, and the uncounted number of rock shelters in the state. California has 4 distinct areas of limestone caves: 1) Sierra Nevada, 2) Trinity-Shasta Lake, 3) Santa Cruz-Monterey, and 4) Mojave Desert, with the greatest majority of the known limestone caves on the western slope of the Sierra Nevada.

Lava tube caves, more than 500 known, are limited, essentially, to the northeastern corner of the state. It is estimated that the area contains more lava tube caves than all the rest of the continental United States. Sea caves, or littoral caves, exist all along the California shoreline wherever rock and shoreline conditions are conducive to their development, and rock shelters, or sandstone caves, are numerous throughout the state. Vandalism of caves is an acute problem. Consequently, speleologists are reluctant to identify cave locations. The accompanying map shows the major commercial caves of the state.

See: Bibliography 63.

Major Commercial Caves

	Cave	County
1	Del Loma Cave	Trinity
2	Lake Shasta Caverns	Shasta
3	Subway Cave	Shasta
4	Masonic Caves	Amador
5	Cataract Gulch Cave	Calaveras
6	Mercer Caverns	Calaveras
7	Moaning Cavern	Calaveras
8	California Caverns	Tuolumne
9	Boyden Caves	Fresno
10	Bear Gulch Cave	Monterey
11	Crystal Cave	Tulare
12	Caverns of Mystery	San Luis Obispo
13	Mitchells Caverns	San Bernardino
14	La Jolla Caves	San Diego

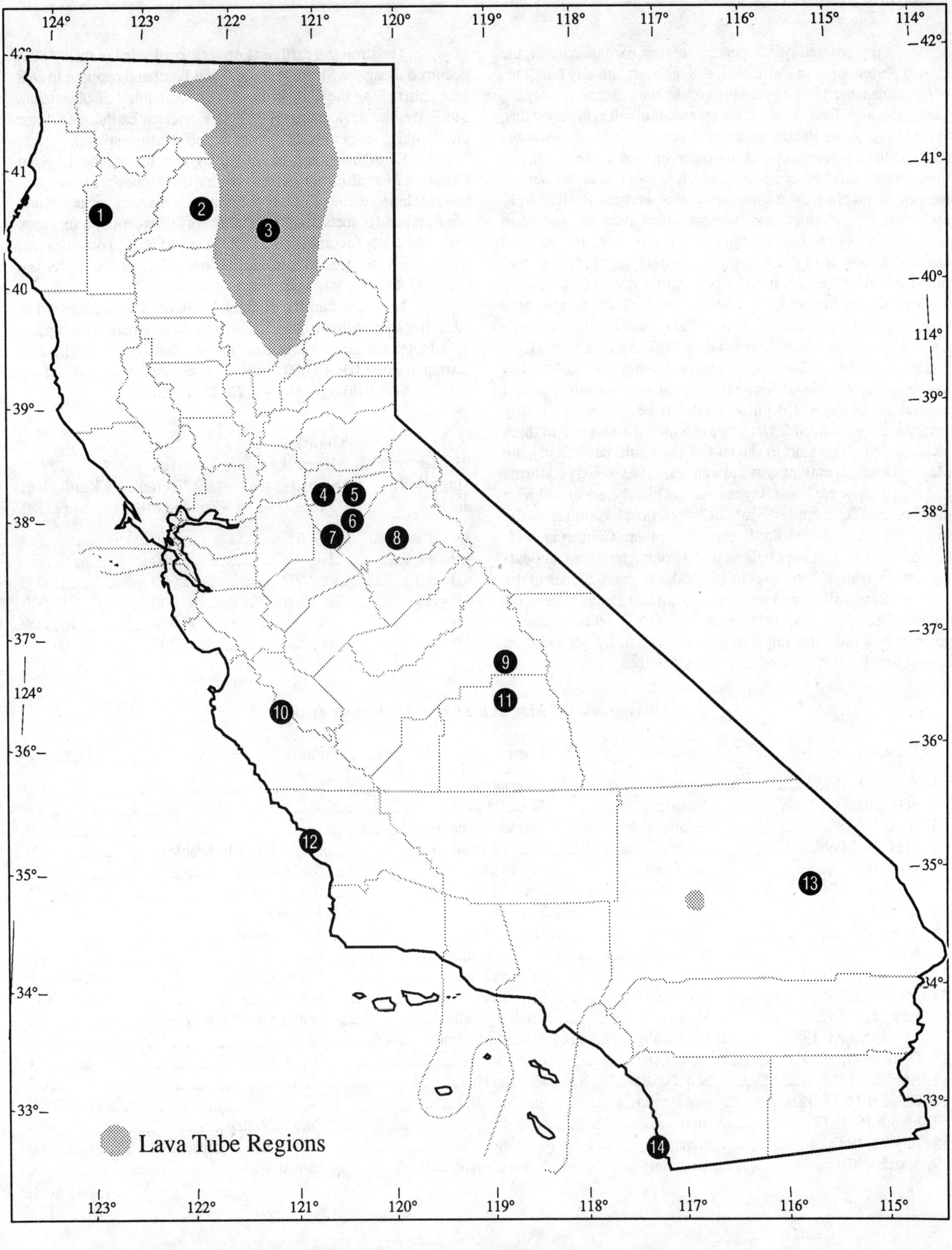

Lava Tube Regions

Approximately 85 percent of the earthquakes in the United States occur in California, and there are literally hundreds of fault lines throughout every area of the state excepting only the Central Valley. This situation, current seismological theory states, is the result of the Pacific Plate on which most of California sits (current theory suggests that the outer surface of the earth, the lithosphere, rests on huge plates, which in turn rest on the molten interior of the planet, the astenosphere) is wedging itself sharply under the North American Plate on which most of the North American continent sits. As the plates move in relation to each other, the outer hard crust responds by buckling, breaking, and fracturing, thus creating fault lines. Fault lines vary from several hundred feet to hundreds of miles in length. Their depths, now being more fully studied, are also of great variance as they run from a few feet to as much as 30 miles deep. The most notorious fault line in the world, the San Andreas Fault, extends over 1,000 miles and descends in an almost vertical line deep into the earth. The land to the west of the San Andreas Fault has been moving steadily northwesterly for millions of years at a rate of about two to three inches a year. The land to the east of the Fault, particularly the Mojave Desert, is moving east and southeasterly. Geology confirms that the rocks of the Point Reyes coast are clearly associated with those of the Tehachapi Mountains, more than 310 miles to the south. In an exactly similar fashion in southern California, rock formations near Taft match those of the Point Arena area, indicating a movement of 350 miles in 150,000,000 years. Shales in the Tembler Range similar to those in the Caliente Range show a 65 mile displacement has occurred in 20,000,000 years. A recent inventory of only the important named faults, by no means a comprehensive list, showed 339 with such titles.

The strongest earthquake ever recorded in the state (1872) occurred along the Sierra Nevada Front Fault and resulted in an 8 foot uplift along the fault and a 20 foot horizontal displacement. Similarly, the devastating 1906 San Francisco earthquake thrust the Point Reyes peninsula almost 20 feet northwestward.

Earthquakes are measured by two scales, the 12 point Modified Mercalli Scale of 1931, based on responses to questionnaires relating to the effects of the quake on humans, property, and life forms and objects. It runs from I: "Not felt except by a very few, under specially favorable circumstances," to XII: "Total damage, waves seen on ground surfaces. Lines of sight and level are distorted. Objects thrown upwards into the air."

The more familiar Richter Magnitude Scale, adopted in 1935, has a logarithmic base in which each increment is multiplied by a factor ten times larger than the previous reading. Thus, an earthquake of 8.0 is 10,000 times more severe than a quake of 4.0

See: Bibliography 46, 73, 110, 145.

Richter Magnitude Scale	Modified Mercalli Intensity Scale	Area Affected (Square Miles)	T.N.T. Equivalent (Tons)
3.0 - 3.9	II - III	750	- -
4.0 - 4.9	IV - V	3,000	- -
5.0 - 5.9	VI - VII	15,000	1,000
6.0 - 6.9	VII - VIII	50,000	10,000
7.0 - 7.9	IX - X	200,000	100,000
8.0 - 8.9	XI - XII	800,000	1,000,000

Earthquakes of Magnitude 6 or over (Richter Scale)

	Date	County	Locale	Fault	Magnitude
1	August 1, 1975	Butte	Palermo	- -	6.0
2	September 12, 1966	Nevada	Boca Reservoir	- -	6.6
3	April 19, 1892	Solano-Yolo	Winters-Vacaville	- -	VIII
4	May 19, 1889	Contra Costa	West of Antioch	Antioch or Mt. Diablo	VIII
5	June, 1838	San Francisco	San Francisco	San Andreas	VIII+
6	April 18, 1906	San Francisco	San Francisco	San Andreas	8.3
7	June 10, 1836	- -	East Bay	Hayward	IX+
8	March 21, 1868	Alameda	Hayward	San Andreas	X
9	May 28, 1980	Mono	East of Mammoth Lakes	Owens Valley	6.1
10	May 25-28, 1980	Mono	Convict Lake	Hilton Creek	6.5(+others)
11	March 8, 1865	San Mateo	Bay Area	San Andreas	VIII+
12	July 21, 1986	Mono	Chalfant Valley	White Mtn. Fault Zone	6.4
13	November 23, 1984	Inyo, Mono	Bishop-Mammoth Lakes	- -	6.1
14	October 17, 1989	Santa Cruz	Loma Prieta	San Andreas	7.1
15	April 24, 1984	Santa Clara	Morgan Hill	Calaveras	6.2
16	October 11-31, 1880	San Benito	San Juan Batista	- -	IX
17	March 26, 1872	Inyo	Lone Pine	Owens Valley	XI+
18	July 22, 1983	Fresno	Coalinga	East of San Andreas	6.0
19	March 1901	Monterey	Stone Canyon	San Andreas	IX

(continued)

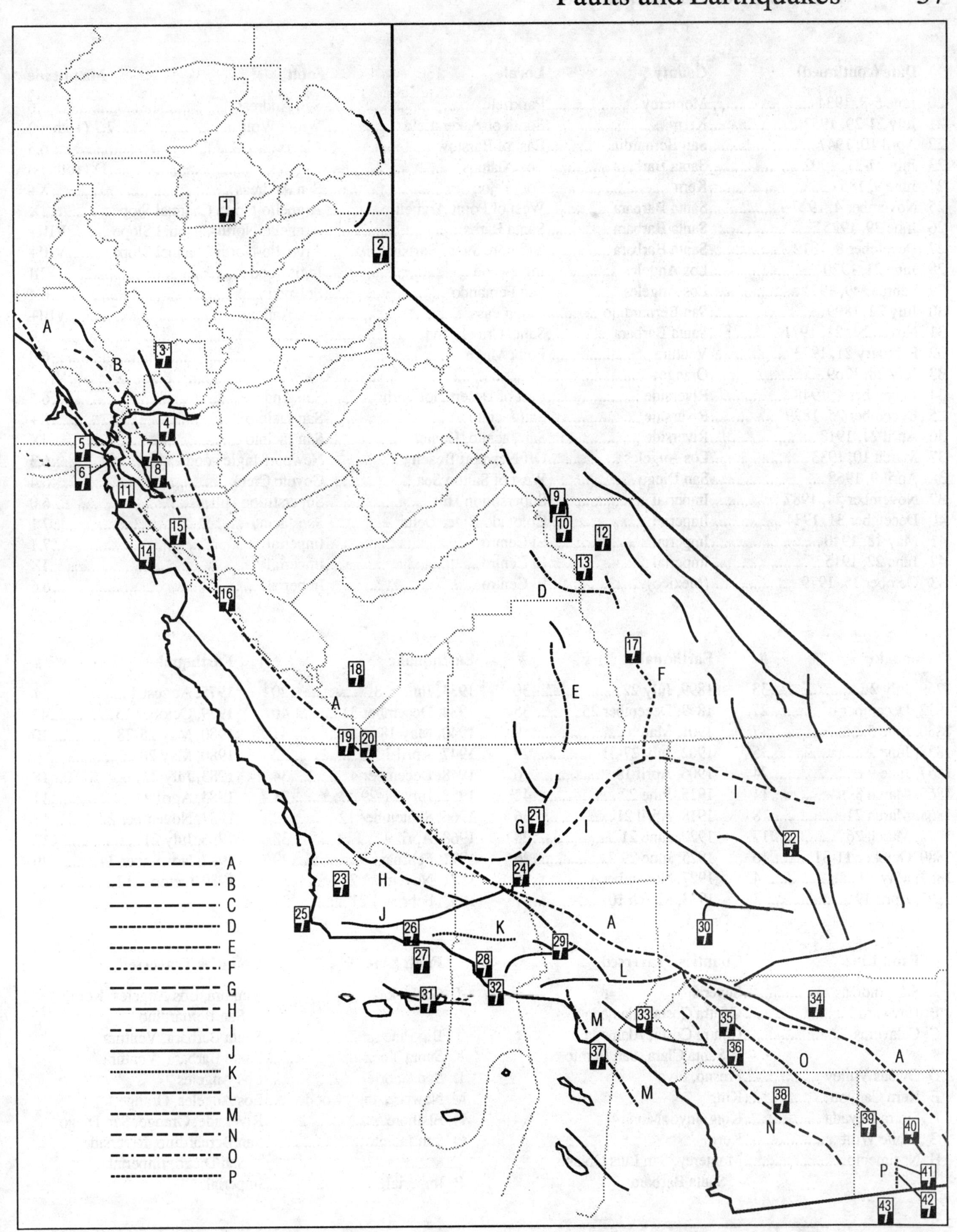
A
B
C
D
E
F
G
H
I
J
K
L
M
N
O
P

	Date (continued)	County	Locale	Fault	Magnitude
20	June 5-8, 1934	Monterey	Parkfield	San Andreas	6
21	July 21-29, 1952	Kern	South of Bakersfield	White Wolf	7.7 (+others)
22	April 10, 1947	San Bernardino	East of Barstow	Manix	6.3
23	July 27-31, 1902	Santa Barbara	Los Alamos	- -	IX(+others)
24	June 9, 1857	Kern	Fort Tejon	San Andreas	X+
25	November 4, 1927	Santa Barbara	West of Point Arguello	Arguello-North Channel Slope	X
26	June 29, 1925	Santa Barbara	Santa Barbara	Arguella-North Channel Slope	VIII+
27	December 8, 1812	Santa Barbara	Offshore Santa Barbara	Arguello-North Channel Slope	VIII+
28	June 21, 1920	Los Angeles	Inglewood	Inglewood	VIII
29	February 9, 1971	Los Angeles	San Fernando	Soledad	6.4
30	July 22, 1899	San Bernardino	Cajon Pass	San Andreas	VIII+
31	November 21, 1971	Santa Barbara	Santa Cruz Island	- -	7.1
32	February 21, 1973	Ventura	Point Mugu	- -	6.0
33	July 28, 1769	Orange	- -	- -	VIII+
34	December 4, 1948	Riverside	East of Desert Hot Springs	San Andreas	6.5
35	December 25, 1899	Riverside	San Jacinto	San Jacinto	IX+
36	April 21, 1918	Riverside	San Jacinto, Hemet	San Jacinto	IX
37	March 10, 1933	Los Angeles	Off Newport Beach	Newport-Inglewood	6.3
38	April 9, 1968	San Diego	West of Salton Sea	Coyote Creek	6.4
39	November 24, 1987	Imperial	Superstition Hill	Superstition Hill	6.0
40	December 31, 1934	Imperial	Colorado River Delta	San Jacinto	7.1
41	May 18, 1940	Imperial	El Centro	Imperial	7.1
42	June 22, 1915	Imperial	El Centro	Imperial	IX
43	October 15, 1979	(Mexico)	El Centro	Imperial	6.6

Earthquake	#
1769, July 2	33
1812, December 8	27
1836, June 5	7
1838, June	5
1857, June 9	24
1865, March 8	11
1868, March 21	8
1872, March 26	17
1880, October 11-31	16
1889, May 19	4
1892, April 19	3
1899, July 22	30
1899, December 25	35
1901, March	19
1902, July 27-31	23
1906, April 18	6
1915, June 22	42
1918, April 21	36
1920, June 21	28
1925, June 29	26
1927, November 4	25
1933, March 10	37
1934, June 5-8	20
1934, December 31	40
1940, May 18	41
1947, April 10	22
1948, December 4	34
1952, July 21-29	21
1966, September 12	2
1968, April 9	38
1971, February 9	29
1971, November 21	31
1973, February 21	32
1975, August 1	1
1979, October 15	43
1980, May 25-28	10
1980, May 28	9
1983, July 22	18
1984, April 24	15
1984, November 22	13
1986, July 21	12
1987, November 24	39
1989, October 17	14

	Fault Line	Counties Traversed
A	San Andreas	Several
B	Hayward	Contra Costa, Alameda
C	Calaveras	Contra Costa, Alameda, Santa Clara, San Benito
D	Owens Valley	Fresno, Inyo
E	Kern Canyon	King
F	Sierra Nevada	King, Inyo, Mono
G	White Wolf	Kern
H	Nacimento	Monterey, San Luis Obispo, Santa Barbara
I	Garlock	Ventura, Los Angeles, Kern, San Bernardino
J	Big Pine	Santa Barbara, Ventura
K	Santa Ynez	Santa Barbara, Ventura
L	San Gabriel	Los Angeles
M	Newport-Inglewood	Los Angeles, Orange
N	Elsinore	Riverside, Orange, San Diego
O	San Jacinto	San Bernardino, Riverside, San Diego, Imperial
P	Imperial	Imperial

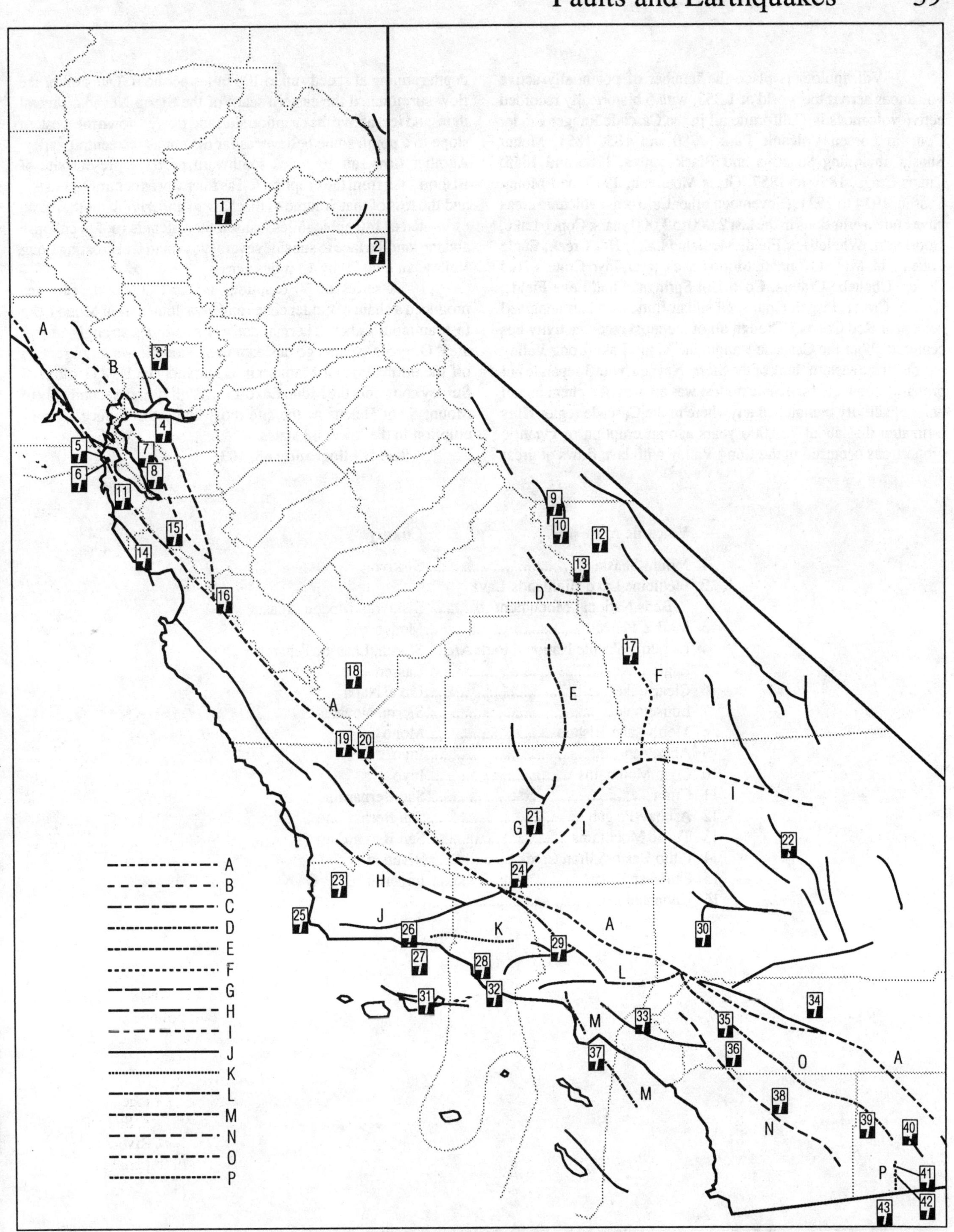
1
2
A
3
B
4
5
7
6
8
11
15
14
16
9
10
12
13
D
17
F
18
E
A
19
20
I
21
G
I
22
23
H
24
25
J
26
K
A
30
27
29
28
L
31
32
34
33
35
M
37
36
O
A
M
38
N
39
40
P
41
42
43
A
B
C
D
E
F
G
H
I
J
K
L
M
N
O
P

Volcanologists place the number of potentially active volcanoes across the world at 1,353, with 5 historically recorded active volcanoes in California, all in the Cascade Range: Cinder Cone in Lassen Volcanic Park, 1780 and 1850-1851; Mount Shasta, including Shastina and Black Buttes, 1786 and 1855; Chaos Crags, 1854 to 1857; Glass Mountain, 1910; and Mount Lassen, 1914 to 1921. (Seventeen other California volcanic areas have known eruptions in the last 2,000 to 3,000 years: Copco Lake, Goosenest, Whaleback Field, Medicine Lake, Hot Creek, Eagle Lake Field, Mount Konocti, Mono Lake Crater, Inyo Craters, Red Cones, Ubehebe Craters, Coso Hot Springs, Cima Lava Fields, Amboy Crater, Pisgah Crater, Obsidian Buttes, and an unnamed peak near Red Cones.) Though all of the most recent activity has centered about the Cascade Range, the Mono Lake-Long Valley region on the eastern flank of the Sierra Nevada, with 3 separate but probably connected volcanic nodes, was an area of a phenomenal burst of activity unmatched anywhere in the Cascade region. It is estimated that about 700,000 years ago an eruption of gigantic proportions occurred in the Long Valley with lava flows of great depth running at speeds up to 100 miles per hour. One arm of the flow surmounted the eastern scarp of the Sierra Nevada, several thousand feet above the eruption site, and flowed down the western slope to a point, some believe, as far distant as the central valley. Another flow ran 50 miles southward past the present site of Bishop. Ash from the eruption fell as far away as central Nebraska, and the loss of that volume of discharged underlying molten rock, estimated at 140 cubic miles, caused the volcanic roof to collapse and the land surface to subside several thousand feet creating Long Valley, an oval 20 by 15 miles across.

A series of 30 eruptions in the last 2,000 years has produced a chain of cinder cones and lava domes from Mono Lake to Mammoth Lakes. Thermal activity continues strongly in that area. Only 500 years ago, a steam explosion blew a large section off the north slope of Mammoth Mountain. The U.S. Geological Survey considers the Mono Lake-Long Valley area second only to Mount Saint Helens as the site most likely to produce a major eruption in the lower 48 states.

See: Bibliography 65, 103.

	Volcanic Area	County
1	Mount Shasta	Siskiyou
2	Medicine Lake Highlands-Lava Beds National Monument	Siskiyou, Modoc, Shasta
3	Modoc Plateau	Modoc
4	Lassen Volcanic National Park Area	Shasta, Lassen, Tehama
5	Eagle Lake	Lassen
6	Clear Lake	Lake, Napa
7	Lousetown	Sierra, Nevada
8	Mono Lake-Bishop	Mono
9	Aberdeen	Inyo
10	Coso Mountains	Inyo
11	Cima	San Bernardino
12	Arubay-Pisgah	San Bernardino
13	Turtle Mountains	San Bernardino
14	Pinto Basin-Salton Creek	Riverside
15	Obsidian Buttes	Imperial
16	Unnamed	- -

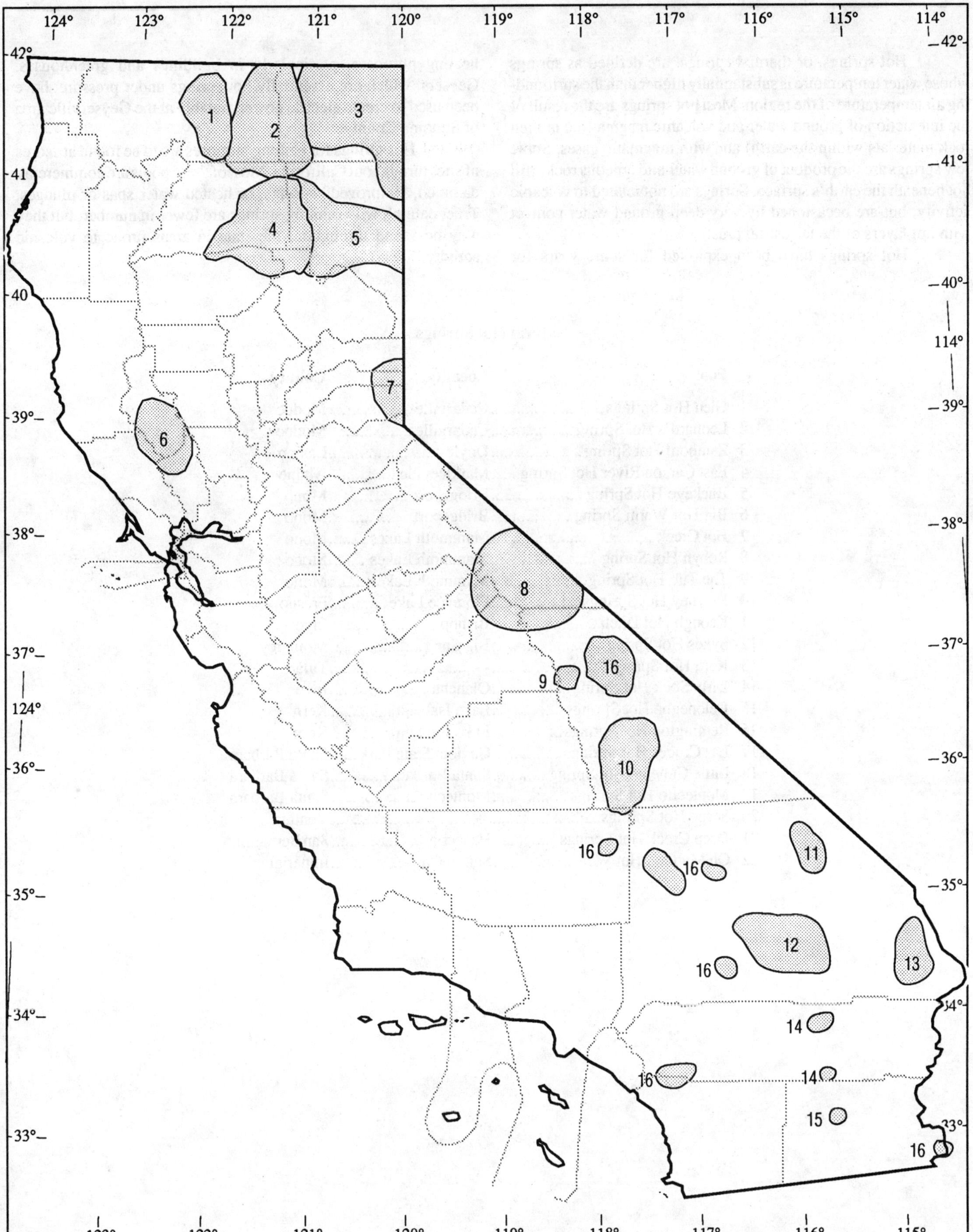
124°
123°
122°
121°
120°
119°
118°
117°
116°
115°
114°
42°
41°
40
39°
38°
37°
36°
35°
34°
33°
1
2
3
4
5
6
7
8
9
10
11
12
13
14
14
15
16
16
16
16
16
16
16

Hot springs, or thermal springs, are defined as springs whose water temperature is substantially higher than the surrounding air temperature of the region. Most hot springs are the result of the interaction of ground water and volcanic magma (the molten rock materials within the earth) and with magmatic gases. Some few springs are the product of ground water and igneous rock, still hot beneath the earth's surface. Springs are not related to volcanic activity, but are occasioned by very deep ground water contact with hot layers of the terrestrial crust.

Hot springs have been exploited for some years for heating purposes, particularly in buildings and greenhouses. Geysers, which are essentially hot springs under pressure, have been used to create electric power, notably at the Geyserville area of Sonoma County.

Hot springs for bathing purposes are to be found at scores of sites throughout California. Most of these pools are commercial, dammed, "improved" springs, or heated water spas or plunges. True, natural, wild running springs are fewer in number, but they may be found throughout the state in areas prone to volcanic activity.

Natural Hot Springs

	Pool	Locality	County
1	Glen Hot Springs	Cedarville	Modoc
2	Leonard's Hot Spring	Cedarville	Modoc
3	Zamboni Hot Spring	Doyle	Lassen
4	East Carson River Hot Spring	Markleeville	Alpine
5	Buckeye Hot Spring	Bridgeport	Mono
6	Big Hot Warm Spring	Bridgeport	Mono
7	Hot Creek	Mammoth Lakes	Mono
8	Robyn Hot Spring	Mammoth Lakes	Mono
9	The Tub Hot Spring	Mammoth Lakes	Mono
10	Blayney Hot Springs	Florence Lake	Fresno
11	Keough Hot Ditch	Bishop	Inyo
12	Sykes Hot Spring	Big Sur	Monterey
13	Kern Hot Spring	- -	Tulare
14	Dirty Sock Hot Spring	Olancha	Inyo
15	Delonegha Hot Springs	Lake Isabella	Kern
16	Remington Hot Springs	Lake Isabella	Kern
17	Las Cruces Hot Springs	Gaviota State Park	Santa Barbara
18	Little Caliente Hot Spring	Santa Barbara	Santa Barbara
19	Montecito Hot Springs	Montecito	Santa Barbara
20	Sespe Hot Springs	- -	Ventura
21	Deep Creek Hot Springs	Hesperia	San Bernardino
22	Oh My God Hot Well	Salton City	Imperial

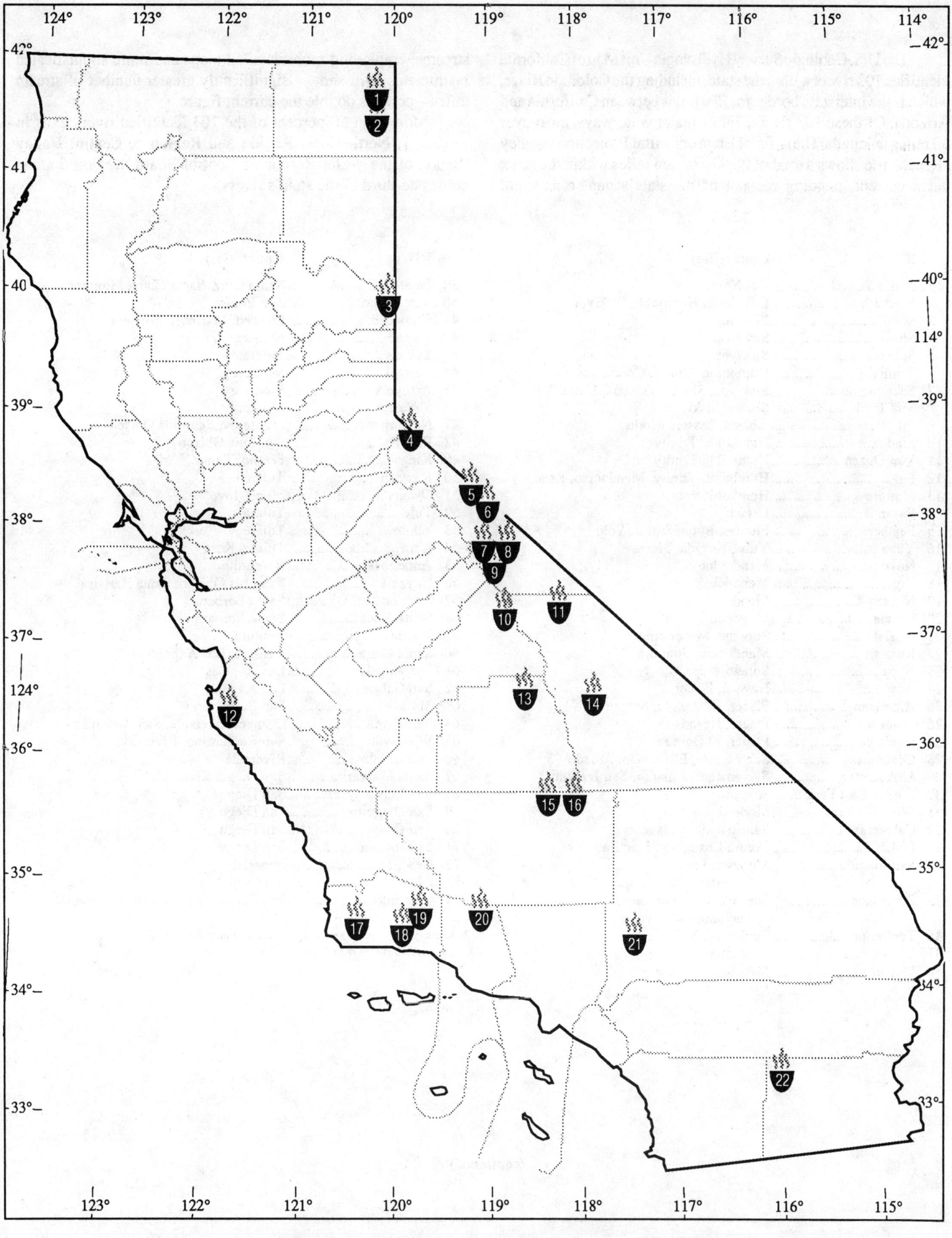
124°
123°
122°
121°
120°
119°
118°
117°
116°
115°
114°
42°
41°
40°
39°
38°
37°
36°
35°
34°
33°
1
2
3
4
5
6
7
8
9
10
11
12
13
14
15
16
17
18
19
20
21
22

The U.S. Geologic Survey Hydrologic Unit Map of California identifies 103 rivers within the state, including the Colorado River, which is the interstate border for 230 miles between California and Arizona. Of these 103 rivers, 74 are major waterways, most over 30 miles in length. The official Environmental Protection Agency "Reach" file shows a total of 26,970 stream miles within the state, but a current, ongoing recount of the state's numerous small streams, employing more detailed maps and more sophisticated instruments, will show a significantly greater number of stream miles—perhaps double the current figure.

More than 50 percent of the 103 identified rivers exist in Region 1, North Coast Region and Region 5, Central Valley Region of the Water Resources Control Board. Region 5 alone holds one-third of the state's rivers.

	River	County(ies)[1]
1	Smith	Del Norte
2	Klamath	Del Norte, Humboldt, Siskiyou
3	Scott	Siskiyou
4	Shasta	Siskiyou
5	Salmon	Siskiyou
6	Trinity	Humboldt, Trinity, Colusa
7	Sacramento	Siskiyou, Shasta, Tehama, Butte, Yolo
8	McCloud	Shasta, Siskiyou
9	Pit	Shasta, Lassen, Modoc
10	Mad	Humboldt, Trinity
11	Van Duzen	Humboldt, Trinity
12	Eel	Humboldt, Trinity, Mendocino, Lake
13	Mattole	Humboldt
14	Susan	Lassen
15	Feather	Plumas, Butte, Sutter, Yolo
16	Yuba	Yuba, Nevada, Sierra
17	Noyo	Mendocino
18	Big	Mendocino
19	Navarro	Mendocino
20	Garcia	Mendocino
21	Gualala	Sonoma, Mendocino
22	Russian	Mendocino, Sonoma
23	Napa	Solano, Napa
24	Bear	Nevada, Placer
25	American	Placer, El Dorado, Sacramento
26	Truckee	Placer, Nevada
27	Rubicon	Placer, El Dorado
28	Cosumnes	Sacramento, El Dorado, Amador
29	Mokelumne	Sacramento, Amador, San Joaquin
30	Carson East Fork	Alpine
31	West Walker	Mono
32	Calaveras	San Joaquin, Calaveras
33	Old	Contra Costa, San Joaquin
34	San Joaquin	Modesto, Fresno, Stanislaus, San Joaquin
35	Stanislaus	San Joaquin, Stanislaus, Tuolomne, Calaveras
36	Tuolomne	Tuolomne, Stanislaus
37	Clavey	Tuolomne
38	Merced	Merced, Mariposa

	River	County(ies)
39	Pajaro	Santa Cruz, Santa Clara, Monterey
40	San Benito	San Benito
41	Chowchilla	Merced, Mariposa, Madera
42	Fresno	Madera
43	Salinas	Monterey
44	Carmel	Monterey
45	Arroyo Seco	Monterey
46	San Antonio	Monterey
47	Nacimiento	Monterey, San Luis Obispo
48	Estrella	San Luis Obispo
49	Kings	Fresno, Tulare, Kings
50	Kaweah	Tulare
51	Owens	Mono, Inyo
52	Tule	Tulare
53	White	Tulare
54	Kern	Tulare, Kern
55	Amargosa	Bernardino
56	Cuyama	San Luis Obispo, Santa Barbara
57	Sisquoc	Santa Barbara
58	Santa Ynez	Santa Barbara
59	Ventura	Ventura
60	Santa Clara	Ventura, Los Angeles
61	Los Angeles	Los Angeles
62	San Gabriel	Los Angeles
63	Mojave	San Bernardino
64	Santa Ana	Orange, Riverside, San Bernardino
65	Whitewater	San Bernardino, Riverside
66	San Jacinto	Riverside
67	Santa Margarita	San Diego, Riverside
68	San Luis Rey	San Diego
69	San Dieguito	San Diego
70	San Diego	San Diego
71	Sweetwater	San Diego
72	New	Imperial
73	Alamo	Imperial
74	Colorado	San Bernardino, Riverside, Imperial

1- In many instances, the river acts as the boundary line between 2 counties. In such a case, both counties are shown.

(continued)

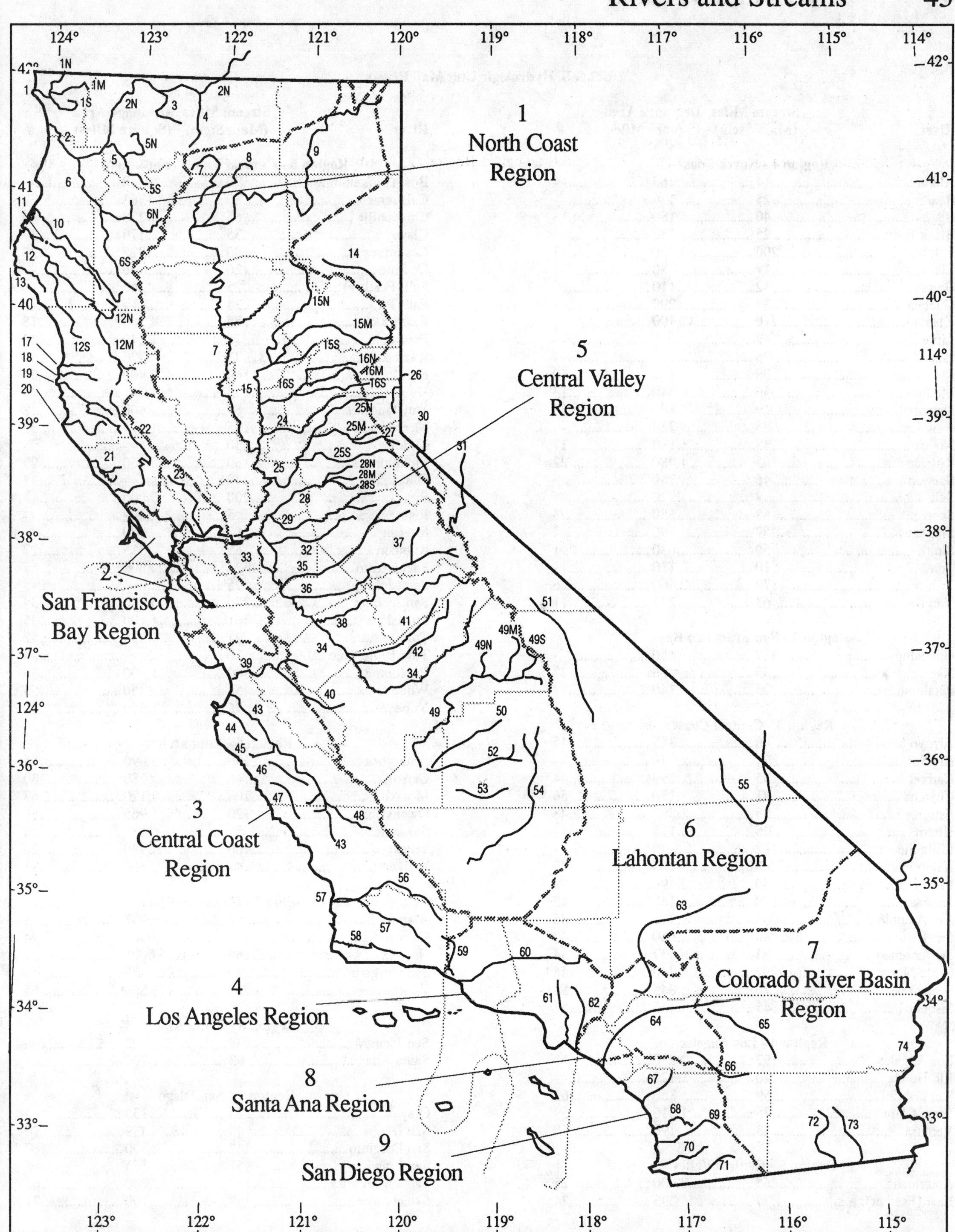
1
North Coast Region
2
San Francisco Bay Region
3
Central Coast Region
4
Los Angeles Region
5
Central Valley Region
6
Lahontan Region
7
Colorado River Basin Region
8
Santa Ana Region
9
San Diego Region

U.S.G.S. Hydrologic Unit Map Rivers

River	Stream Miles (Main Stem)	Drainage Area (Square Miles)	#
Region 1 - North Coast			
Albion	14	65	- -
Bear	25	120	- -
Big	40	180	18
Black Butte	25	130	- -
Eel	200	3,120	12
Elk	17	80	- -
Garcia	32	110	20
Gualala	35	290	21
Klamath	210	12,100	2
Little	17	40	- -
Lost	26	- -	- -
Mad	90	490	10
Mattole	56	340	13
Navarro	19	300	19
New	25	220	- -
Noyo	35	130	17
Russian	105	1,480	22
Salmon	46	750	5
Salt	8	20	- -
Scott	68	650	3
Shasta	52	790	4
Smith	50	630	1
Ten Mile	10	110	- -
Trinity	170	2,860	6
Van Duzen	63	275	11
Region 2 - San Francisco Bay			
Guadalupe	12?	150	- -
Napa	55	426	23
Petaluma	25	140	- -
Region 3 - Central Coast			
Arroyo Seco	40	385	45
Big Sur	21	70	- -
Carmel	35	250	44
Cuyama	91	1,130	56
Estrella	55	800	48
Huasna	25	115	- -
Little Sur	12	45	- -
Nacimiento	65	325	47
Pajaro	40	1,190	39
Salinas	180	4,160	43
San Antonio	60	310	46
San Benito	80	540	40
San Lorenzo	25	137	34
Santa Maria	20?	1,740	46
Santa Ynez	70	845	58
Sisquoc	45	445	57
Region 4 - Los Angeles			
Los Angeles	97	830	61
Rio Honda	20	125	- -
San Gabriel	59	350	62
Santa Clara	75	1,616	59
Ventura	33	190	60
Region 5 - Central Valley			
American	265	2,000	25
Bear (Feather)	77	295	24

River	Stream Miles (Main Stem)	Drainage Area (Square Miles)	#
Region 5 - Central Valley (continued)			
Bear (Mokelumne)	20	60	- -
Calaveras	80	365	32
Chowchilla	65	250	41
Clavey	35	170	37
Cosumnes	80	725	28
Downie	20	40	- -
Fall (Feather)	25	40	- -
Fall (Pit)	25	- -	- -
Feather	175	4,580	15
Fresno	75	240	42
Kaweah	77	720	50
Kern	164	2,400	54
Kings	133	1,745	49
McCloud	60	600	8
Merced	135	1,275	38
Middle	30	- -	- -
Mokelumne	160	660	29
Old	48	- -	33
Pit	200	5,000	9
Rising	5	- -	- -
Roaring	17	80	- -
Rubicon	65	315	27
Sacramento	327	24,000	7
Saint Johns	25	- -	- -
San Joaquin	330	13,540	34
Stanislaus	161	1,100	35
Tule	91	395	52
Tule-Little Tule	10	- -	- -
Tuolumne	149	1,900	36
White	55	850	53
Yuba	96	1,350	16
Region 6 - Lahontan			
Amargosa	198	3,090	55
Carson	46	280	30
Mojave	100	2,120	63
Owens	120	1,965	51
Susan	59	185	14
Truckee	60	930	26
Walker	47	360	31
Region 7 - Colorado River			
Alamo	52	695*	73
Colorado	230	3,950*	74
New	60	1,000*	72
San Gorgonio	30	155	- -
Whitewater	25?	1,500	65
Region 8 - Santa Ana			
San Jacinto	38	725	66
Santa Ana	93	1,700	64
Region 9 - San Diego			
Otay	25	135	- -
San Diego	45	439	70
San Dieguito	11?	300	69
San Luis Rey	51	575	68
Santa Margarita	11?	740	67
Sweetwater	11?	190	71

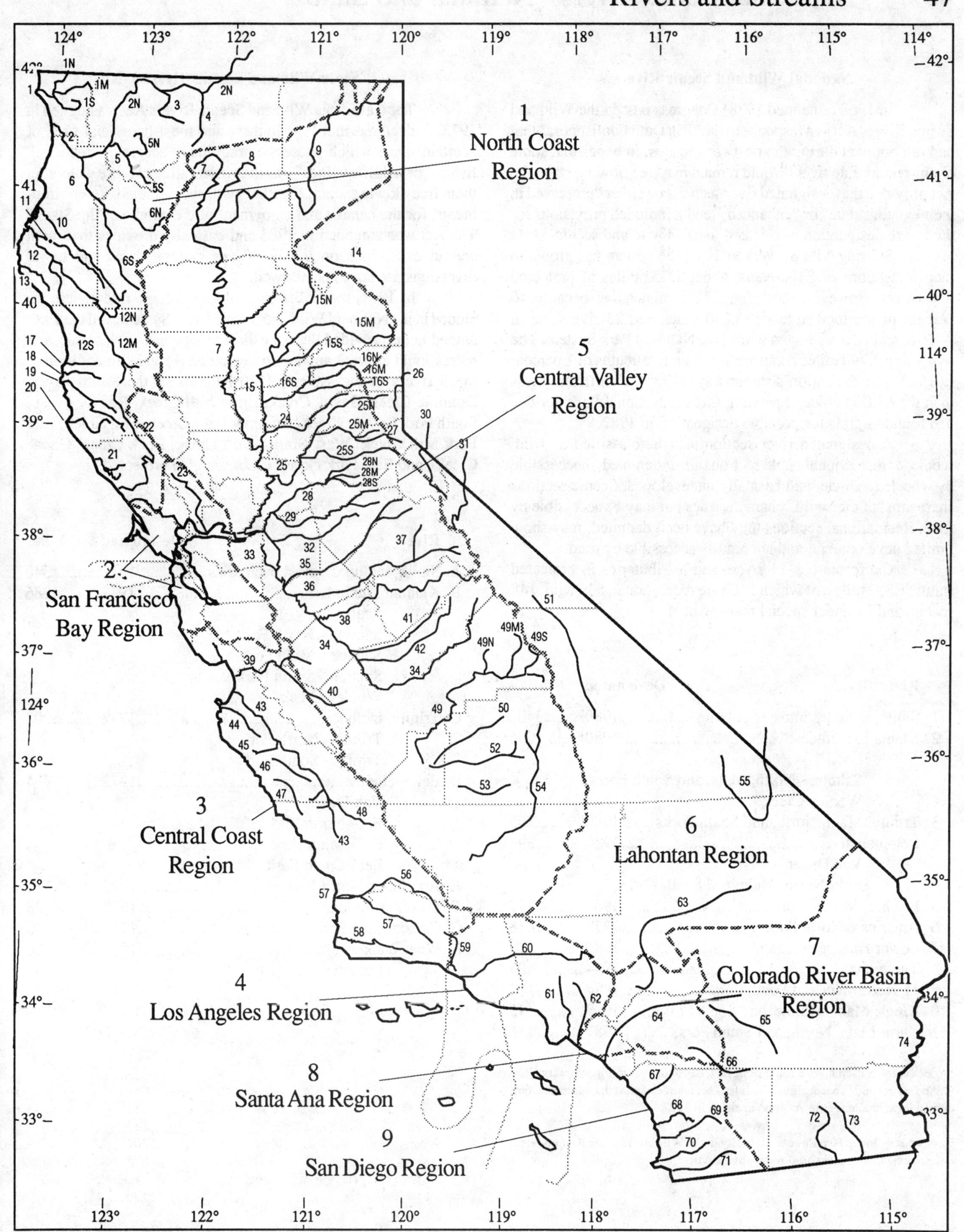
1
North Coast
Region
2
San Francisco
Bay Region
3
Central Coast
Region
4
Los Angeles Region
5
Central Valley
Region
6
Lahontan Region
7
Colorado River Basin
Region
8
Santa Ana Region
9
San Diego Region

National Wild and Scenic Rivers

In 1968 (amended 1978) Congress passed the Wild and Scenic Rivers Act as a response to the "Big Dam Foolishness" that had run rampant during the postwar decades. In hopes that some of our remarkable rivers would remain in a free flowing state, the Act provided that designated river sections be either "preserved in their existing state" or "enhanced," and it included provisions for the future designation of selected units as wild and scenic.

Starting with 8 rivers in 1968, the system has grown to include sections of 83 streams, about 8,000 miles of protected waterways. However, those figures are misleading because 40 percent of the total mileage, 3,210 miles and 25 rivers, lie in Alaska, and 18 are situated within the National Park System. The remaining 4,790 miles comprise less than two-tenths of 1 percent (.0157 %) of the nation's streamways (1989), significantly less than the 61,000 miles, 2 percent, that the National Park Service had found eligible for possible designation in 1982.

A designated river section may be classified as wild, scenic, or recreational. Wild sections are undammed, inaccessible by wheeled vehicle, and basically undeveloped. Scenic sections share most of the "wild" characteristics but may be accessible by road. Recreational sections may have been dammed, may show limited development, and are readily accessible by road.

California has 11 rivers and 5 tributaries in protected status, 1,803 miles, of which 1108 are recreational, 536 wild, 148 scenic, and 11 under special management.

	River	Designated	Miles
1	Smith, includes: most tributaries	1980[1]	3402
2	Klamath, includes:	1980[1]	286
	Scott		
	Salmon - North, Main, and South Forks		
	Wooley Creek		
3	Trinity: Main, North, and South Forks	1980[1]	203
4	Eel, includes:	1980[1]	394
	Van Duzen		
	Eel - North, Main, and South Forks		
5	Feather: Middle Fork	1968	77
6	American: North Fork	1979	38
7	Lower American	1980[1]	23
8	Tuolomne	1984	83
9	Merced: Main and South Forks	1987	114
10	Kings: Main, Middle, and South Forks	1987	95[2]
11	Kern: Main, North, and South Forks	1987	151

[1] Added to the National Wild And Scenic River System by Interior Secretary Cecil C. Andrus in 1980. That designation did not become final until the Supreme Court denied legal challenges to Andrus' decision in 1982.

[2] Does not includes 10 miles not officially in the Wild And Scenic River System but protected from dams amd water diversions.

State Wild and Scenic Rivers

The California Wild and Scenic Rivers Act was signed in 1972. It declared that "the highest and most beneficial use" of "certain rivers which possess extraordinary scenic, recreational, fishery, or wildlife values" was the preservation of those waters "in their free-flowing state, together with their immediate environments, for the benefit and enjoyment of the people of the State." The Act was amended in 1982 and basically provides that, with certain exceptions on the Eel River system, dams and major diversions are not to be allowed.

In 1981, the California wild and scenic rivers were included in the National Wild and Scenic River System, and the 1982 amendments provided the same three categories: wild, scenic, and recreational as were applied to the national rivers. In addition to the 5 rivers in the system, 12 tributaries of the Smith River: Dominie Creek, Savoy Creek, Little Mill Creek, Rowdy Creek, South Fork Rowdy Creek, Brimmer Lake Creek, Mill Creek, East Fork Mill Creek, West Branch Mill Creek, Rock Creek, Goose Creek , and East Fork Goose Creek are also protected.

	River	Designated	Miles
A	Smith, includes: most tributaries	1972	340
B	Klamath, includes:	1972	286
	Scott		
	Salmon - North Fork		
	Salmon - Main Fork		
	Salmon - South Fork		
	Wooley Creek		
C	Trinity, includes:	1972	203
	Trinity - North Fork		
	Trinity - South Fork		
D	Eel, includes:	1972	394
	Van Duzen		
	Eel - North Fork		
	Eel - Main Fork		
	Eel - South Fork		
E	American: North Fork	1972	38
F	Lower American	1972	23
G	Carson: East Fork	1989	10
H	West Walker, Includes:	1989	38
	Leavitt Creek (1 mile)		

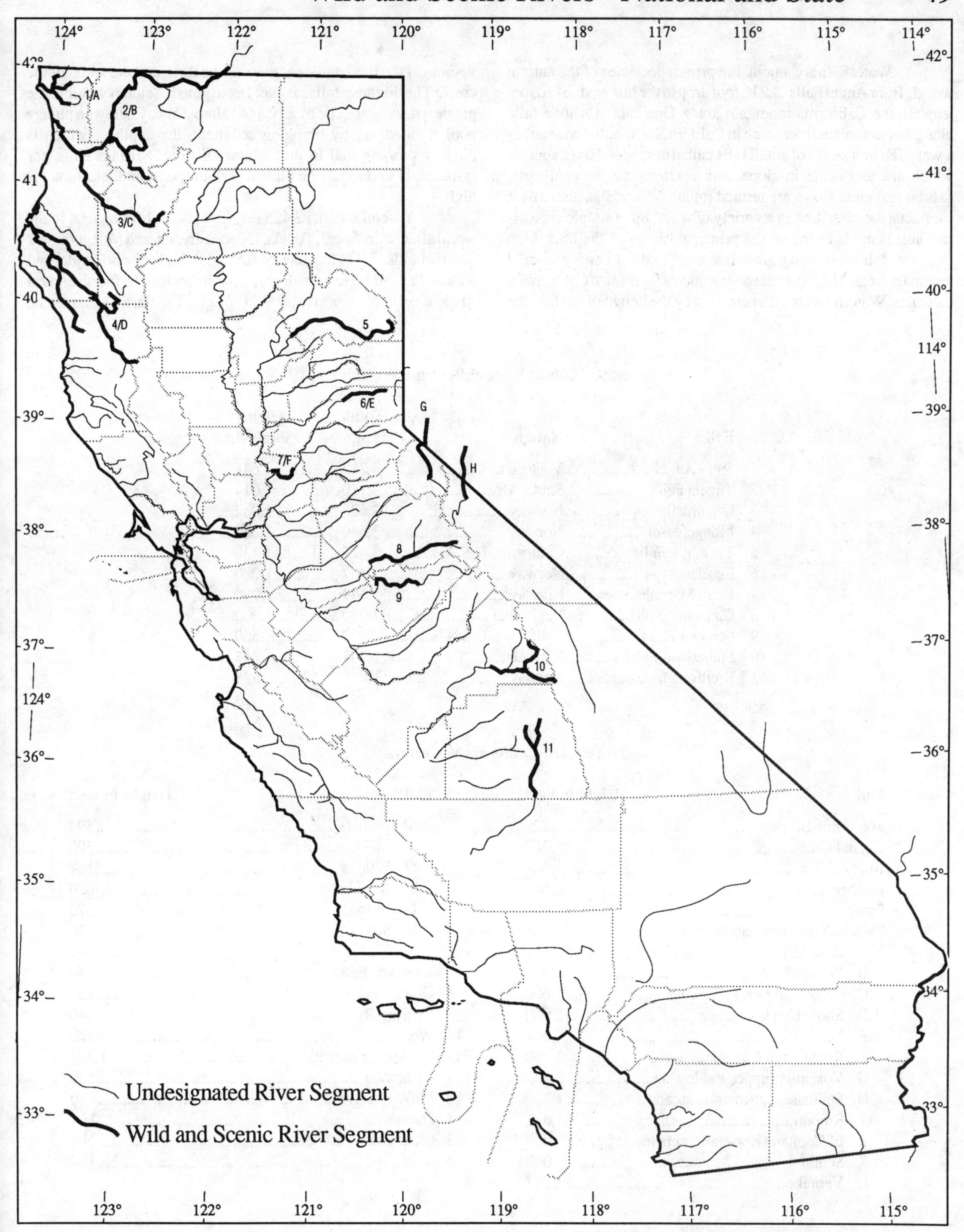
1/A
2/B
3/C
4/D
5
6/E
G
H
7/F
8
9
10
11
Undesignated River Segment
Wild and Scenic River Segment

Waterfalls are among the prime spectacles of the natural world, from Angel Falls' 3,212 foot drop to the hundreds of 10 foot leaps in the California mountain areas. One cannot walk a mile along any mountain river reach in California without encountering a waterfall or a series of small falls called cascades. River courses which are less acute in slope than cascades but, nevertheless, exhibit turbulent flows are termed rapids. Waterfalls, also called cataracts, are classified in a variety of ways, but a simple classification scheme is based on the principal region of the falls: high plateaus, fall lines along great fractures, and formerly glaciated mountain areas. The latter category accounts for most of California's cataracts. Waterfalls are characterized by the height of the fall, the volume of the flow, the steadiness of the flow, and the width of the crest. The highest falls, it has been noted, seldom exhibit the greatest discharge. The base of a fall almost always shows a plunge pool scoured out by plunging sediment, the depth of which is directly proportional to the volume of flow. Niagara Falls, for instance, is said to have a plunge pool deeper than the cataract is high.

Yosemite Falls, 2,425 feet high, is the fifth highest in the world after Salto Angel, Tugela, Utigordsfossen, and Mongefossen. Sentinel falls, 2,000 feet, is tied for eighth highest worldwide, and Ribbon Falls' 1600 feet single leap establishes it as the fourth highest single drop in the world after Salto Angel, Tugela, and Cuquenan.

Note: Highest Waterfalls of the World

	Fall	Nation	Total Height	Highest Single Fall
1	Salto Angel	Venezuela	3,212	2,648
2	Tugela Falls	South Africa	3,110	2,014
3	Utigordsfossen	Norway	2,625	1,062
4	Mongefossen	Norway	2,540	1,251
5	Yosemite Falls	California, U.S.A.	2,425	1,430
6	Espelandsfoss	Norway	2,307	1,080
7	Østre Mardalsfossen	Norway	2,149	990
8	Cuquenan Falls	Venezuela	2,000	2,000
9	Sentinel Falls	California, U.S.A.	2,000	500
10	Sutherland Falls	New Zeland	1,904	815
11	Kjellfossen	Norway	1,841	828

Selected California Waterfalls

	Fall	Height in Feet
1	McArthur-Burney	129
2	Crystal Creek	200
3	Brady Creek	NA
4	Feather	640
5	Eagle	NA
6	Falls in Yosemite National Park	
A	Rancheria	NA
B	Waterwheel	NA
C	Tenaya (Pyweak), cascade	600
D	Snow Creek, cascade	2,000
E	Lehamite	NA
F	Royal Arch Cascade	1,250
G	Yosemite (upper and lower)	2,425
H	Staircase, ephemeral cascade	1,300
I	Ribbon	1,612
J	El Capitan (Horsetail), ephemeral	1,000
K	Sentinel	2,000
L	Vernal	317
M	Nevada	594
N	The Cascades	500
O	Wildcat	NA
P	Bridalveil	620
Q	Illilouette	370
R	Silver Strand	1,170
S	Chilnualna	NA
7	Rainbow Falls	140
8	Grizzly Falls	100
9	Roaring River	40
10	West	100
11	Tokopah, cascade	1,200
12	Chagoopa	500
13	McWay Cove	70
14	Darwin	NA
15	Clear Lake Cascade	NA
16	Nojoqui	164

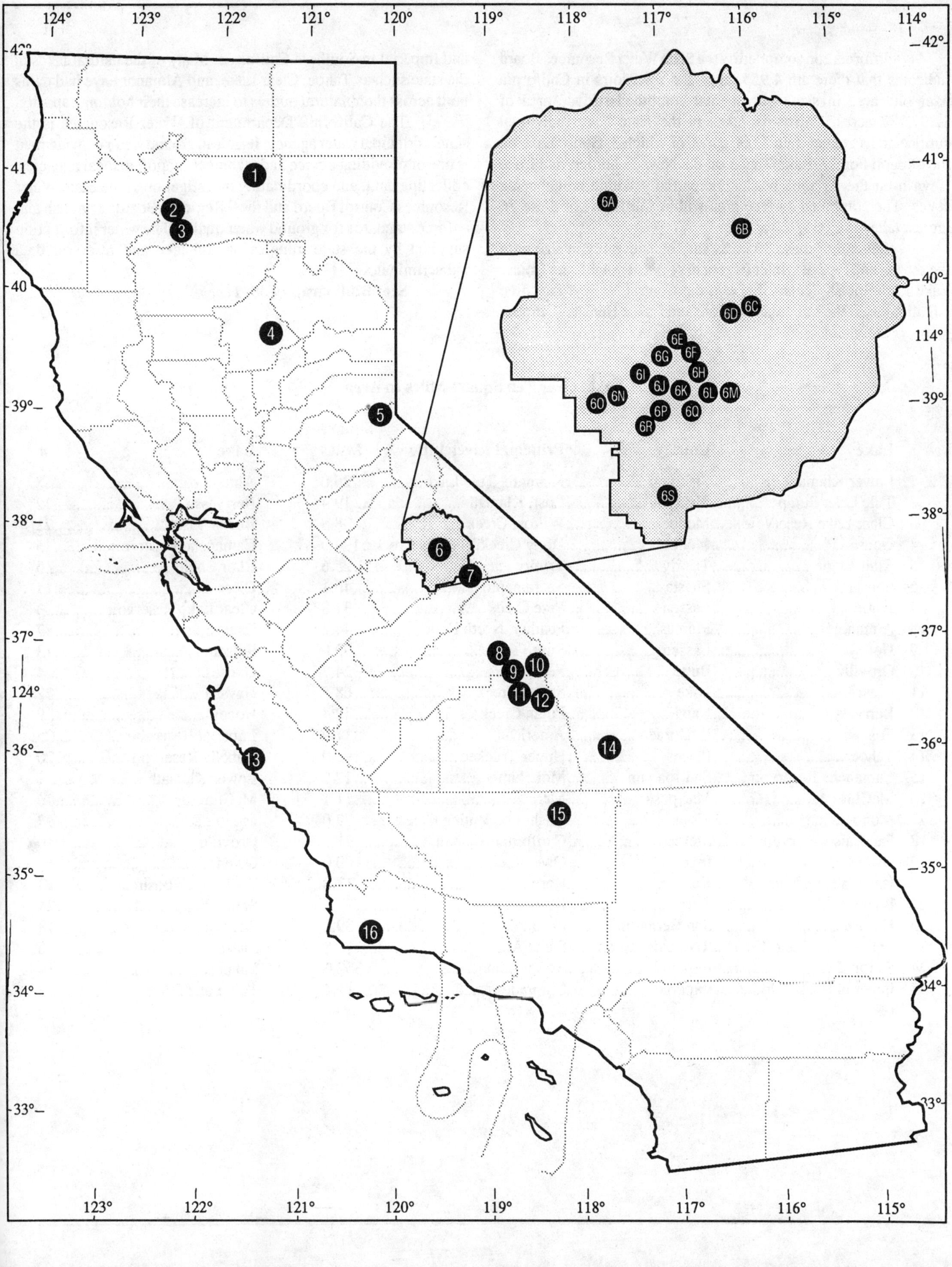
124°
123°
122°
121°
120°
119°
118°
117°
116°
115°
114°
42°
41°
40°
39°
38°
37°
36°
35°
34°
33°
1
2
3
4
5
6
7
8
9
10
11
12
13
14
15
16
6A
6B
6C
6D
6E
6F
6G
6H
6I
6J
6K
6L
6M
6N
6O
6P
6Q
6R
6S

Information compiled by the State Water Resources Board indicates that there are 4,955 lakes and reservoirs in California over one acre in size, with a total combined surface area of 1,397,137 acres. Thirty-six lakes in the state have individual surface areas greater than 5,000 acres, of which 3, Goose Lake on the Oregon border, Lake Tahoe on the Nevada border, and Lake Havasu on the Arizona border, lie partially within neighboring states. The combined surface area within California of these 36 largest lakes is 1,025,781 acres.

Several of these 36 lakes vary considerably in size with the season and type of winter (wet or dry). Most notable are Honey Lake in the north; Tulare Lake in the Central Valley; the three Alkali lakes in the Lahontan Region; and Lakes Henshaw, Prado, and Imperial in Southern California. Many of the natural lakes in the state such as Tahoe, Clear Lake, and Almanor have had dams built across their natural outlets to increase their holding capacity.

The California Department of Water Resources is the state's principal water agency. Its role in ground water management is one of providing advice and technical support to local agencies, collecting data, and coordinating investigations. The State Water Resources Control Board and the 9 Regional Boards establish and enforce standards for ground water quality.The water information supplied by the state mingles natural lake and reservoir data indiscriminately.

See: Bibliography 38, 113.

Lakes Over Ten Square Miles In Area

	Lake	County	Principal River Inflow	Square Miles
1	Lower Klamath	Siskiyou	Klamath, Tule Lake	26.0
2	Tule Lake Sump	Siskiyou	Lost, Klamath	19.4
3	Clear Lake Reservoir	Modoc	Willow Creek	38.8
4	Goose	Modoc	Drew Creek,Thomas Creek	194.0
5	Clair Engle	Trinity	Trinity	25.6
6	Shasta	Shasta	Sacramento	46.1
7	Eagle	Lassen	Pine Creek	31.8
8	Almanor	Plumas	Feather, North Fork	44.2
9	Honey	Lassen	Susan	90.0
10	Oroville	Butte	Feather	24.7
11	Clear Lake	Lake	Scotts Creek	68.5
12	Berryessa	Napa	Putah Creek	32.4
13	Folsom	El Dorado	American	17.9
14	Tahoe	Placer	Upper Truckee	191.0
15	Camanche Reservoir	San Joaquin	Mokelumne	12.0
16	McClure	Mariposa	Merced	11.1
17	Mono	Mono	Rush, Lee Vining Creek	77.0
18	San Luis Reservoir	Merced	California Aqueduct	21.6
19	Owens	Inyo	Owens	100.0
20	Isabella Reservoir	Kern	Kern	17.8
21	Buena Vista	Kern	Kern	37.5
22	Havasu	San Bernardino	Colorado	39.2
23	Prado F. C. Basin	Riverside	Santa Ana	13.8
24	Salton Sea	Imperial	New, Alamo	352.0
25	Imperial Reservoir	Imperial	Colorado	11.4

Lake	#
Almanor	8
Berryessa	12
Buena Vista	21
Camanche Reservoir	15
Clair Engle	5
Clear	11
Clear Lake Reservoir	3
Eagle	7
Folsom	13
Goose	4
Havasu	22
Honey	9
Imperial Reservoir	25
Isabella Reservoir	20
Lower Klamath	1
McClure	16
Mono	17
Oroville	10
Owens	19
Prado F. C. Basin	23
Salton Sea	24
San Luis Reservoir	18
Shasta	6
Tahoe	14
Tule Lake Sump	2

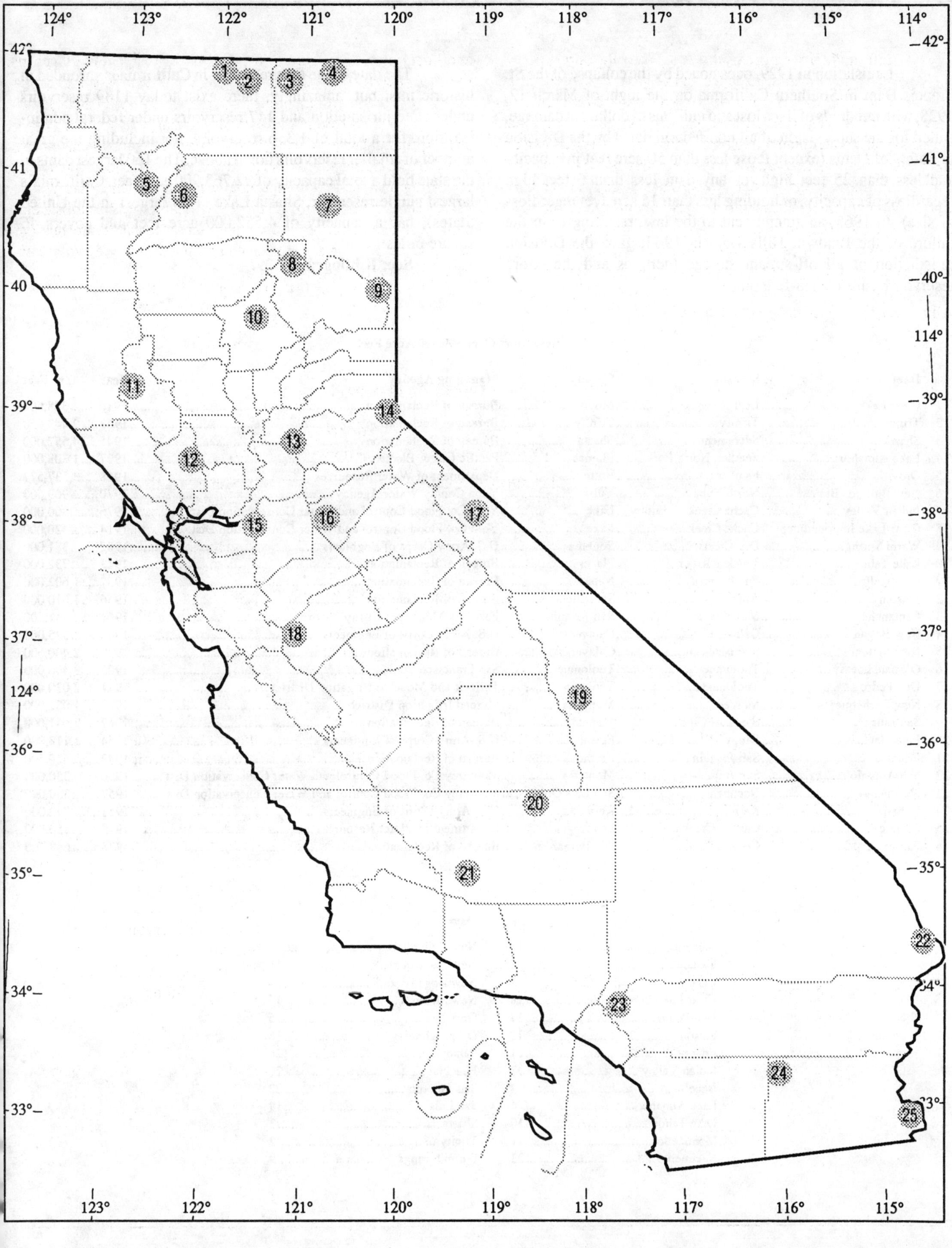
124°
123°
122°
121°
120°
119°
118°
117°
116°
115°
114°
42°
41°
40°
39°
38°
37°
36°
35°
34°
33°
1
2
3
4
5
6
7
8
9
10
11
12
13
14
15
16
17
18
19
20
21
22
23
24
25

Legislation in 1929, occasioned by the collapse of the St. Francis Dam in Southern California on the night of March 12, 1928, with hundreds of lives lost and millions of dollars in damage, called for the supervision of all non-federal dams by the Division of Safety of Dams (except those less than 50 acre feet in capacity and less than 25 feet high, or any dam less than 6 feet high regardless of capacity, or holding less than 15 acre feet regardless of size). In 1965, an amendment to the law, resulting from the failure of the Baldwin Hills Dam in 1963, gave the Division jurisdiction of all off-stream storage facilities and the entire reservoir basins of non-federal dams.

The date of the first reservoir in California is shrouded in historic mist but, amazingly, there exist today 1189 reservoirs under state jurisdiction and 147 reservoirs under federal administration, for a total of 1,336 reservoirs, not including the great number of smaller reservoirs noted above. The 160 largest dams in the state hold a total capacity of 92,763,900 acre feet. California's largest single reservoir, Shasta Lake (14th largest in the United States), has a capacity of 4,552,000 acre feet and covers 46 square miles.

See: Bibliography 39.

Reservoirs Over 300,000 Acre Feet

	Dam	Stream	County	Managing Agency	Year	Acre Feet
1	Clear Lake	Lost River	Modoc	Bureau of Reclamation	1910	388,500
2	Trinity	Trinity	Trinity	Bureau of Reclamation	1962	2,447,650
3	Shasta	Sacramento	Shasta	Bureau of Reclamation	1945	4,552,000
4	Lake Almanor	Feather, North Fork	Plumas	Pacific Gas & Electric Co.	1927	1,308,000
5	Oroville	Feather	Butte	Department of Water Resources	1968	3,537,577
6	New Bullards Bar	North Yuba	Yuba	Yuba County Water Agency	1970	969,600
7	Indian Valley	Cache Creek Tributary	Lake	Yolo Co. Flood Control and Water Conservation Dist.	1976	300,000
8	Clear Lake Impoundment	Cache Creek	Lake	Yolo Co. Flood Control and Water Conservation Dist.	1914	420,000
9	Warm Springs	Dry Creek	Sonoma	U.S. Army Corps of Engineers	1982	381,000
10	Lake Tahoe	Truckee River	Placer	Bureau of Reclamation	1913	732,000
11	Monticello	Putah Creek	Napa	Bureau of Reclamation	1957	1,602,000
12	Folsom	American	Sacramento	Bureau of Reclamation	1956	1,010,000
13	Camanche	Mokelumne	San Joaquin	East Bay Municipal Utility District	1963	431,500
14	New Hogan	Calaveras	Calaveras	U.S. Army Corps of Engineers	1963	325,000
15	New Melones	Stanislaus	Calaveras	Bureau of Reclamation	1979	2,400,000
16	O'Shaughnessy	Tuolumne	Tuolomne	San Francisco	1923	3,400,000
17	Don Pedro	Tuolumne	Tuolumne	Turlock and Modesto Irrigation District	1971	2,030,000
18	New Exchequer	Merced	Mariposa	Merced Irrigation District	1967	1,032,000
19	San Luis	San Luis Creek	Merced	Bureau of Reclamation	1967	2,041,000
20	Pine Flat	Kings	Fresno	U.S. Army Corps of Engineers	1954	1,113,000
21	Friant	San Joaquin	Fresno	Bureau of Reclamation	1942	520,500
22	San Antonio	San Antonio	Monterey	Monterey Co. Flood Control and Water Conservation Dist.	1956	350,000
23	Nacimento	Nacimento	San Luis Obispo	Monterey Co. Flood Control and Water Conservation Dist.	1957	350,000
24	Isabella	Kern	Kern	U.S. Army Corps of Engineers	1953	842,000
25	Castaic	Castaic Creek	Los Angeles	Department of Water Resources	1973	350,000
26	Parker	Colorado	San Bernardino	Bureau of Reclamation	1938	648,000

Dam	#	Dam	#
Camanche	13	New Bullards Bar	6
Castaic	25	New Exchequer	18
Clear Lake	1	New Hogan	14
Clear Lake Impoundment	8	New Melones	15
Don Pedro	17	Oroville	5
Folsom	12	O'Shaughnessy	16
Friant	21	Parker	26
Indian Valley	7	Pine Flat	20
Isabella	24	San Antonio	22
Lake Almanor	4	San Luis	19
Lake Tahoe	10	Shasta	3
Monticello	11	Trinity	2
Nacimento	23	Warm Springs	9

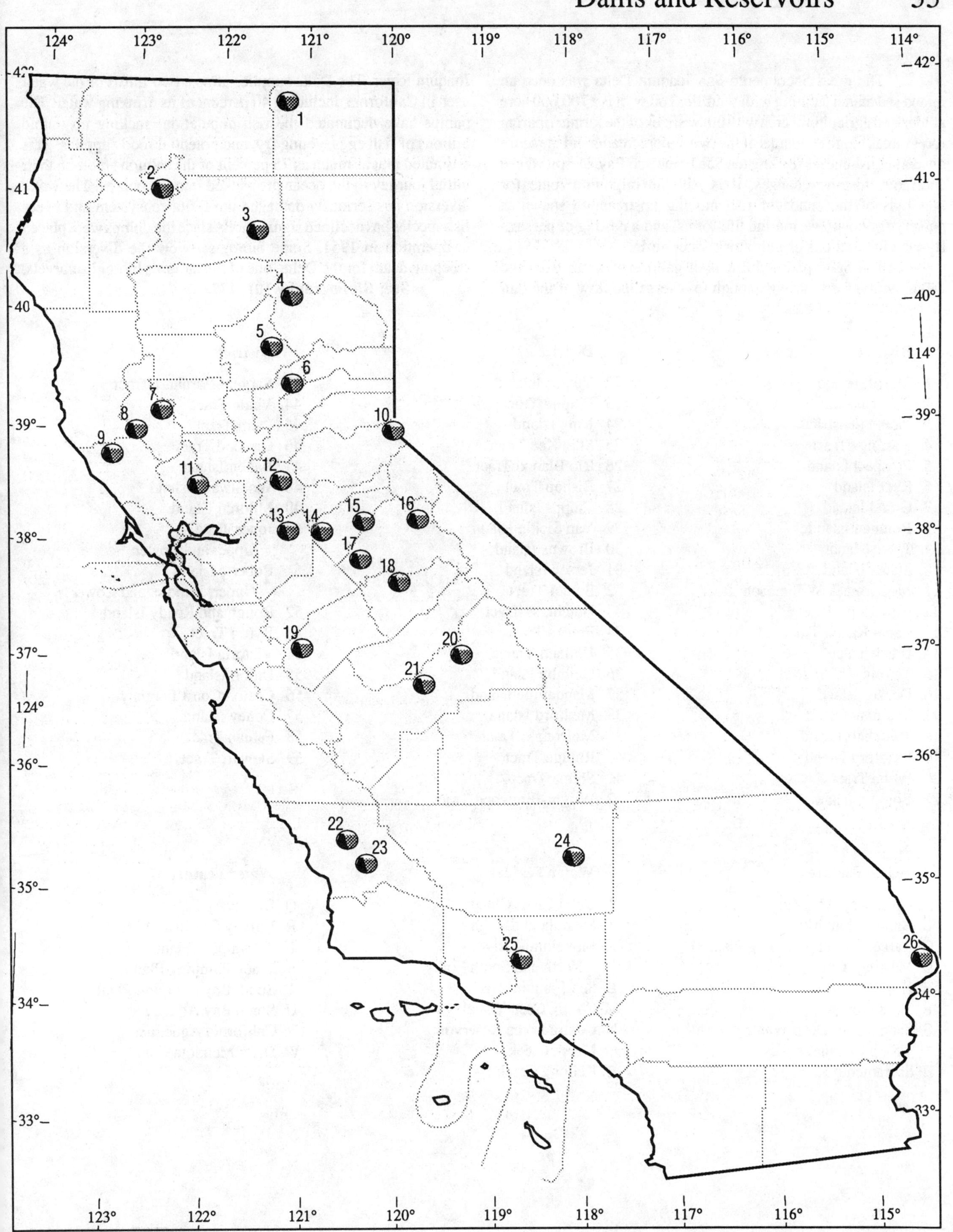
124°
123°
122°
121°
120°
119°
118°
117°
116°
115°
114°
42°
41°
40°
39°
38°
37°
36°
35°
34°
33°
1
2
3
4
5
6
7
8
9
10
11
12
13
14
15
16
17
18
19
20
21
22
23
24
25
26

The great Sacramento-San Joaquin Delta was once an inland sea marsh teeming with wildlife. Today, it is a 700,000 acre polderized agricultural area with but vestiges of the former riparian forest areas along the banks of the two major streams and the score of smaller tributaries feeding the San Francisco Bay. Despite these enormous historic changes, it is still the migration route for hundreds of thousands of fish moving upstream to spawn, a nursery for countless marine life forms, and a nesting or passage fly-way for millions of migratory water birds.

The State pumps billions of gallons of water from the Delta, with a force strong enough to reverse the flow of the San Joaquin River. The Delta supplies almost two-thirds of the water used in California, including 40 percent of its drinking water. The pumps have decimated the fish population, sucking up untold billions of fish eggs, young fry, and potential food supplies. It is estimated that as much as 75 percent of the salmon smolt on their initial journey to the ocean are sucked into the pumps. The water diversion has seriously damaged the Delta ecosystem and every fish species has declined significantly since the pumps were placed in operation in 1951. Some biologists view the diversion as a creeping death for the Delta, one of the nation's biologic marvels.

See: Bibliography 101, 177.

District

1 Merritt Island
2 Sutter Island
3 Pierson District
4 Hastings Tract
5 Prospect Island
6 Ryer island
7 Grand Island
8 Brannan Island
9 Tyler Island
10 Staten Island
11 McCormack Williamson Tract
12 New Hope Tract
13 Canal Ranch Tract
14 Brack Tract
15 Terminous Tract
16 Decker Island
17 Sherman Island
18 Twitchell Island
19 Bradford Island
20 Webb Tract
21 Bouldin Island

District

22 Venice Island
23 Empire Tract
24 King Island
25 Shin Kee Tract
26 Rio Blanco Tract
27 Bishop Tract
28 Chipps Island
29 Van Sickle Island
30 Browns Island
31 Jersey Island
32 Bethel Tract
33 Hotchkiss Tract
34 Franks Tract
35 Holland Tract
36 Quimby Island
37 Mandeville Island
38 Medford Island
39 McDonald Tract
40 Rindge Tract
41 Shima Tract
42 Wright-Elmwood Tract

District

43 Sargent Barnhart Tract
44 Veale Tract
45 Palm Tract
46 Orwood Tract
47 Bacon Island
48 Woodward Island
49 Mildred Island
50 Jones Tract - Upper and Lower
51 Roberts Island - Upper, Middle, and Lower
52 Rough and Ready Island
53 Byron Tract
54 Victoria Island
55 Union Island
56 Clifton Court Forebay
57 Coney Island
58 Fabian Tract
59 Stewart Tract

Water Feature

A North Bay Aqueduct
B Barker Slough
C Barker Slough Pumping Plant
D Mastings Cut
E Hass Slough
F Skag Slough
G Sacramento Deep Water Ship Channel
H Sacramento River

Water Feature

I Delta Cross Channel
J Steamboat Slough
K Mokelumne River - North and South Forks
L San Joaquin River
M Contra Costa Canal
N Contra Loma Reservoir
O Marsh Creek
P Kellogg Creek

Water Feature

Q Discovery Bay
R Harvey O. Banks Delta Pumping Plant
S Tracy Pumping Pland
T South Bay Pumping Plant
U South Bay Aqueduct
V California Aqueduct
W Delta Mendota Canal

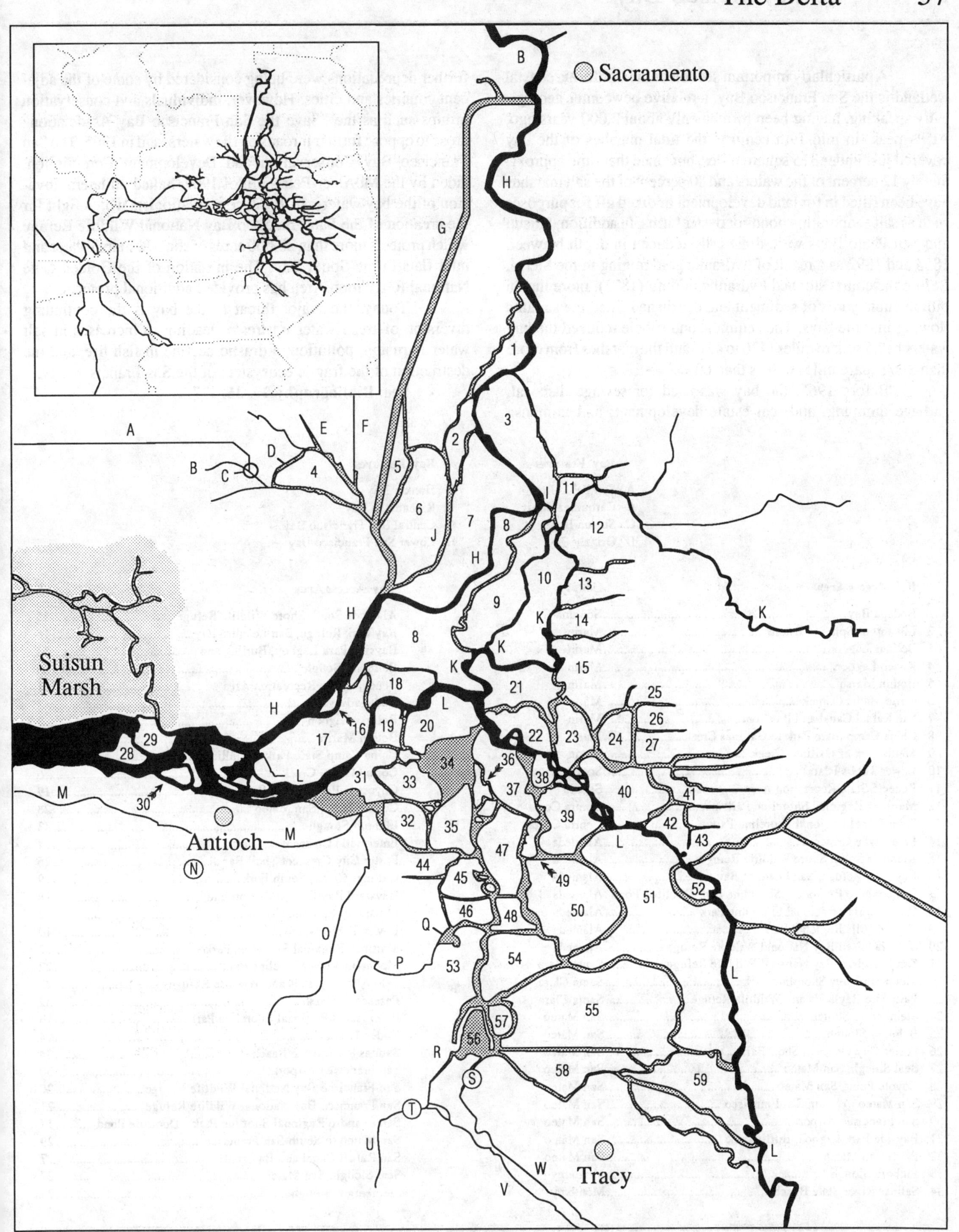
Sacramento
Suisun
Marsh
Antioch
Tracy
A
B
C
D
E
F
G
H
I
J
K
L
M
N
O
P
Q
R
S
T
U
V
W
1
2
3
4
5
6
7
8
9
10
11
12
13
14
15
16
17
18
19
20
21
22
23
24
25
26
27
28
29
30
31
32
33
34
35
36
37
38
39
40
41
42
43
44
45
46
47
48
49
50
51
52
53
54
55
56
57
58
59

A particularly important section of the California coastal wetland is the San Francisco Bay, a relative newcomer geologically speaking, having been formed only about 2,000 years ago. At its peak, in mid-19th century, the tidal marshes of the bay covered just under 315 square miles, but since that time approximately 12 percent of the waters and 80 percent of the salt marshes have been filled in for land development or diked off for purposes such as salt evaporation ponds or oyster farms. In addition, Suisun and San Pablo Bays were drastically reduced in depth between 1853 and 1892 as a result of hydraulic gold mining in the Sierra. Before the courts stopped hydraulic mining (1893), more than a billion cubic yards of sediment and debris had filled the streams flowing into the bays. The sediment and rubble reduced the bay waters by 55 square miles (476 to 421) and the marshes from more than 300 square miles to less than 60.

Before 1965, the bay was used for sewage disposal, garbage dumping, and real estate development; and immense further depredations were being considered by some of the adjacent counties and cities. However, individuals and conservation groups such as the "Save the San Francisco Bay Association" arose to oppose further inroads on the waters, and in 1965, The San Francisco Bay Conservation and Development Commission, aided by the McAteer-Petris bill of 1969, halted unilateral invasion of the bay waters. In 1972, conservationists won a fight for the creation of the San Francisco Bay National Wildlife Refuge which protects more than 20,000 acres of shallow marsh water and mud flats in the South Bay. The initiation of the Golden Gate National Recreation Area has provided additional protection.

Today, the major threat to the bay is the continuing diversion of fresh water upstream, leading to increases in salt water intrusion, pollution, a drastic decline in fish life, and the destruction of the fragile ecosystem of the San Francisco Bay.

See: Bibliography 29, 35, 177.

Bay Features

	Bay Features
A	San Pablo Bay
B	Carquinez Strait
C	Suisun Bay
D	Grizzly Bay
E	Honker Bay
F	Richardson Bay
G	Central San Francisco Bay
H	Lower San Francisco Bay

	Bay Access Area	County
1	Bodega Bay	Sonoma
2	Limantour Spit and Estero	Marin
3	Bolinas Lagoon	Marin
4	Rodeo Lagoon	Marin
5	Bothin Marsh	Marin
6	Corte Madera Creek	Marin
7	San Rafael Canal and Bayfront	Marin
8	China Camp State Park to Gallinas Creek	Marin
9	South Fork of Gallinas Creek	Marin
10	Lower Tubbs Island	Sonoma
11	Benecia State Recreation Area	Solano
12	Martinez Regional Shoreline Park	Contra Costa
13	Point Isabel Regional Shoreline Park	Contra Costa
14	Emeryville Crescent	Alameda
15	Alameda South Shore Wildlife Refuge	Alameda
16	Bay Park Refuge, San Leandro Bay	Alameda
17	San Leandro Regional Shoreline Park-Doolittle Pond	Alameda
18	Hayward Bay Regional Shoreline Park	Alameda
19	Coyote Hills Regional Park	Alameda
20	San Francisco Bay National Wildlife Refuge	Alameda
21	San Francisco Bay National Wildlife Refuge	Santa Clara
22	Mountain View Shoreline Park	Santa Clara
23	Palo Alto Baylands and Wildlife Refuge	Santa Clara
24	Steinberger Slough	San Mateo
25	Belmont Slough	San Mateo
26	Foster City Crescent Shell Bar	San Mateo
27	Seal Slough, San Mateo	San Mateo
28	Coyote Point, San Mateo	San Mateo
29	San Mateo to South San Francisco	San Mateo
30	San Francisco Airport	San Mateo
31	Bayside Park Lagoon, Burlingame	San Mateo
32	Pescadero Marsh	San Mateo
33	Elkhorn Slough	Monterey
34	Salinas River State Beach	Monterey

Bay Access Area	#
Alameda South Shore Wildlife Refuge	15
Bay Park Refuge, San Leandro Bay	16
Bayside Park Lagoon, Burlingame	31
Belmont Slough	25
Benecia State Recreation Area	11
Bodega Bay	1
Bolinas Lagoon	3
Bothin Marsh	5
China Camp State Park to Gallinas Creek	8
Corte Madera Creek	6
Coyote Hills Regional Park	19
Coyote Point, San Mateo	28
Elkhorn Slough	33
Emeryville Crescent	14
Foster City Crescent Shell Bar	26
Gallinas Creek, South Fork	9
Hayward Bay Regional Shoreline Park	18
Limantour Spit and Estero	2
Lower Tubbs Island	10
Martinez Regional Shoreline Park	12
Mountain View Shoreline Park	22
Palo Alto Baylands and Wildlife Refuge	23
Pescadero Marsh	32
Point Isabel Regional Shoreline Park	13
Rodeo Lagoon	4
Salinas River State Beach	34
San Francisco Airport	30
San Francisco Bay National Wildlife Refuge	20
San Francisco Bay National Wildlife Refuge	21
San Leandro Regional Shorline Park - Doolittle Pond	17
San Mateo to South San Francisco	29
San Rafael Canal and Bayfront	7
Seal Slough, San Mateo	27
Steinberger Slough	24

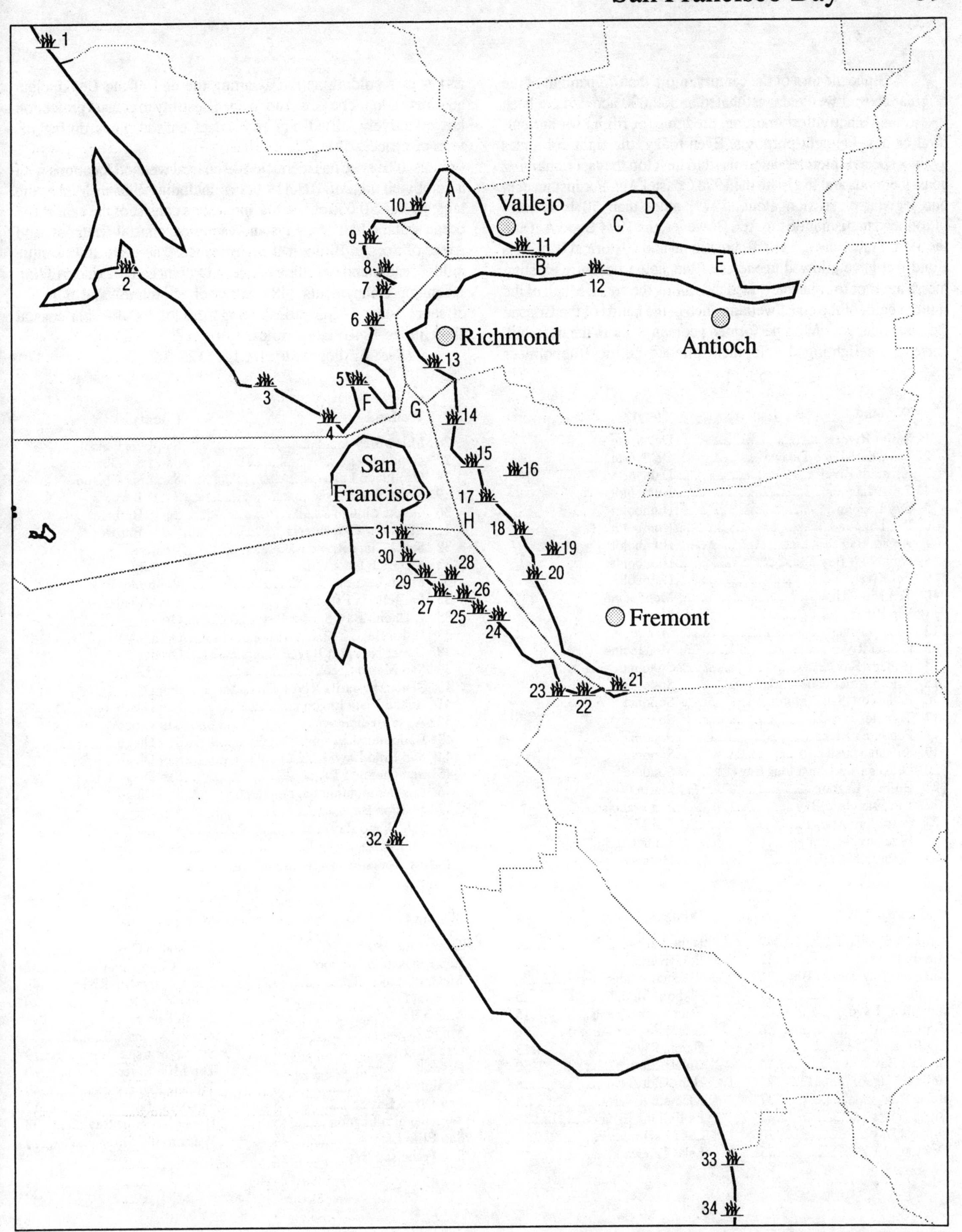
Vallejo
Richmond
Antioch
San Francisco
Fremont
A
B
C
D
E
F
G
H
1
2
3
4
5
6
7
8
9
10
11
12
13
14
15
16
17
18
19
20
21
22
23
24
25
26
27
28
29
30
31
32
33
34

Since the turn of the century, more than 75 percent of the original coastal wetlands, estimated at 382,000 acres, have been lost to man's activities: draining, dredging, or filling for agricultural or development purposes. Even today, the fight continues against special interests determined to develop the last remaining coast wetlands and to vitiate the 1976 Coastal Act. For instance, in one legislative session alone, 1979, more than 50 bills were introduced to dismantle the Act. However, the 1974 Bond Act and the 1976 State Beach, Park, Recreational and Historical Facilities Bond Act have allowed agencies of the state to acquire existing marsh areas or to create new wetlands along the coast. Much of the future health of the coast wetlands lies in the hands of the Coastal Commission, which has performed yeoman work in the past, but, currently is fighting development minded political appointees and is in a holding action awaiting the end of the Deukmejian administration. The Governor's open hostility to coastal protection has effectively stifled any significant coastal gains during his term of office.

The state has identified 49 coastal wetland locations with a combined area of 107,419 acres, including Suisun Marsh, the largest, with 50,000 acres. Marine waters consist of the near shore ocean waters, harbors, bays and estuaries, coastal wetlands, and Areas of Special Biological Significance. There are 1073 statute miles of mainland coastline between Oregon and Mexico. Adding 10 major embayments (485 miles of shoreline) and 9 major offshore islands (282 miles) brings the total California coastal shoreline to 1840 statute miles.

See: Bibliography 18, 116, 125, 169, 173.

	Wetland *	County
1	Smith River	Del Norte
2	Lake Earl-Lake Talawa	Del Norte
3	Klamath River	Del Norte
4	Stone Lagoon	Humboldt
5	Dry Lagoon	Humboldt
6	Big Lagoon	Humboldt
7	Arcata Bay	Humboldt
8	Humboldt Bay	Humboldt
9	Eel River	Humboldt
10	Ten Mile River	Mendocino
11	Big River	Mendocino
12	Albion River	Mendocino
13	Garcia River	Mendocino
14	Bodega Bay	Sonoma
15	Estero Americano	Sonoma
16	Estero de San Antonio	Sonoma
17	Tomales Bay	Sonoma
18	Petaluma River	Sonoma
19	Suisun Marsh	Sonoma
20	Drakes Bay-Limantour Bay	Sonoma
21	Bolinas Lagoon	Marin
22	San Francisco Bay	San Francisco
23	Pescadero Marsh	San Mateo
24	Watsonville Slough	Santa Cruz
25	Elkhorn Slough	Monterey
26	Morro Bay	San Luis Obispo
27	Dune Lakes	San Luis Obispo
28	Oso Flaco Lake	San Luis Obispo
29	Santa Ynez River	Santa Barbara
30	Goleta Slough	Santa Barbara
31	El Estero-Carpenteria	Santa Barbara
32	Santa Clara River	Ventura
33	McGrath Lake	Ventura
34	Mugu Lagoon	Ventura
35	La Bellona Estuary	Los Angeles
36	Anaheim Bay-Sunset Bay	Orange
37	Bolsa Bay	Orange
38	Upper Newport Bay	Orange
39	San Mateo Creek	San Diego
40	Santa Margarita River	San Diego
41	Buena Vista Lagoon	San Diego
42	Agua Hedionda	San Diego
43	Batiquitos Lagoon	San Diego
44	San Elijo Lagoon	San Diego
45	San Dieguito Lagoon	San Diego
46	Los Penasquitos Lagoon	San Diego
47	Mission Bay	San Diego
48	San Diego Bay	San Diego
49	Tijuana Estuary	San Diego

* These 49 named areas represent 86 discrete coastal sites.

Wetland	#
Agua Hedionda	42
Albion River	12
Anaheim Bay-Sunset Bay	36
Arcata Bay	7
Batiquitos Lagoon	43
Big Lagoon	6
Big River	11
Bodega Bay	14
Bolinas Lagoon	21
Bolsa Bay	37
Buena Vista	41
Drake's Bay	20
Dry Lagoon	5
Dune Lakes	27
Eel River	9
El Estero-Carpenteria	31
Elkhorn Slough	25
Estero Americano	15
Estero de San Antonio	16
Garcia River	13
Goleta Slough	30
Humboldt Bay	8
Klamath River	3
La Bellona Estuary	35
Lake Earl	2
Lake Talawa	2
Limantour Bay	20
Los Penasquitos Lagoon	46
McGrath Lake	33
Mission Bay	47
Morro Bay	26
Mugu Lagoon	34
Oso Flaco	28
Pescadero Marsh	23
Petaluma River	18
San Diego Bay	48
San Dieguito Lagoon	45
San Elijo Lagoon	44
San Francisco Bay	22
San Mateo Creek	39
Santa Clara River	32
Santa Margarita River	40
Santa Ynez River	29
Smith River	1
Stone Lagoon	4
Suisun Marsh	19
Ten Mile River	10
Tijuana Estuary	49
Tomales Bay	17
Upper Newport Bay	38
Watsonville Slough	24

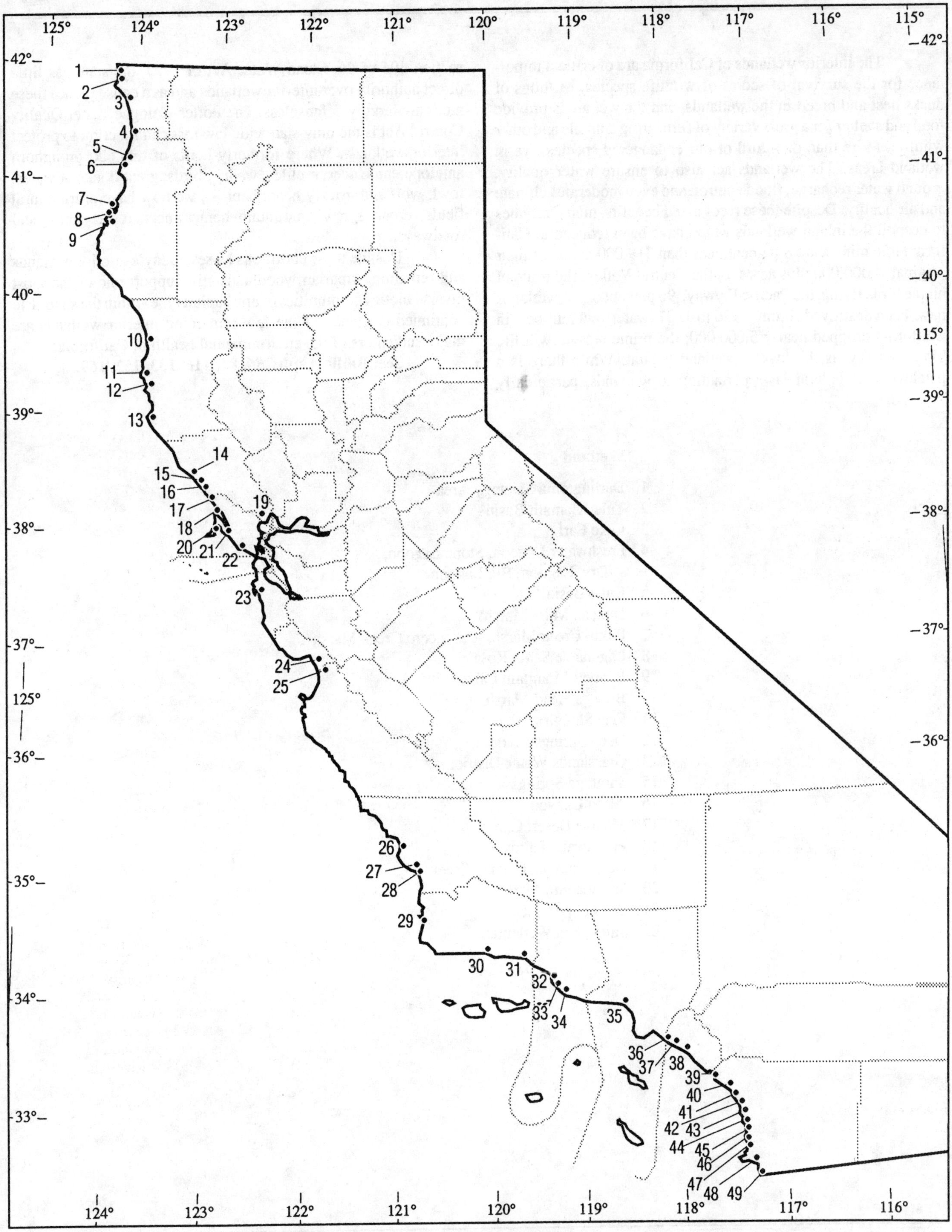
125°
124°
123°
122°
121°
120°
119°
118°
117°
116°
115°
42°
41°
40°
39°
38°
37°
36°
35°
34°
33°
1
2
3
4
5
6
7
8
9
10
11
12
13
14
15
16
17
18
19
20
21
22
23
24
25
26
27
28
29
30
31
32
33
34
35
36
37
38
39
40
41
42
43
44
45
46
47
48
49

The interior wetlands of California are of critical importance for the survival of scores of wildlife species. Millions of ducks nest and breed in the wetlands, and the wet areas provide food and shelter for a great variety of furbearing animals and other wildlife. More than one-third of our endangered species live in wetland areas. The wetlands act also to ensure water quality, ground water recharge, flood control, and even moderates climate and air quality. Despite these recognized benefits, man continues to despoil the inland wetlands which have been reduced in California to a minuscule 8 percent, less than 319,000 acres, of their original 4,000,000 plus acres. In the Central Valley, the route of all the birds flying the Pacific Flyway, 96 percent of the wetlands have been destroyed. From 1976 to 1987, waterfowl numbers in California dropped nearly 5,000,000; the prime reason, wildlife biologists say, is the loss of wetland habitat. While there is a patchwork of federal laws pertaining to wetlands, particularly section 404 of the Clean Water Act of 1972, the state has little direct authority over interior wetlands and as a consequence these areas are largely defenseless. The Porter-Cologne Water Quality Control Act is the only statewide law that can function to protect interior wetlands. Where formerly herds of tule elk, pronghorn antelope, mule deer, reptiles, wading birds, great flocks of waterfowl, wolf and grizzly bears habited, we now have agricultural fields, urban sprawl, mountain home tracts, resort areas, and roadways.

Despite these enormous losses, today's interior wetlands and remaining riparian woodlands still support one of the most diverse biotic communities of any habitat type within the state. The continued existence of the few remaining interior wetlands are major indicators of the environmental health of California.

See: Bibliography 56, 84, 116, 133, 152, 169.

Wetland

1 Darlingtonia Swampy Area
2 Tule-Klamath Basin
3 Lake Earl
4 Freshwater Lagoon, Stone Lagoon,
  Dry Lagoon, Big Lagoon
5 Butte Basin
6 Duncan Mills Marsh
7 Pitkin Creek Marsh, Atascadero Creek Marsh
8 Laguna de Santa Rosa
9 Bennett Mountain Lake
10 Bodega Head Marsh
11 Fish Slough
12 Deep Springs Marsh
13 Grasslands Water District
15 Saratoga Springs
16 Afton Canyon
17 Mojave Desert Camp
18 Pushawalla Palms
19 Deep Canyon, Carrizo Creek
20 San Joaquin Marsh
21 San Felipe Creek
22 Salton Sea Wetlands

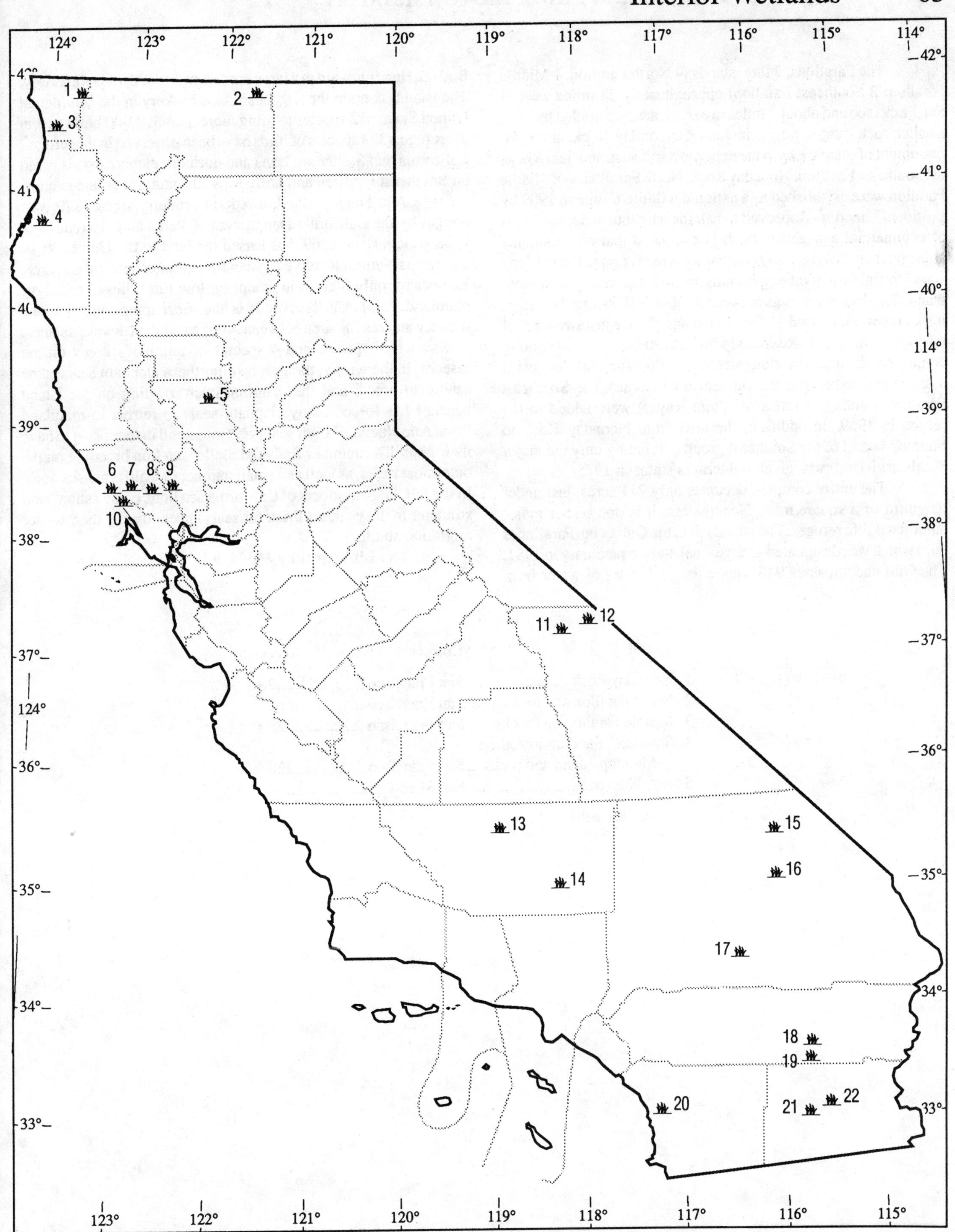
124°
123°
122°
121°
120°
119°
118°
117°
116°
115°
114°
42°
41°
40°
39°
38°
37°
36°
35°
34°
33°
1
2
3
4
5
6
7
8
9
10
11
12
13
14
15
16
17
18
19
20
21
22

The Farallons, 7 tiny islands (4 North Farallon; 1 Middle Farallon; 2 Southeast Farallon) approximately 28 miles west of San Francisco and about 7 miles in extent, are surrounded by even smaller rock outcroppings, including Noonday Rock, the most prominent of these rocks in the extreme northwest, and Seal Rocks off Southeast Farallon. Noonday Rock, North Farallon, and Middle Farallon were established as a national wildlife refuge in 1909 by President Theodore Roosevelt to halt the gargantuan depredations of commercial egg gatherers. It is estimated that one company alone (and there were many companies which engaged in the "egg wars" to fight off rival egg hunting troops) reportedly sold more than 4,000,000 Murre eggs between 1850 and 1856. One boatload, it was reported, carried "1,000 dozen eggs" after just two days of thievery. (Earlier, the Russians, who had established a fur hunting colony on the islands, slaughtered 200,000 fur seals in just 3 seasons and wiped out the fur seal population.) The Southeast Farallon islands (18 miles off Point Reyes) were added to the refuge in 1969. In addition, the area from Noonday Rock to Maintop Island of the Southeast group (excluding only the major Southeast island) was given wilderness status in 1975.

The entire complex occupies only 211 acres, just under one-third of a square mile. Nonetheless, it is one of our major natural wildlife refuges. The islands sit in the Gulf of the Farallones (sic) which was designated as a national marine sanctuary in 1981. The Gulf encompasses 948 square nautical miles of water from Bodega Head outward in a large arc to a point south of Bolinas Bay. The islands contain the largest sea bird rookery in the continental United States; 12 species totaling more than 350,000 birds nest in the refuge. (328 species of birds have been observed in the refuge.) California and Steller sea lions and northern elephant seals breed on Southeast Farallon and northern seals haul out on the islands.

Año Nuevo Island, a federal property since 1870, was bought by the California Department of Parks and Recreation in 1958 which, since 1969, has leased the land to the University of California Natural Reserve System for management and research. The island, only 8 acres in extent (at low tide), lies .3 miles off mainland Point Año Nuevo. It is the most important pinniped rookery and resting area between Alaska and the Channel Islands. A dozen bird species and 4 species of pinnipeds breed on the reserve. In the winter, thousands of northern elephant seals come ashore to breed, and the resulting overcrowding on the island beaches has forced many elephant seals to retreat to mainland Point Año Nuevo. Harbor seals visit the island in the spring to have their pups. The summer finds the Steller sea lion breeding on the outermost rocks, as well as a small number of California sea lions. In the fall, large numbers of California sea lions come ashore, and still later in the year, northern fur seals appear during their winter migration south.

See: Bibliography 66, 96, 97.

	Island	**County**	**Acres**
1	Noonday Rock	San Francisco	2 *
2	North Farallon and rocks	San Francisco	61
3	Middle Farallon and rocks	San Francisco	28 *
4	Southeast Farallon including Maintop Island and rocks	San Francisco	120
5	Año Nuevo	San Mateo	8

* awash seasonally

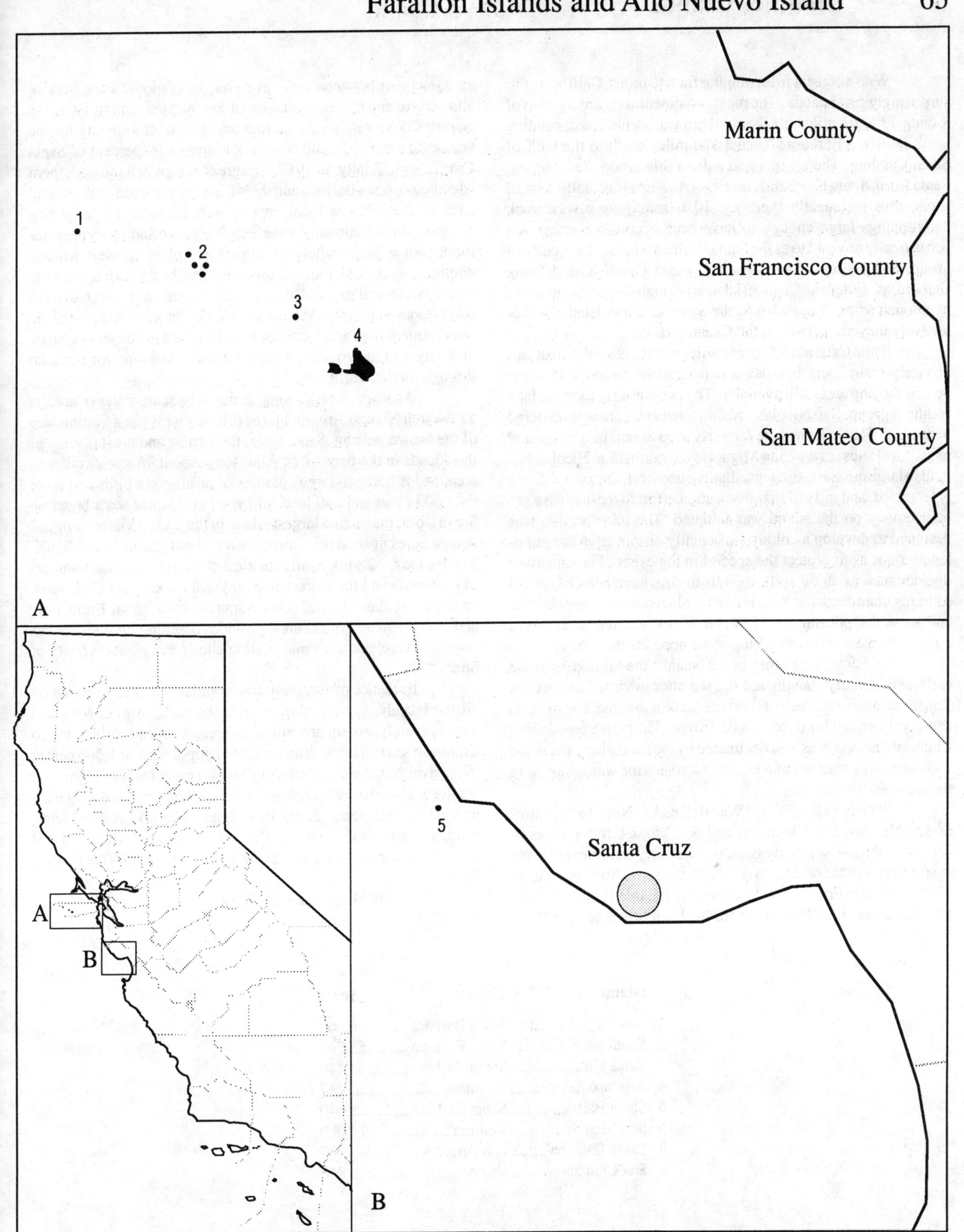
Marin County
San Francisco County
San Mateo County
1
2
3
4
5
A
Santa Cruz
A
B
B

With a coastal front running for 840 miles, California has surprisingly few islands. The most prominent isles are a group of 8 lying 13 to 61 miles off the southern California coast, running from the Santa Barbara Channel 160 miles south to the Gulf of Santa Catalina. The northern islands of this group, San Miguel, Santa Rosa, Santa Cruz, and Anacapa (Anacapa is actually 3 small islets; thus technically there are 10 islands plus several rock outcroppings large enough to have been accorded names), are occasionally referred to as the Santa Barbara Islands. The southern group, San Nicolas, San Clemente, Santa Catalina, and Santa Barbara, are called the Channel Islands (though they are sometimes mentioned as the San Pedro or the Santa Catalina Islands). Collectively they are known as the Channel Islands.

Santa Catalina, 75 square miles in area, is for the most part privately owned and is today a major tourist mecca with some beach camping areas still available. The two westerly islands of the southern group, San Nicolas and San Clemente, are administered by the U.S. Navy. They were formerly used as bombing range and missile test sites (as was San Miguel in the north). San Nicolas still holds a large missile testing installation; however, the local wildlife is protected, and in 1987, a translocation effort to reestablish a sea otter colony on the island was initiated. The translocation was designed to develop a colony sufficiently distant from the mainland groups as to protect the species in the event of an unnatural disaster such as an oil spill. Results to date have been mixed — some resultant deaths, several returns of older otters to the mainland, and some disappearances. However, a few younger otters have remained on San Nicolas giving some hope for the future.

Earlier in the history of the islands, the fur traders of the early 19th century slaughtered the sea otter to virtual extinction. The same fate befell the seals and sea lions of the area, and by mid-century the island fur trade was destroyed. The native population, Chumash Indians, was also decimated by the fur traders, and in the 1830s a pitiful remnant of a once numerous tribe was removed to the mainland.

Shortly before World War II, the U.S. Navy took control of San Nicolas, San Clemente, and San Miguel. It continues in control of the two southerly isles, but the Navy's former presence is still very visible on San Miguel and on Santa Rosa as well. In 1938, President Franklin D. Roosevelt proclaimed Santa Barbara and Anacapa as the Channel Islands National Monument. In 1976, an agreement between the Navy and the National Park Service allowed for supervised visitation of San Miguel, and in 1978, the Nature Conservancy entered into an agreement allowing for the protection, research, and educational use of 90 percent of Santa Cruz Island. Finally, in 1980, Congress designated the 4 northern islands and Santa Barbara and the waters for 1 nautical mile around each as the 40th national park. Later that year the area was designated as a National Marine Sanctuary. The Sanctuary provides shelter for a large variety of animal and plant species: whales, dolphins, great colonies of breeding sea birds, extensive kelp forest communities, and one of the largest and most diverse populations of pinnipeds in the world. The park is a Reserve of the International Man and the Biosphere Program to conserve genetic diversity and to provide an environmental baseline for research throughout the world.

Anacapa, 5 miles long, about 1.1 square miles in area, is a Research Natural Area and is the primary west coast nesting area of the brown pelican. Santa Cruz, the largest and most diverse of the islands in the park, is 24 miles long, about 96 square miles in area, and is home to large colonies of nesting sea birds and more than 600 plant and 140 land bird species. Harbor seals breed on Santa Rosa, the second largest island in the park, 15 miles long, 83 square miles in area; it is surrounded by kelp forests. San Miguel, 8 miles long, 32 square miles in area, shelters enormous numbers of pinnipeds and the largest elephant seal rookery off California, estimated at about 10,000 seals. Santa Barbara, about 1 mile long and only 1 square mile in area, provides breeding grounds for sea lions and elephant seals and nesting sites for a great diversity of birds.

Evidence of former native populations is to be found on all the islands. San Nicolas, outside the park, 9 miles long and approximately 36 square miles in area, was formerly a major Chumash settlement with an estimated population of 1,500 before the arrival of the fur traders. San Miguel contains more than 600 fragile undisturbed archaeological sites, including one dating back at least 10,000 years. Santa Rosa holds 160 documented sites, some at least 8,000 years old. Some archaeologists maintain that Native Americans occupied Santa Rosa as long as 40,000 years back.

See: Bibliography 27, 187.

	Island	**County**	**Acres**
1	San Miguel	Santa Barbara	8,960
2	Santa Rosa	Santa Barbara	53,760
3	Santa Cruz	Santa Barbara	61,440
4	Anacapa	Ventura	717
5	Santa Barbara	Santa Barbara	640
6	San Nicolas	Ventura	14,080
7	Santa Catalina	Los Angeles	48,000
8	San Clemente	Los Angeles	35,840

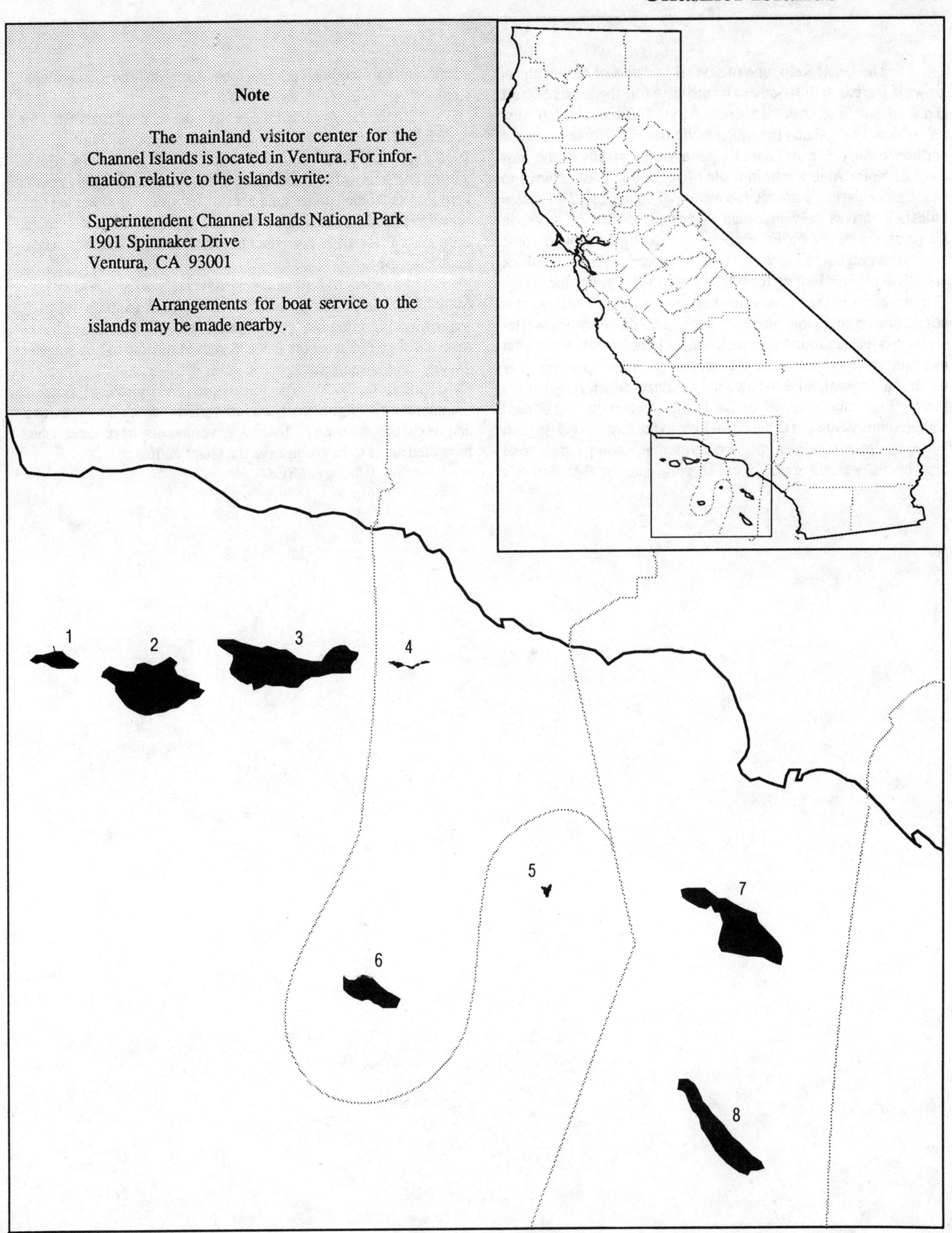
Note
The mainland visitor center for the Channel Islands is located in Ventura. For information relative to the islands write:
Superintendent Channel Islands National Park
1901 Spinnaker Drive
Ventura, CA 93001
Arrangements for boat service to the islands may be made nearby.
1
2
3
4
5
6
7
8

The Giant Kelp, one of several varieties of brown algae, grows in 9 areas in the southern hemisphere (off the coasts of west and southeast South America, South Africa, Tasmania, south Australia, New Zealand and the subantarctic islands). However, in the northern hemisphere the Giant Kelp is to be found only on the west coast of North America from a site offshore Point Año Nuevo to Baja California, in communities covering 27,181 acres(42.47 square miles). It thrives only in ocean depths of 65 feet or more, in temperatures below 65°Fahrenheit (18° Centigrade), with some specimens ranging to 215 feet. The floating surface canopy is thick enough to provide footing for birds as large as the great blue heron. The forests provide a nursery, feeding grounds, and shelter, so it is not surprising that large numbers and a great diversity of invertebrates and fish are found in association with the forests. Somewhat more surprising is the number of mammals (California sea lion, gray whale, harbor seal, killer whale, and sea otter) which frequent the forests. Even more unusual are the 13 bird species (among others) which commonly resort to the Giant Kelp as feeding ground (pigeon guillemot, brown pelican, pelagic cormorant, snowy egret, great blue heron, western grebe, western gull, eared grebe, Brandt's cormorant, surf scooter, common loon, common murre, and elegant tern).

In the early years of the century the kelp was harvested for potash, and as much as 400,000 wet tons were gathered in 1917 and 1918 to provide potash for gunpowder. However, the unrestricted cutting and removal of whole plants was severely destructive of the kelp forests. Today, under state control, about 150,000 wet tons are harvested annually for a sale exceeding $35,000,000, not so much for potash as for algin, a product used by the food and pharmaceutical industries.

A variety of influences can adversely affect the great kelp (which are estimated to grow for 100 to 800 years to attain full size), but sewage pollution is a specific peril, as was demonstrated by the loss of a large forest off the Palos Verdes Peninsula in Los Angeles County and disappearances off several other sites in southern California in the 1950s. The forests are directly related to major commercial fishing interests and to varied recreational activities, and both the state and federal governments have established protected areas to help conserve the Giant Kelp.

See: Bibliography 54.

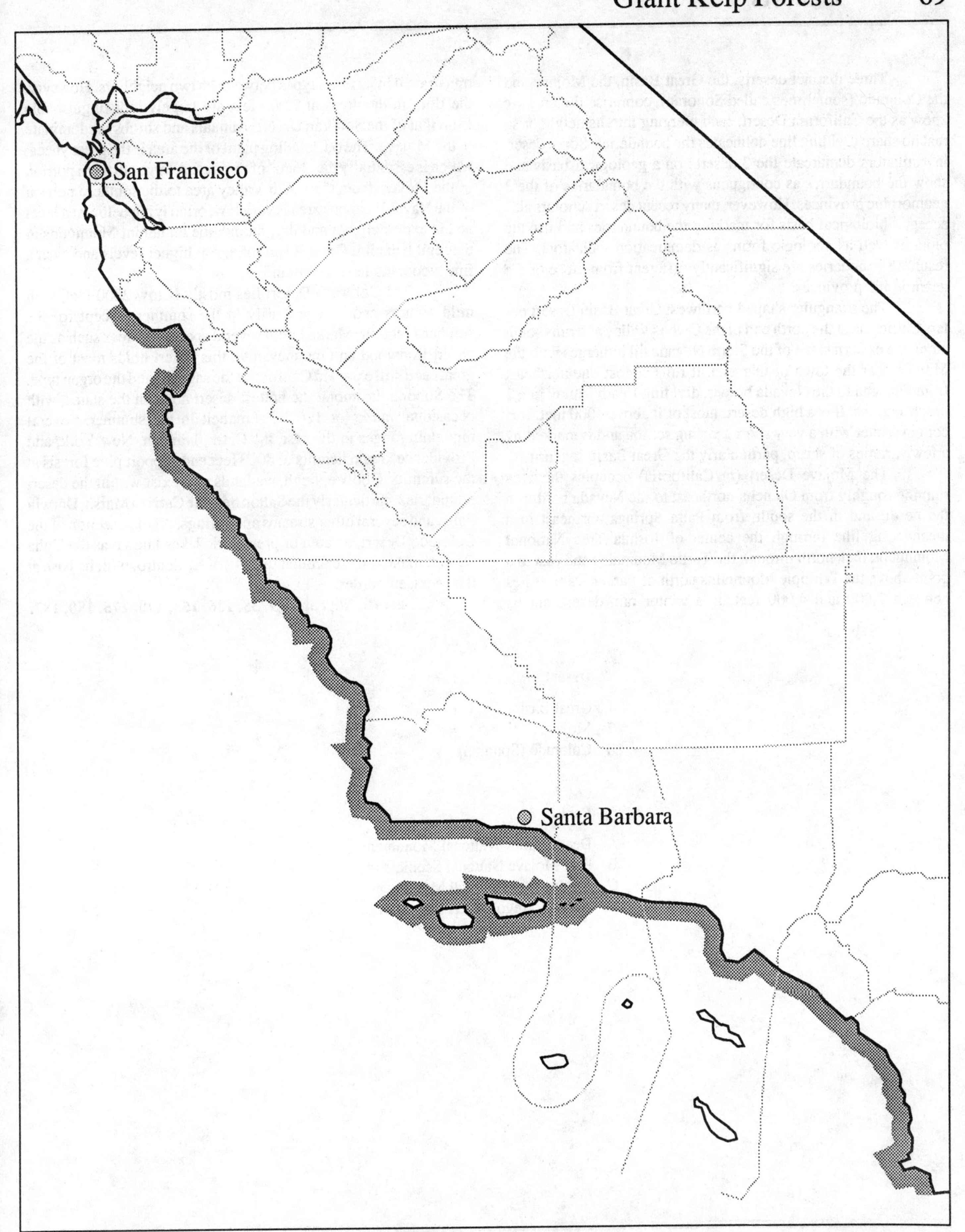
San Francisco
Santa Barbara

Three distinct deserts, the Great Basin, the Mojave and the Colorado (sometimes called Sonoran), comprise the area we know as the California Desert, each merging into its neighbor so that no sharp dividing line delineates the boundaries. Some desert investigators demarcate the 3 deserts on a geological basis and show the boundaries as contiguous with the boundaries of the 3 geomorphic provinces. However, many recent desert scholars also accept a biological basis for marking the boundaries and use the biota as well as geologic forms as demarcation indicators. The resulting boundaries are significantly different from those of the geomorphic provinces.

The triangular shaped northwest Great Basin Desert enters California at the north end of the Owens Valley and runs south along the eastern slope of the Sierra Nevada till it merges with the Mojave near the town of Olancha. It runs almost due northeast from Olancha to the Nevada border, dividing Death Valley into 2 desert regions. It is a high desert, most of it above 4000 feet. It is cold in winter with a very short growing season and is marked by a few varieties of shrub, particularly the Great Basin sagebrush.

The Mojave Desert (in California) occupies the area running roughly from Olancha northeast to the Nevada border in the north, and in the south, from Palm Springs northeast in a meandering line through the center of Joshua Tree National Monument, then north through the Turtle Mountains and east to a point above the Whipple Mountains north of Parker Dam. It lies between 2,000 and 4,000 feet, is a winter rain desert, and is transitional in its floral types between its two neighbors. However, the flora in the areas at 2,000 feet are largely indistinguishable from that of the Sonoran Desert. Annuals and shrubs predominate in the Mojave. The outstanding plant of the area is the great yucca, which is essentially the "logo" of the area. A large northern portion of the Mojave from the Death Valley area to the southern portion of the Naval Weapons area is very low, often lying below sea level and is extremely hot and dry, though the Panamint Mountains in the region exhibit Great Basin foliage at higher levels and even a fine woodland near the summit.

The Colorado Desert lies mostly below 2000 feet, with mild winters and rain primarily in the summer, except for the southern section. Marked with several species of trees such as the desert ironwood and the mesquite, this desert holds most of the great cacti still extant in California: the saguaro and the organ pipe. The Sonoran is among the hottest desert areas in the states, with occasional marks of 120 ° Fahrenheit in the summer. Several mountain ranges in the east, the Clark, Ivanpah, New York, and Providence, attain heights of 8000 feet and support pine forests at the summits. Oddly enough, wetlands also exist within the desert boundaries, particularly the Salton Sea, the Carrizo Marsh, Darwin Falls, and several other streams and springs. A small section of the Colorado Desert, an area of prehistoric lakes known as the Yuha Desert, makes an arc around the city of El Centro, with its base at the Mexican border.

See: Bibliography 9, 35, 136, 153, 170, 175, 180, 182.

Desert

1 Great Basin
2 Mojave
3 Colorado (Sonoran)

Park

A Death Valley National Monument
B East Mojave National Scenic Area
C Joshua Tree National Monument
D Anza-Borrego Desert State Park

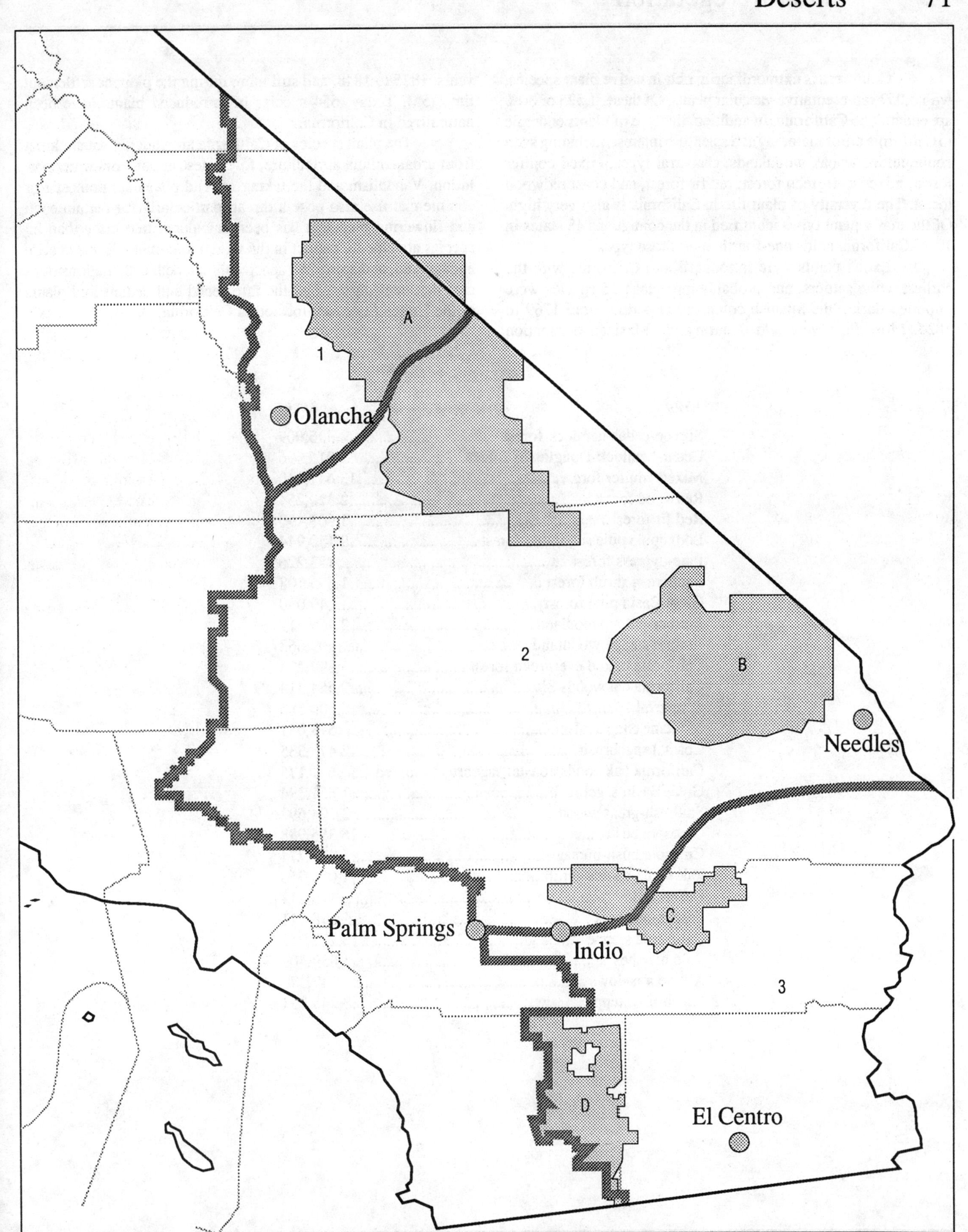
A
1
Olancha
2
B
Needles
C
Palm Springs
Indio
3
D
El Centro

California is extraordinarily rich in native plant species, with 5,057 representative vascular plants. Of these, 1,525 or 30% are endemic to California. In addition, the type of plants endemic to California are of major lay and scientific interest, including such communities as oak woodlands, chaparral types, mixed conifer forest, mixed evergreen forest, red fir forest, and coast redwood forest. The diversity of plant life in California is also very high. Of the major plant types identified in the contiguous 48 states in 1964, California holds one-fourth of all those types.

Exotic plants were introduced into California with the earliest white settlers, and probably more than 16 species were imported during the Spanish colonization period, from 1769 to 1824. More (63) were added during the Mexican occupation years, 1825 to 1848, and still more during the pioneer settlement time (55). Today, 654 species of introduced plants have been naturalized in California.

The plant species of California are under constant threat from urbanization, agriculture, forest destruction, and water pollution. Vandalism and the taking of wild plants for domestic or commercial use also pose a threat, particularly for certain cacti and flowering plants. It has been calculated that more than 25 species are already extinct in the wild, more than 150 species are endangered, and some 280 species threatened. Calculations indicate that one-fourth of all the threatened and endangered plants of the United States are located in California.

Type	Acres
Spruce-cedar-hemlock forest	5,009
Cedar-hemlock-Douglas fir forest	2,015,696
Mixed conifer forest	13,641,010
Redwood forest	2,320,254
Red fir forest	1,903,490
Lodgepole pine subalpine forest	2,150,944
Pine-cypress forest	123,226
Ponderosa shrub forest	1,695,108
Great Basin pine forest	49,090
Juniper-piñon woodland	2,463,517
Juniper steppe woodland	909,668
California mixed evergreen forest	3,399,232
California oakwoods	9,554,518
Chaparral	8,500,585
Montane chaparral	573,051
Coastal sagebrush	2,473,535
California oakwoods/coastal sagebrush - mixed	641,175
Great Basin sagebrush	1,851,394
Saltbush-greasewood	3,104,692
Creosote bush	16,355,988
Creosote bush-bursage	5,330,774
Paloverde-cactus shrub	1,052,950
Desert sparse vegetation	115,211
Fescue-oatgrass	878,711
California steppe	13,222,242
Tule marshes	1,859,409
Alpine meadows	747,370
Sagebrush steppe	3,245,951

Coniferous Forest
Oak Woodland
Coniferous Woodland
California Prairie
Chaparral
Sagebrush
Desert Shrub
Marsh-grass
Redwood

The California Wildlife Habitat Relationships System is an information system employing advanced technology to describe "the management status, distribution, life history and habitat requirements of California's wildlife species." The system currently provides: "1) a wildlife species list; 2) species notes; 3)species distribution maps; 4) computer data base (species-habitat relationship models); and 5) habitat classification and vegetation descriptions." It lists, under 5 major headings: Tree Dominated Habitats, Shrub Dominated Habitats, Herbaceous Dominated Habitats, Aquatic Habitats, and Developed Habitats, 52 different habitat types. (In 1975, The University of California Natural Reserve System attempted a similar classification scheme showing 168 types under 10 major headings.) The "species distribution maps" indicate that most of the prominent species of California wildlife are dispersed across the state's wild areas so as to make impractical any mapping of their habitat sites in this *Atlas*. However, several species reside in discrete locations. Most of these animals are reduced to these areas as a result of man's depredation, habitat invasion, or livestock grazing. Wild creatures with readily identifiable habitat sites are:

Sea Otters: Worldwide, almost a million sea otters were slaughtered for their fur, and by the early twentieth century, the California sea otter was considered extinct. In 1938, a small group of these otters were discovered and protection has permitted their numbers to grow, but not yet to a measure of safety. About 1,700 otters, including 200 pups, were reported in 1989. To further protect them, a translocation scheme has been effected in an effort to establish a colony on their former breeding ground on San Nicolas Island.

Pronghorn Antelope: Not a true antelope, the Pronghorn was almost wiped out by hunting and grazing. Today, Pronghorn reside in 14 areas across California.

California Condor: The Condors, devastated by illegal hunting and poisonous herbicides, exist today only in 2 zoos in Los Angeles and San Diego. Recently,8 chicks were hatched, raising the California Condor population to 40 (27 origional birds; 1 hatched in 1988; 4 in 1989, 8 in 1990). Hopefully, they may some day return to their former habitat.

Tule Elk: Killed for their meat and hides, the original 500,000 Tule Elk in California were severely reduced in number, and limited to a few very small sites in the Tulare and Buena Vista Lakes areas by 1860. Both these marsh areas were drained later in the century, and the Tule Elk were reduced to 28 individuals by 1895, 22 years after supposedly protective legislation had been enacted in 1873. Translocation of the elk has been attempted and several such efforts have met with success. In December 1988, a survey indicated that between 2,300 and 2,700 Tule Elk were living in 19 locations.

Bighorn Sheep: Hunting and diseases carried by domestic animals in California almost eradicated the Bighorn Sheep, which were reduced to scattered small groups and 1 remaining herd on Mount Baxter by the mid-twentieth century. In 1986, a portion of the Mount Baxter herd was translocated to the Lee Vining Canyon where they are faring reasonably well but are not yet fully established.

Monarch Butterfly: Monarchs residing west of the Rocky Mountains cluster in dense concentrations, between October and March in approximately 100 coastal groves along the California Coast. Their winter range extends from the Monterey peninsula south to a point just north of Santa Barbara. (Monarchs resident east of the Rocky Mountains roost in the transvolcanic mountain ranges of central Mexico.) These sites provide winter refuge for an estimated 100,000,000,000 butterflies. After mating in March, the Monarchs leave, have their young inland all across the western U.S. and southern Canada, and die shortly thereafter. The next generation then makes the long flight to its western winter home.

Elephant Seal: Ranging historically from Prince of Wales Island in Alaska to Cape San Lazaro in Baja California, the seals were hunted to virtual extermination for their fine quality oil. In 1892, scientists found what they thought were the last 9 individuals on Guadalupe Island off northern Baja California. They then took 7 of those individuals as specimens! However, other elephant seals had found refuge out of human sight and when discovered, were protected by the Mexican government in 1911. By 1930, enjoying full protection by Mexico and the United States, the herd grew to approximately 1,500 seals. By 1978, breeding colonies had been reestablished on San Miguel, San Nicolas, Santa Barbara, Año Nuevo and Southeast Farallon Islands, in addition to a colony on mainland Point Año Nuevo. Today, more than 50,000 seals haul out off the coast of Baja California and California.

See: Bibliography 108, 135, 149, 163, 165, 201.

(continued)

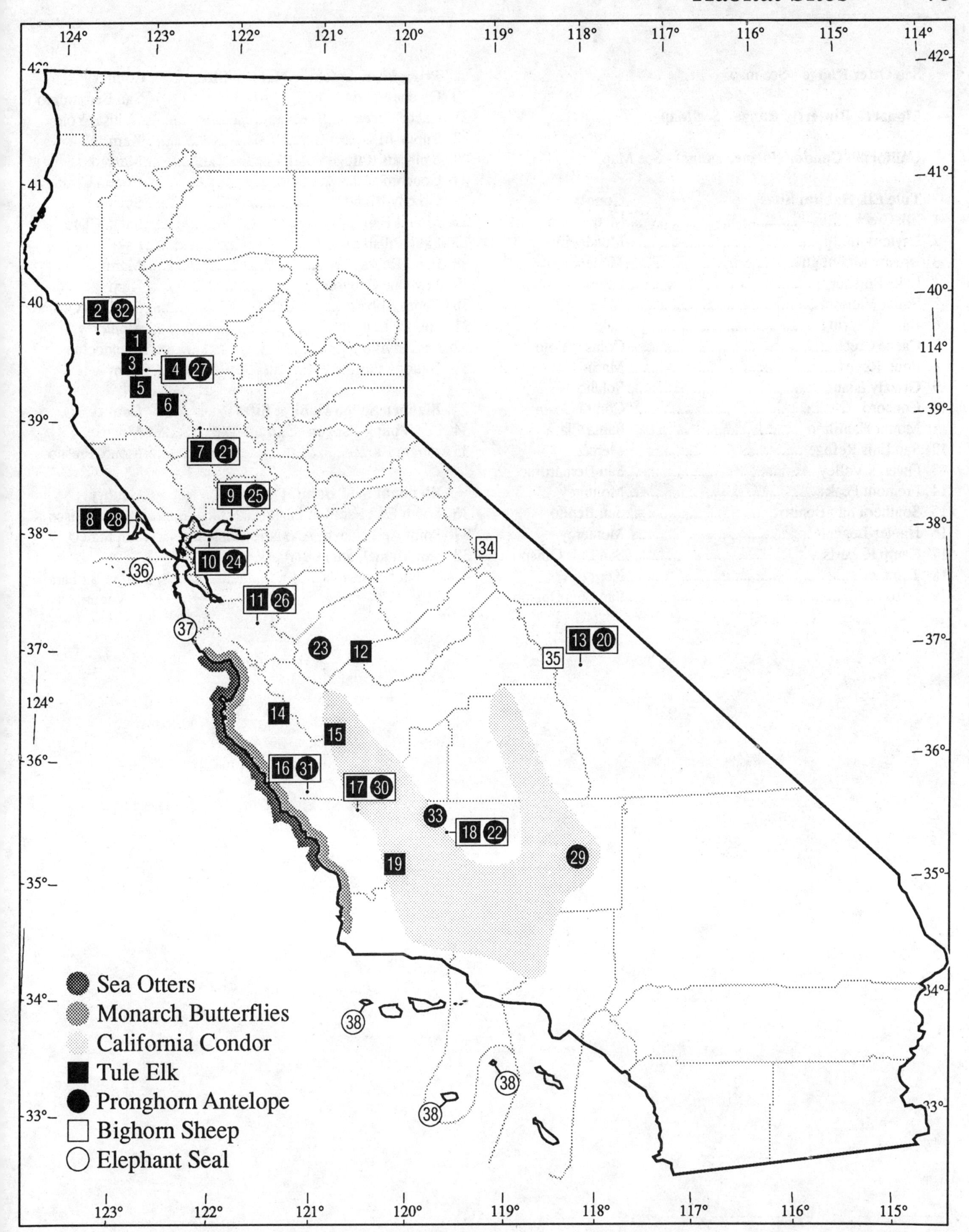
Sea Otters
Monarch Butterflies
California Condor
Tule Elk
Pronghorn Antelope
Bighorn Sheep
Elephant Seal

Sea Otter Range - See map.

Monarch Butterfly Range - See Map.

California Condor (former range) - See Map.

	Tule Elk Habitat Sites	**County**
1	Elk Creek	Mendocino
2	Laytonville	Mendocino
3	Brushy Mountain	Mendocino
4	Lake Pillsbury	Lake
5	Potter Valley	Lake
6	Bartlett Spring	Lake
7	Cache Creek	Colusa, Yolo
8	Point Reyes	Marin
9	Grizzly Island	Solano
10	Concord	Contra Costa
11	Mount Hamilton	Santa Clara
12	San Luis Refuge	Merced
13	Owen's Valley	San Bernardino
14	Fremont Peak	Monterey
15	Southern San Benito	San Benito
16	Hunter-Leggett	Monterey
17	Camp Roberts	San Luis Obispo
18	Tupman	Kern
19	Pozo	San Luis Obispo

	Pronghorn Antelope Habitat Sites	**County**
20	Owens Valley	San Bernardino
21	Cache Creek	Colusa, Yolo
22	Tupman	Kern
23	San Luis Refuge	Merced
24	Concord	Contra Costa
25	Grizzly Island	Solano
26	Mount Hamilton	Santa Clara
27	Lake Pillsbury	Lake
28	Point Reyes	Marin
29	Jawbone Canyon	Kern
30	Camp Roberts	San Luis Obispo
31	Hunter-Leggett	Monterey
32	Laytonville	Mendocino
33	Buttonwillow	Kern

	Bighorn Sheep Habitat Sites	**County**
34	Lee Vining Canyon	Mono
35	Mount Baxter	Mono, Fresno

	Elephant Seal Habitat Sites	**County**
36	South East Farallon Island	San Francisco
37	Point Año Nuevo, Año Nuevo Island	San Mateo
38	San Miguel, Santa Barbara, and San Nicolas Islands	Santa Barbara, Ventura

124°
123°
122°
121°
120°
119°
118°
117°
116°
115°
114°
42°
41°
40°
39°
38°
37°
36°
35°
34°
33°
1
2
3
4
5
6
7
8
9
10
11
12
13
14
15
16
17
18
19
20
21
22
23
24
25
26
27
28
29
30
31
32
33
34
35
36
37
38
Sea Otters
Monarch Butterflies
California Condor
Tule Elk
Pronghorn Antelope
Bighorn Sheep
Elephant Seal

In order to identify relict areas of major floral or faunal value in California, the U.S. Fish and Wildlife Service, in cooperation with other federal, state, and local agencies; private conservation groups, and concerned individuals, compiled, in 1980, an inventory of the important fish and wildlife habitat areas in the state. Working from a lengthy list of nominations, the Service applied a systematized evaluation check which ranked and scored each suggested area under 3 major headings: Biological values - 8 subdivisions provided for a possible maximum score of 58; Habitat Values - 2 subdivisions for a possible maximum score of 21; and Management - with a possible maximum score of 6. The resultant scores enabled the Service to name the sites they deemed the 49 most important habitat areas within California.

See: Bibliography 24, 37, 117, 184.

	Area	County	Score	Rank	Acres
1	Goose Lake	Modoc	24	44	124,160
2	Humboldt Lagoons	Humboldt	23	45 (tie)	4,221
3	Pit River Canyon	Modoc	23	45 (tie)	- -
4	Cuckoo Island Corridor	Tehama	58	53 (tie)	3,500
5	Deer Creek	Tehama	26	40 (tie)	32,000
6	Big River Estuary	Mendocino	38	25 (tie)	1,500
7	Sierra Valley Marsh	Plumas	40	21 (tie)	1,600
8	Valley Vernal Pools	Placer	33	32 (tie)	- -
9	Putah Creek Riparian Corridor	Yolo, Solano	46	14	900
10	Dozier Grasslands and Vernal Pools	Solano	34	28 (tie)	2,800
11	West Marin Island	Marin	35	27	160
12	Emeryville Crescent	Alameda	51	10 (tie)	500
13	San Bruno Mountain	San Mateo	42	20	3,500
14	McNamee's Cave	Tuolomne	20	48	3
15	Mono Lake	Mono	55	7	54,500
16	Hickman Vernal Pools	Stanislaus	40	21 (tie)	2,000
17	San Luis Island	Merced	34	28 (tie)	21,000
18	Fish Slough	Mono	45	15 (tie)	200
19	Watsonville Slough	Santa Cruz	44	17 (tie)	1,800
20	Saline Valley Salt Marsh	Inyo	45	15 (tie)	200
21	Creighton Ranch	Tulare	34	28 (tie)	- -
22	China Ranch	Inyo	50	12	320
23	Fiscalini Ranch	San Luis Obispo	40	26 (tie)	380
24	San Joaquin Natural Areas	Kern	48	13	12,980
25	Kelso Creek	Kern	40	21 (tie)	700
26	Meadow Creek	San Luis Obispo	26	40 (tie)	- -
27	Nipomo Dunes	San Luis Obispo	72	1	3,000-10,000
28	Soda Lake	San Luis Obispo	58	3 (tie)	2,360
29	Desert Tortoise Natural Area	Kern	53	9	8,320
30	Afton Canyon	San Bernardino	33	32 (tie)	200
31	Gorman Post Road	Kern	37	36 (tie)	200
32	South Fork Kern River	Kern	57	6	5,000
33	Las Tunas	Santa Barbara	21	47	12
34	Scorpion Rock	Santa Barbara	33	32 (tie)	1
35	Upper Santa Clara River	Los Angeles	29	38	312
36	El Segundo Dunes	Los Angeles	32	35	245
37	Brea-Olinda Wilderness	Orange	34	28 (tie)	1,100
38	Lake Norconian	Riverside	30	36 (tie)	450
39	Coachella Valley Fringe-Toed Lizard Site	Riverside	44	17 (tie)	2,900
40	Rancho Dos Palmos	Riverside	58	3 (tie)	1,371
41	Bolsa Chica Marsh	Orange	54	8	900
42	Weir Canyon	Orange	43	19	2,570
43	Rubber Boa Habitat	Riverside	28	39	20,000
44	Garver Valley	Riverside	38	25 (tie)	25,000
45	Whitewater River Marsh	Riverside	61	2	150
46	Volcan Mountains	San Diego	26	40	12,000
47	San Sebastian Marsh	Imperial	51	10 (tie)	7,680
48	Pine Creek	San Diego	40	21 (tie)	6,000
*	Heater Cave	Calaveras	19	49	- -

* not shown

Area	#
Afton Canyon	30
Big River Estuary	6
Bolsa Chica Marsh	41
Brea-Olinda Wilderness	37
China Ranch	22
Coachella Valley Fringe-Toed Lizard Site	39
Creighton Ranch	21
Cuckoo Island Corridor	4
Deer Creek	5
Desert Tortoise Natural Area	29
Dozier Grassland and Vernal Pools	10
El Segundo Dunes	36
Emeryville Crescent	12
Fiscalini Ranch	25
Fish Slough	18
Garner Valley	44
Goose Lake	1
Gorman Post Road	31
Heater Cave	*
Hickman Vernal Pools	16
Humboldt Lagoons	2
Kelso Creek	25
Lake Norconian	38
Las Tunas Grasslands	33
McNamee's Cave	14
Meadow Creek	26
Mono Lake	15
Nipomo Dunes	27
Pine Creek	48
Pit River Canyon	3
Putah Creek Riparian Corridor	9
Rancho Dos Palmos	40
Rubber Boa Habitat	43
Saline Valley Salt Marsh	20
San Bruno Mountain	13
San Joaquin Natural Areas	24
San Sebastian Marsh	47
San Luis Island	17
Scorpion Rock	34
Sierra Valley Marsh	7
Soda Lake	28
South Fork Kern River	32
Upper Santa Clara River	35
Valley Vernal Pools	8
Volcan Mountains	46
Watsonville Slough	19
Weir Canyon	42
West Marin Island	11
Whitewater River Marsh	45

*not shown

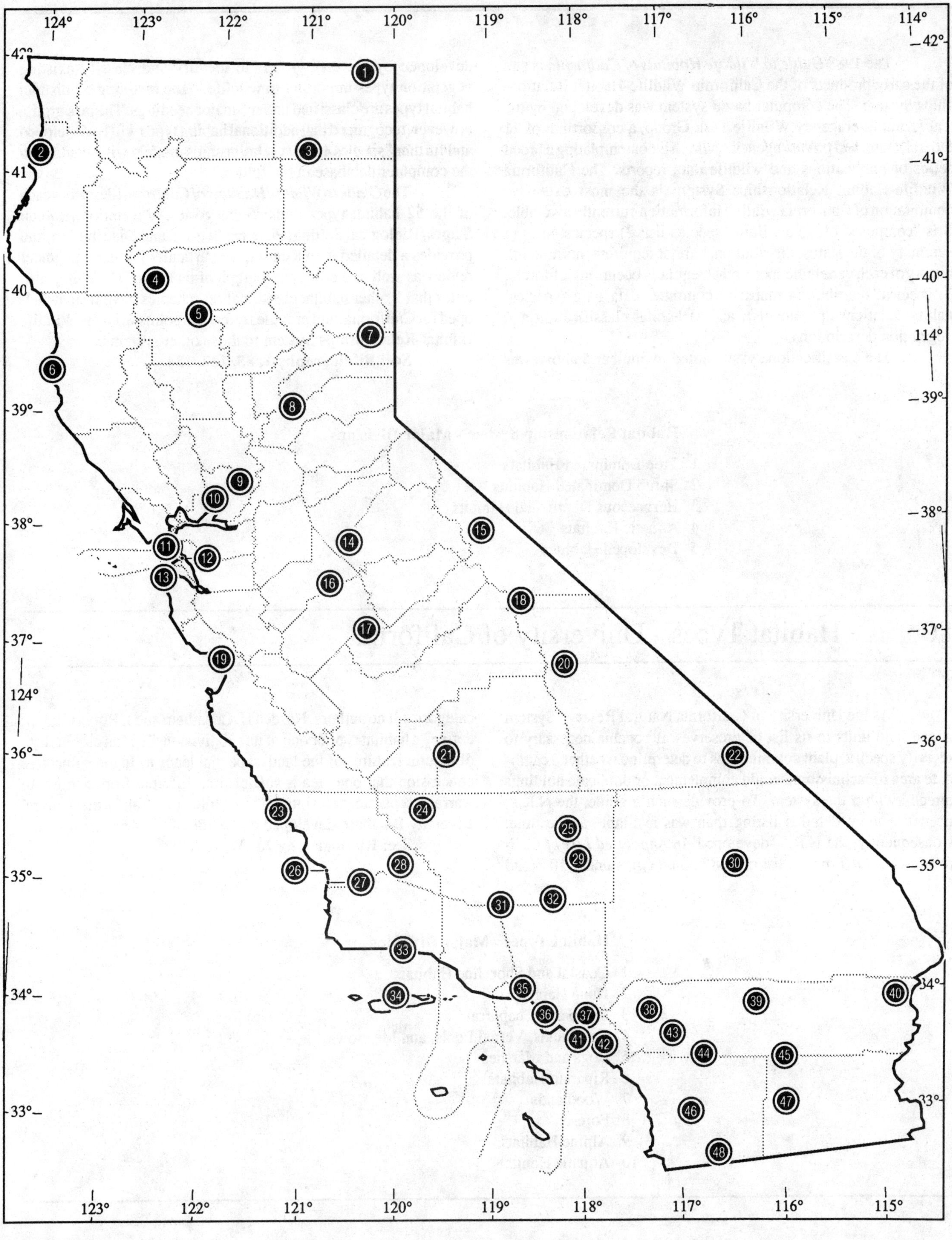

124° 123° 122° 121° 120° 119° 118° 117° 116° 115° 114°

42° 41° 40° 39° 38° 37° 36° 35° 34° 33°

1 2 3 4 5 6 7 8 9 10 11 12 13 14 15 16 17 18 19 20 21 22 23 24 25 26 27 28 29 30 31 32 33 34 35 36 37 38 39 40 41 42 43 44 45 46 47 48

123° 122° 121° 120° 119° 118° 117° 116° 115°

80a Habitat Types - California Interagency Wildlife Task Group

The 1988 *Guide to Wildlife Habitats of California* is one of the early products of the California Wildlife-Habitat Relationships System. The computer based system was developed by the California Interagency Wildlife Task Group, a consortium of 21 federal, state, and private agencies, which is contemplating a broad series of publications and wildlife data reports. The California Wildlife-Habitat Relationships System is the most extensive compilation of California wildlife information currently available. It is "composed of: 1) a wildlife species list; 2) species notes (a summary of the status, distribution, habitat requirements, and life history of each vertebrate species that regularly occurs in California; 3) species' distribution maps; 4) computer data base (species' habitat relationships models); and 5) habitat classification and vegetation descriptions."

The classification system noted in number 5 above was developed by the task group "to identify and classify existing vegetation types important to wildlife." The resulting 52 distinct habitat types are classified under 5 major headings. The task group, however, recognized that additional habitat types will be developed and further "species-habitat relationships models will be added to the computer database in the future."

The *Guide to Wildlife Habitats of California* describes each of the 52 habitat types under 5 categories: Vegetation, Habitat Stages, Biological Setting, Physical Setting, and Distribution, and provides a detailed map locating the particular habitat type under review as well as a color photograph of that type. The *Guide* also notes that 9 other habitat classification schemes have been developed for California, and provides reference points from the Wildlife Habitat-Relationships System to those other systems.

See: Bibliography 1, 85, 122, 184

Habitat Relationship System - Major Divisions

1 Tree Dominated Habitats
2 Shrub Dominated Habitats
3 Herbaceous Dominated Habitats
4 Aquatic Habitats
5 Developed Habitats

80b Habitat Types - University of California

As the University of California Natural Reserve System added land units to its list of preserves, it became necessary to identify specific plant communities to determine whether a candidate area for acquisition would complement or duplicate holdings already within the system. To provide such a guide, the N.R.S. needed a more detailed listing than was available at the time. Consequently, the N.R.S. developed *An Annotated List of California Habitat Types*, a list of 168 habitat types under 10 broad categories. The authors, Norden H. Cheatham and J. Robert Haller, define "a habitat type or one of its subdivisions" as "an assemblage of natural features of the landscape that leads us to the subjective conclusion that one area is sufficiently different from another to warrant separate description." In 1986, the California Natural Diversity Database developed a revision of the *Annotated List*.

See: Bibliography 28, 37, 137.

Habitat Types - Major Divisions

1 Coastal and Shoreline Habitats
2 Dune Habitats
3 Scrub and Chaparral
4 Grasslands, Vernal Pools, and Meadows
5 Bogs and Marshes
6 Riparian Habitats
7 Woodlands
8 Forests
9 Alpine Habitats
10 Aquatic Habitats

Wildlife is dependent on habitat, and wildlife losses in California are due, in the main, to the enormous decrements of natural habitat since the mid 1800s.Oak woodlands, for instance, have declined precipitously in the last century under the pressures that affect all natural habitat: agricultural conversion, development and urbanization, grazing, road construction, and cutting for fuel. Hardest hit has been the Valley Oak which formerly ran in miles wide swaths, but is reduced today to individual trees and tiny stands and is nearing endangered status. Other major habitat areas have suffered similar or even greater losses.

Coastal wetlands, recognized as some of the most biologically productive systems in the state, offer fish and wildlife habitat and act as natural water purifiers and shoreline protection systems. More than 80 percent of these habitats have been lost to development.

Interior wetlands of the Great Central Valley once offered forage and security for countless numbers of pronghorn antelope, tule elk, mule deer, grizzly bear and other fur bearing mammals, and reptiles, and provided nesting , breeding, and foraging sites for millions of migrating water fowl.Today, only 4 percent of those wetlands remain.

Riparian woodland along the Sacramento and San Joaquin valleys once flourished across 1,000,000 acres and the wildlife along the banks was equally numerous. Currently, just 11 percent still stands and much of that area is degraded.

Valley grasslands covered a quarter of the state in the early 1900s, and reports of that period tell of grassland and attendant wildflowers stretching to the horizon. A pitiful 1 percent of our natural grasslands remain.

Vernal pools are unique to California, and many of these remarkable ponds have ecosystems unique unto themselves. Efforts to recreate vernal pools have been largely unsuccessful, yet hundreds of acres of vernal pools are lost to the bulldozer annually. Sixty-six percent of all the vernal pools in the central valley are gone.

Habitat Losses Since 1850

Habitat	Acres Lost	Percentage Lost
Coastal Wetland	202,000	80%
Interior Wetland	3,750,000	96%
Riparian Woodland	819,000	89%
Valley Grassland	21,978,000	99%
Vernal Pools	2,770,000	66%

Note: Native Animals Now Extinct in California.

Mammals- Bison*, Grey Wolf*, Grizzly Bear*, Jaguar*, Long Eared Kit Fox, White-Tailed Deer.

Birds- San Clemente Bewick's Wren, Santa Barbara Song Sparrow, Sharp-Tailed Grouse*.

Fish- Bull Trout*, Clear Lake Splittail, Shoshone Pupfish, Tecopa Pupfish, Thick-Tailed Chub.

Invertebrates- Antioch Dunes Katydid, Antioch Robber Fly, Antioch Sphecid Wasp, Antioch Weevil, Atossa Fritillary Butterfly, Castle Lake Rhyacophilan Caddisfly, Fort RossTrigonoscuta Weevil, Mono Lake Hygrotus Diving Beetle, Pasadena Freshwater Shrimp*, Pheres Blue Butterfly, Sooty Crayfish*, Sthenele Wood Nymph Butterfly, Strohbeen's Parnassian Butterfly, Valley Mydas Fly, Voluntine Stonemyian Tabanid Fly, Xerxes Blue Butterfly, Yellow-Banded Andrenid Bee, Yorba Linda Trigonoscuta Weevil.

*- Exists outside of California

The first state forest, Las Posadas, was acquired by gift in 1930, and 3 further gifts followed over the next 14 years. These were all small tracts, and the fourth donation, Loughry, only 68 acres, is no longer administered as a state forest. Extraordinary opposition to the purchase of forest lands succeeded in denying appropriations for that purpose until 1945, when funding was legislated for the purchase of Latour, (January 8, 1946) and Mountain Home (January 9, 1946).

Today (December 1990), the California Department of Forestry and Fire Protection manages 7 relatively small Demonstration State Forests totaling 68,664 acres, and contracts with the State Lands Commission to manage 18 State School Lands Areas for an additional 3,836 acres. The primary purpose of the state forest program, in addition to commercial timber sales, is the performance of demonstrations, research, and education in forest management for the production of cut lumber. Consequently, timber for harvest is a major component of state management, and 41,893,000 board feet were cut in 1987, enough to build 3,793 single family homes for a total value of $4,905,718. The funds raised were used to provide cost-share incentives for reforestation and similar forest improvements. However, in an effort to demonstrate that timber cutting and recreation are compatible uses, the department does provide some facilities for camping and picnicking.

Very little of the optimistic ideas and ideals of the 1940s regarding state forest development, in which it was contemplated that the state would manage great tracts of forest land, have come to fruition. Today, the state forest system plays a minor role in recreation and wildlife habitat programs. California's state forests are slight, particularly in comparison with Oregon's state forests, 800,000 acres, and Washington's forests 2,116,000 acres.

	State Forest	County	Date Acquired	Acres
1	Ellen Pickett	Trinity	1939 (gift)	160
2	Latour	Shasta	1946	9,013
3	Jackson	Mendocino	1947-51, 1968	50,505
4	Boggs Mountain	Lake	1949, 1972	3,464
5	Las Posadas	Napa	1929 (gift)	796
6	Mount Zion	Amador	1932 (gift)	164
7	Mountain Home	Tulare	1946	4,562
				Total: 68,664

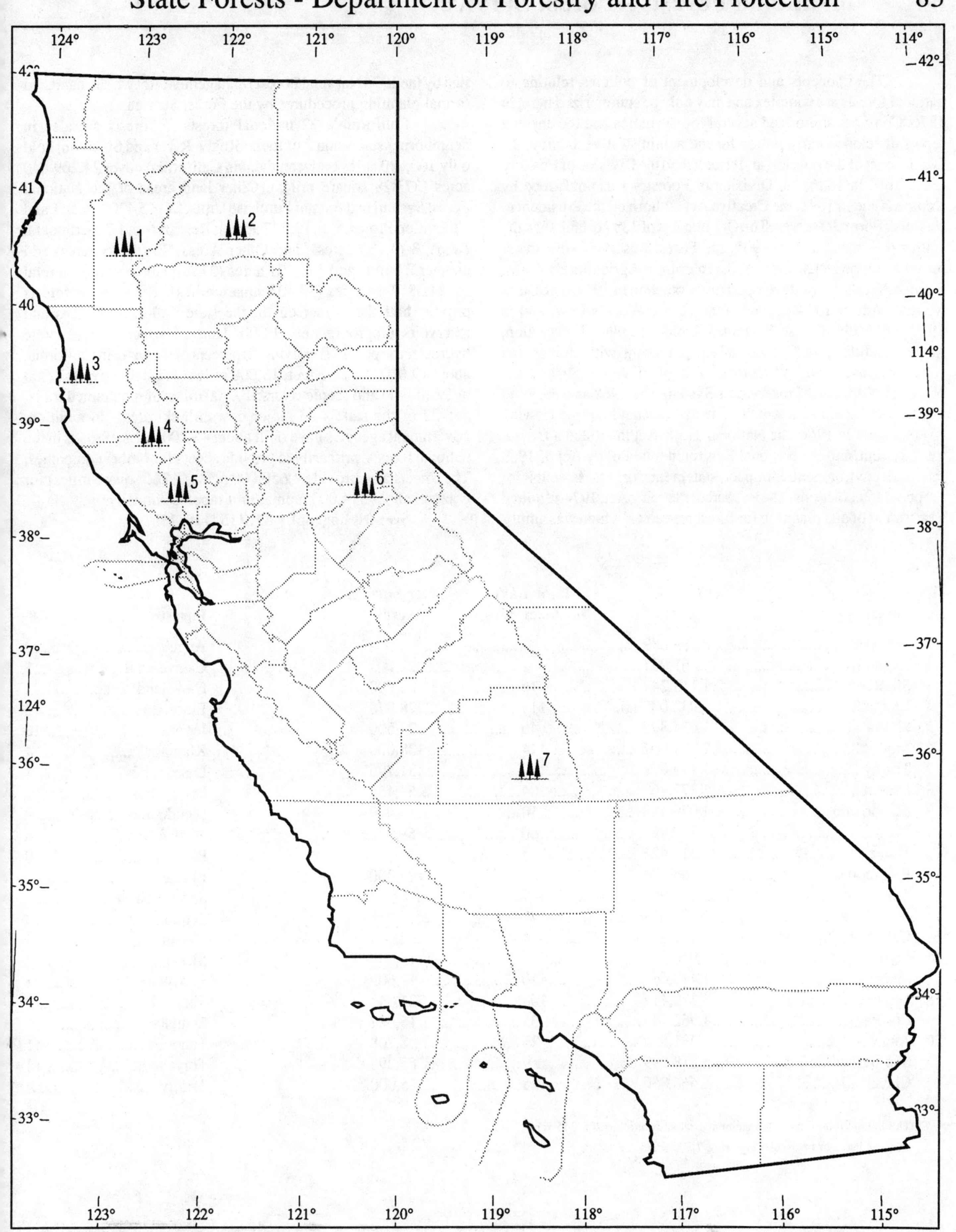
124°
123°
122°
121°
120°
119°
118°
117°
116°
115°
114°
42°
41°
40°
39°
38°
37°
36°
35°
34°
33°
1
2
3
4
5
6
7

The concepts and development of policies relating to national forests are complex and may only be summarized here. In 1876, Congress authorized several forest studies and the ensuing report developed early policy for the administrative agency, the Department of the Interior, and established the Division of Forestry (Fort Bill). In 1886, the Division of Forestry was confirmed by Congress and in 1891, the Creative Act authorized the President to establish Forest Reserves. The Organic Act of 1897 established the National Forests, and in 1905, the Forest Reserves were transferred from the Department of the Interior to Agriculture's Forest Service. Week's Law developed fire protection in 1911. The Clark McNary Act of 1924 expanded the earlier Week's Law, and in 1960, the Multiple Use Sustained Yield Act placed recreation, range, wildlife, and fish on an equal footing with timber and watershed uses. The Wilderness Act of 1964 established the National Wilderness Preservation System, and 1968 saw the Wild and Scenic Rivers Act, which protected outstanding free flowing rivers. Again in 1968, the National Trails Act instituted a federal trails system, and the National Environmental Policy Act of 1969 required environmental impact statements or assessments for proposed forest actions. The Resources Planning Act, 1974, required assessment of and planning for forest resources, which was amplified by the 1976 National Forest Management Act which mandated formal planning procedures by the Forest Service.

California's 22 national forests (5 forests are also in neighboring states and 2 of them, Rogue River and Siskiyou, hold only 100,720 acres between them in California) cover 24,209,840 acres (37,828 square miles). Other land areas of the National Forest System in the state: Purchase Units, 2 (145,470 acres); Land Utilization Projects, 2 (19,222 acres); Research and Experimental Areas, 3 (4,783 acres); and Other Areas, 20 (6,646 acres) add another 27 units and 176,121 acres (275 square miles) for a total of 24,385,961 acres (38,103 square miles). The national forests provide half the timber cut in the state; hold 2,467 lakes and reservoirs (approximately 1,200 dams of which 167 generate hydroelectric power) supplying 50 percent of the water in California, about 19,000,000 acre feet; 13,000 miles of fishable rivers; 1,800 miles of wild and scenic rivers; 10,500 miles of maintained trails; and 22 of the state's 33 major downhill ski areas. In addition, government agencies have built about 45,200 miles of roads in the national forests, primarily to provide access for timber companies. The forests also provide 4,650,000 acres (7,266 square miles) for grazing, about 514,000 animal unit months annually.

See: Bibliography 87, 115, 118, 150.

	Forest	Acres	Major Lakes or Reservoirs	Water Surface Acres
1	Siskiyou	39,689	- -	- -
2	Rogue River	61,031	- -	- -
3	Six Rivers	1,118,247	16	500
4	Klamath	1,932,000	111	228,878
5	Shasta	1,634,896	46	27,500
6	Modoc	1,979,407	114	138,000
7	Trinity	1,179,098	49	31,795
8	Lassen	1,377,969	104	63,436
9	Mendocino	1,079,483	10	35,000
10	Plumas	1,400,895	60	58,370
11	Tahoe	1,211,425	115	21,739
12	Eldorado	786,994	170	127,300
13	Toiyabe	694,988	- -	- -
14	Stanislaus	1,090,039	77	6,711
15	Calaveras Big Trees	380	- -	- -
16	Inyo	2,046,346	600	315,000
17	Sierra	1,412,641	440	17,609
18	Sequoia	1,178,417	108	721,045
19	Los Padres	1,962,743	35	15,627
20	Angeles	691,539	25	8,708
21	San Bernadino	818,999	9	85,395
22	Cleveland	566,850	25	15,000

Forest	#
Angeles	20
Calaveras Big Trees	15
Cleveland	22
Eldorado	12
Inyo	16
Klamath	4
Lassen	8
Los Padres	19
Mendocino	9
Modoc	6
Plumas	10
Rogue River	2
San Bernardino	21
Sequoia	18
Shasta	5
Sierra	17
Siskiyou	1
Six Rivers	3
Stanislaus	14
Tahoe	11
Toiyabe	13
Trinity	7

The Lake Tahoe Basin Management Unit adds another 208,945 acres, and 362 lakes or reservoirs covering 7,300 acres.

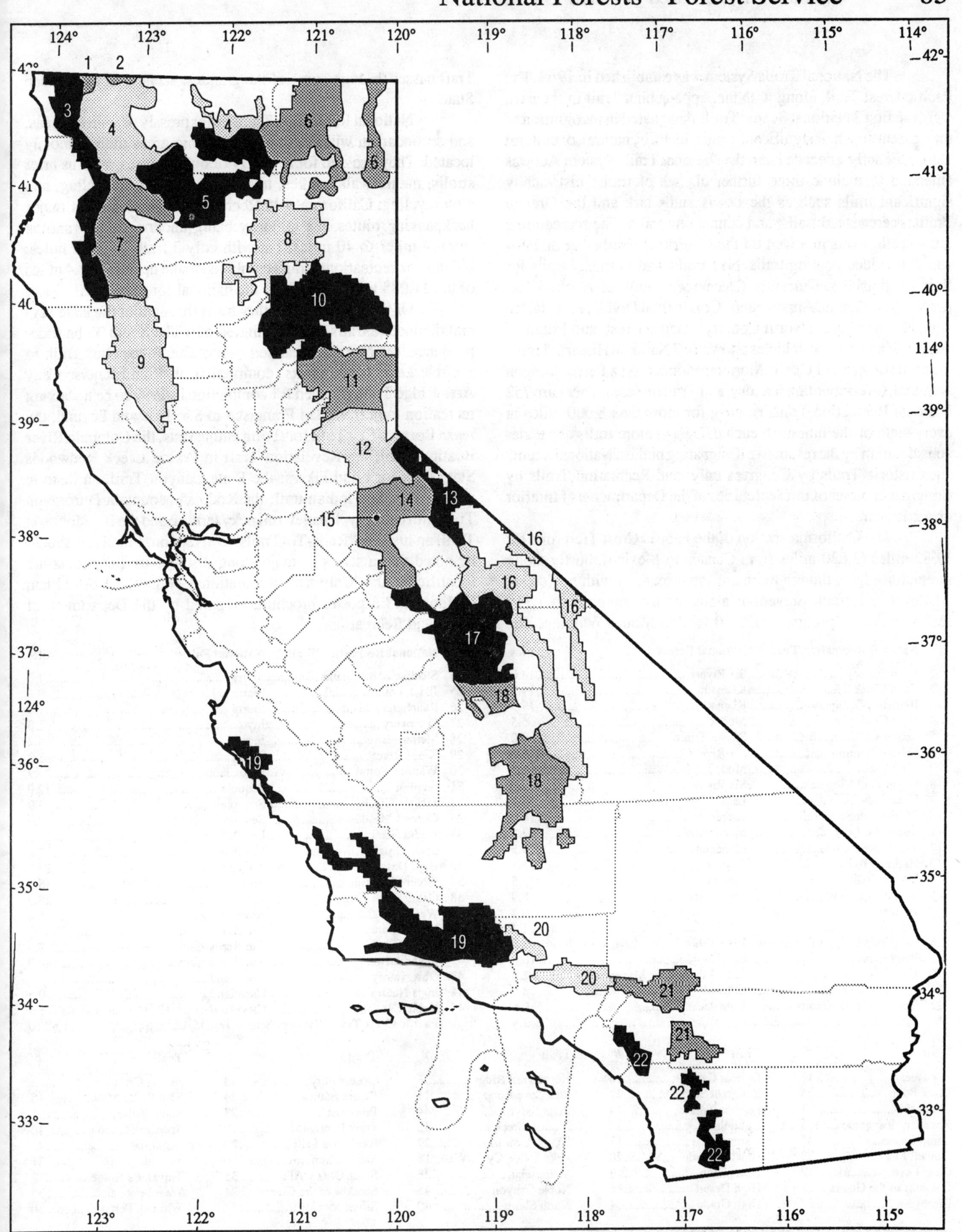
124°
123°
122°
121°
120°
119°
118°
117°
116°
115°
114°
42°
41°
40°
39°
38°
37°
36°
35°
34°
33°
1
2
3
4
4
5
6
6
7
8
9
10
11
12
13
14
15
16
16
16
17
18
18
19
19
20
20
21
21
22
22
22
22

The National Trails System was established in 1968. The Pacific Crest Trail, along with the Appalachian Trail in the east, were the first 2 National Scenic Trails designated in recognition of their extent and their significant scenic, historic, natural, or cultural values. Exactly a decade later, the National Trails System Act was amended to include three further classes of trails: historically significant trails such as the Lewis and Clark and the Oregon Trails; recreational trails; and connecting trails. The recreational trails section was modeled on the California Trails Act of 1974 which included boating trails, bike trails, and even sky trails for flying enthusiasts. Currently (December 1990), there are 8 National Scenic Trails: Appalachian, Continental Divide, Florida, Ice Age, Natchez Trace, North Country, Pacific Crest, and Potomac Heritage (no wheeled vehicles allowed); 7 National Historic Trails: Iditarod, Lewis and Clark, Mormon Pioneer, Nez Perce, Oregon National, Overmountain Victory, and Trail of Tears. There are 752 National Recreation Trails running for more than 8,000 miles in every state of the union. In each category, more trails are under consideration by the respective designating bodies: National Scenic and Historic Trails by Congress only, and Recreation Trails by Congress or either of the Secretaries of the Departments of Interior or Agriculture.

The California section of the Pacific Crest Trail runs for 1,682 miles (2,620 miles from Canada to Mexico) mostly along mountain ridges, though there are some sections with no "crest" whatever underfoot. Elevations along the trail range from 1,110 feet near Palm Springs, to 13,180 feet on Mount Whitney. The Trail passes through some of the grandest scenery in the Golden State.

National Recreation Trails are expressly near urban areas, and secondarily, within established scenic areas more remotely located. They provide for a variety of outdoor uses, such as brief strolls, nature walks, hikes, horse travel, skiing, bicycling, and motorcycling. California's 45 Recreational Trails are not major backpacking routes, the majority being under 6 miles, running from .4 miles to 40 miles, but with only 3 trails over 20 miles. Indeed, the recreational trails combined make up only 377.4 miles of the 11,085 miles of trails in the national forests alone.

Of the hundreds of other trails throughout the state several distinctive trails may be singled out: The South Yuba Independence Trail for handicapped users; The Tahoe Rim Trail, to encircle Lake Tahoe on its completion; the San Francisco Bay Area Ridge Trail, planned for completion in 1992 to be a walking recreation loop from San Francisco to San Jose and beyond; the Anza Borrego Sky Trail for flying enthusiasts, the Colorado River Boating Trail; the Revelation Trail in Prairie Creek Redwoods State Park for blind hikers; the Piute Canyon Trail, an historic wagon route and Indian trail; the Rodman Mountain Petroglyph Trail with petroglyph panels on rocky terrain; and the Bizz Johnson Trail, an historic "Rails-To-Trails" conversion. Finally, it should be noted that a series of major long distance trail corridors are identified in the "California Recreational Trails and Hostel Plan; Preliminary" a poster brochure prepared by the Department of Parks and Recreation.

	National Recreation Trail	National Forest	Miles
1	South Kelsey	Six Rivers	14.0
2	Clear Creek	Klamath	19.3
3	Boundary-Kangaroo	Klamath	19.5
4	High Grade	Modoc	5.5
5	Sisson-Callahan	Shasta-Trinity	9.0
6	Salmon Summit	Six Rivers	5.2
7	Blue Lake	Modoc	1.5
8	Horse Trail Ridge	Six Rivers	15.0
9	Heart Lake	Lassen	3.5
10	Spencer Meadow	Lassen	6.0
11	Ides Cove Loop	Mendocino	7.8
12	Traveler's Home	Mendocino	9.5
13	Hartman Bar	Plumas	3.5
14	Feather Falls	Plumas	3.5
15	Sled Ridge Motorcycle	Mendocino	8.9
16	Donner Camp	Tahoe	0.4
17	Big Trees	Tahoe	1.4
18	Lake Tahoe Cycle/Pedestrian	Lake Tahoe Basin Management Unit	3.5
19	Pony Express	Eldorado	10.0
20	Hawley Grade	Lake Tahoe Basin Management Unit	2.0
21	Emigrant Summit	Eldorado	18.0
22	Columns of the Giants	Stanislaus	0.6
23	Pinecrest	Stanislaus	3.6
24	Shadow of the Giants	Sierra	1.0
25	Black Point	Sierra	0.6
26	Rancheria Falls	Sierra	1.0
27	Discovery	Inyo	1.0
28	Methuselah	Inyo	4.2
29	Kings River	Sierra	3.0
30	Whitney Portal	Inyo	3.9
31	Summit	Sequoia	12.0
32	Jackass Creek	Sequoia	3.0
33	Cannell Meadow	Sequoia	9.0
34	Piedra Blanca	Los Padres	18.2
35	Santa Cruz y Aliso	Los Padres	12.6
36	High Desert	Angeles	27.0
37	Silver Mocassin	Angeles	15.5
38	Gabrielino	Angeles	28.0
39	West Fork	Angeles	6.8
40	North Shore	San Bernardino	5.2
41	Camp Creek	San Bernardino	3.6
42	Sugarloaf	San Bernardino	40.0
43	Observatory	Cleveland	2.1
44	Inaja Nature	Cleveland	0.5
45	Noble Canyon	Cleveland	8.0
A	**Pacific Crest Trail** (National Scenic Trail)		1,682.0

Trail	#	Trail	#	Trail	#	Trail	#	Trail	#
Big Trees	17	Donner Camp	16	Horse Trail Ridge	8	Observatory	43	Sisson-Callahan	5
Black Point	25	Emigrant Summit	21	Ides Cove Loop	11	Piedra Blanca	34	Sled Ridge Motorcycle	15
Blue Lake	7	Feather Falls	14	Inaja Nature	44	Pinecrest	23	South Kelsey	1
Boundary-Kangaroo	3	Gabrielino	38	Jackass Creek	32	Pony Express	19	Spencer Meadow	10
Camp Creek	41	Hartman Bar	13	Kings River	29	Rancheria Falls	26	Sugarloaf	42
Cannell Meadow	33	Hawley Grade	20	Lake Tahoe Cycle/Walk	18	Salmon Summit	6	Summit	31
Clear Lake	2	Heart Lake	9	Methuselah	28	Santa Cruz y Aliso	35	Traveler's Home	12
Columns of the Giants	22	High Desert	36	Noble Canyon	45	Shadow of the Giants	24	West Fork	39
Discovery	27	High Grade	4	North Shore	40	Silver Mocassin	37	Whitney Portal	30

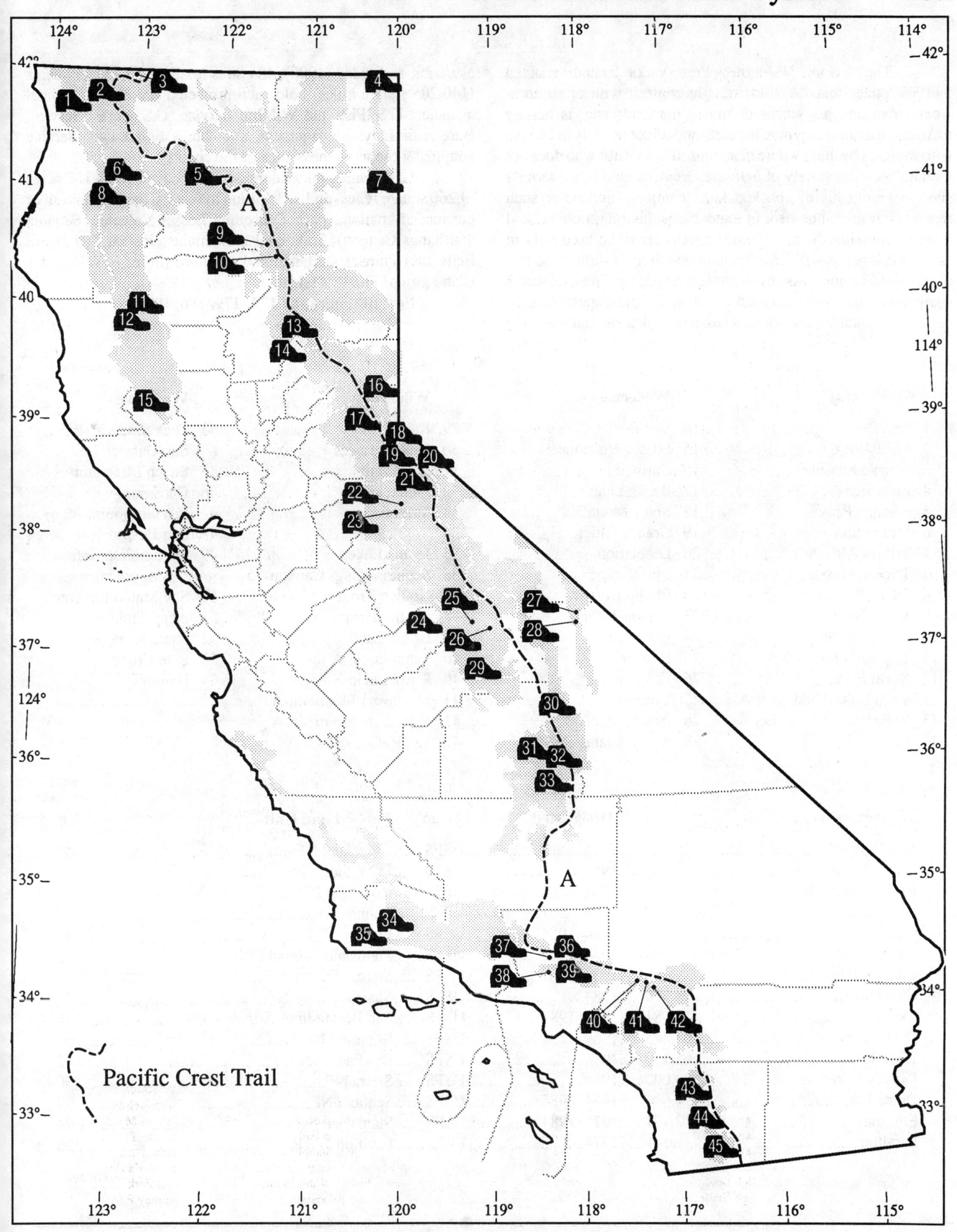
124°
123°
122°
121°
120°
119°
118°
117°
116°
115°
114°
42°
41°
40°
39°
38°
37°
36°
35°
34°
33°
A
A
1
2
3
4
5
6
7
8
9
10
11
12
13
14
15
16
17
18
19
20
21
22
23
24
25
26
27
28
29
30
31
32
33
34
35
36
37
38
39
40
41
42
43
44
45
Pacific Crest Trail

The National Wilderness Preservation System, enacted in 1964, states that "A wilderness in contrast with those areas where man and his works dominate the landscape, is hereby recognized as an area where the earth and its community of life are untrammeled by man, where man himself is a visitor who does not remain." A wide variety of activities are permitted in wilderness sites, not only hiking, backpacking, camping, and other such recreational uses, but also, in many cases, hunting, mining, and livestock grazing. Nonetheless, the areas are to be used only in "such a manner as will leave them unimpaired for future use and enjoyment as wilderness" by the American people. The wilderness designation has been applied to diverse landscapes: forests, mountains, islands, seashores, lakes, rivers, deserts, canyons, and estuaries, and today (1990), 484 sites totalling 89,731,147 acres (140,205 square miles) hold such protected status. Four federal agencies: U.S. Fish and Wildlife Service, U.S. Forest Service, Bureau of Land Management, and the National Park Service administer the wilderness sites of the country.

California's 60 wilderness areas cover 5,926,158 acres (9,260 square miles) and, in addition to forests, they include all or portions of 4 national parks (Yosemite, Lassen Volcanic, Sequoia and Kings Canyon), 3 national monuments (Joshua Tree, Lava Beds, and Pinnacles), 1 national seashore (Point Reyes) and 1 island group (Farallon Islands).

See: Bibliography 172, 179, 186, 199.

	Wilderness
1	Red Buttes
2	Siskiyou
3	Marble Mountain
4	Lava Beds
5	Russian Peak
6	Mount Shasta
7	Trinity Alps-A
7	Trinity Alps-B
8	Castle Crags
9	South Warner
10	Chanchelulla
11	Thousand Lakes
12	North Fork
13	Yolla Bolla-Middle Eel-A
13	Yolla Bolly-Middle Eel-B
14	Ishi-A
14	Ishi-B
15	Lassen Volcanic
16	Caribou
17	Bucks Lake
18	Snow Mountain
19	Granite Chief
20	Desolation
21	Mokelumne
22	Phillip Burton
23	Carson-Iceberg
24	Emigrant
25	Hoover
26	Farallon
27	Yosemite
28	Ansel Adams-A
28	Ansel Adams-B
29	Kaiser
30	Dinkey Lakes
31	John Muir
32	Pinnacles
33	Monarch
34	Ventana
35	Jennie Lakes
36	Sequoia-Kings Canyon
37	Golden Trout
38	South Sierra
39	Dome Land
40	Santa Lucia-A
40	Santa Lucia-B
41	Machesna Mountain-A
41	Machesna Mountain-A
42	San Rafael
43	Dick Smith
44	San Gabriel
45	Sheep Mountain
46	Cucamonga
47	San Gorgonio
48	San Jacinto
49	Joshua Tree National Monument
50	San Mateo Canyon
51	Agua Tibia
52	Santa Rosa
53	Pine Creek
54	Hauser

Wilderness Area	Acres	Designated	Agency	Public Land Unit	#
Agua Tibia	15,933	1975	USFS	Cleveland NF	51
Ansel Adams (A)	228,669	1964	USFS	Inyo, Sierra NFs	28
Ansel Adams (B)	665	1984	NPS	Yosemite NP	28
Bucks Lake	21,000	1984	USFS	Plumas NF	17
Caribou	20,625	1964, 1984	USFS	Lassen NF	16
Carson-Iceberg	160,000	1984	USFS	Stanislaus, Toiyabe NFs	23
Castle Crags	7,300	1984	USFS	Shasta-Trinity NF	8
Chanchelulla	8,200	1984	USFS	Shasta-Trinity NF	10
Cucamonga	12,981	1964, 1984	USFS	San Bernardino, Angeles NFs	46
Desolation	63,475	1969	USFS	Eldorado NF	20
Dick Smith	65,130	1984	USFS	Los Padres NF	43
Dinkey Lakes	30,000	1984	USFS	Sierra NF	30
Dome Land	94,686	1964, 1984	USFS	Sequoia NF	39
Emigrant	112,191	1975, 1984	USFS	Stanislaus NF	24
Farallon	141	1974	FWS	Farallon NWR	26

(continued)

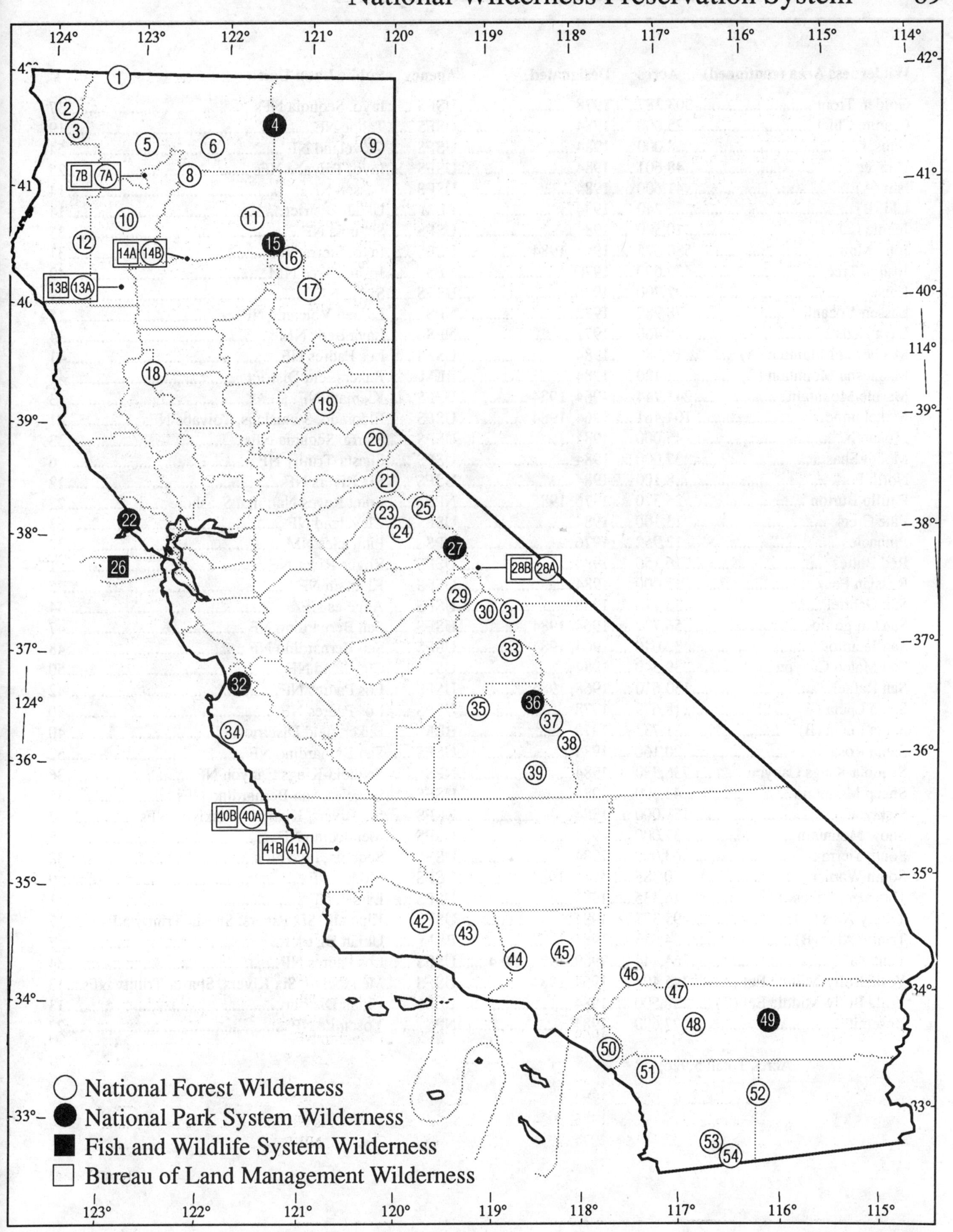

National Forest Wilderness
National Park System Wilderness
Fish and Wildlife System Wilderness
Bureau of Land Management Wilderness

Wilderness Area (continued)	Acres	Designated	Agency	Public Land Unit	#
Golden Trout	303,287	1978	USFS	Inyo, Sequoia NFs	37
Granite Chief	25,000	1984	USFS	Tahoe NF	19
Hauser	8,000	1984	USFS	Cleveland NF	54
Hoover	48,601	1984	USFS	Inyo, Toiyabe NFs	25
Ishi (A)	41,600	1984	USFS	Lassen NF	14
Ishi (B)	240	1984	BLM	Ukiah District	14
Jennie Lakes	10,500	1984	USFS	Sequoia NF	35
John Muir	580,675	1964, 1984	USFS	Inyo, Sierra NFs	31
Joshua Tree	429,690	1976	NPS	Joshua Tree NM	49
Kaiser	22,700	1976	USFS	Sierra NF	29
Lassen Vocanic	78,982	1972	NPS	Lassen Volcanic NP	15
Lava Beds	28,460	1972	NPS	Lava Beds NM	4
Machesna Mountain (A)	19,880	1984	USFS	Los Padres NF	41
Machesna Mountain (B)	120	1984	BLM	Bakersfield District	41
Marble Mountain	241,744	1964, 1984	USFS	Klamath NF	3
Mokelumne	104,461	1964, 1984	USFS	Eldorado, Stanislaus, Toiyabe NFs	21
Monarch	45,000	1984	USFS	Sierra, Sequoia NFs	33
Mount Shasta	37,000	1984	USFS	Shasta-Trinity NF	6
North Fork	8,100	1984	USFS	Six Rivers NF	12
Phillip Burton	25,370	1976, 1985	NPS	Point Reyes National Seashore	22
Pine Creek	13,100	1984	USFS	Cleveland NF	53
Pinnacles	12,952	1976	NPS	Pinnacles NM	32
Red Buttes	16,150	1984	USFS	Rogue River NF	1
Russian Peak	12,000	1984	USFS	Klamath NF	5
San Gabriel	36,118	1968	USFS	Angeles NF	44
San Gorgonio	56,722	1964, 1984	USFS	San Bernardino NF	47
San Jacinto	32,040	1964, 1984	USFS	San Bernardino NF	48
San Mateo Canyon	39,540	1984	USFS	Cleveland NF	50
San Rafael	150,610	1968, 1984	USFS	Los Padres NF	42
Santa Lucia (A)	18,679	1978	USFS	Los Padres NF	40
Santa Lucia (B)	1,733	1978	BLM	Bakersfield Disctrict	40
Santa Rosa	20,160	1984	USFS	San Bernardino NF	52
Sequoia-Kings Canyon	736,980	1984	NPS	Sequoia-Kings Canyon NP	36
Sheep Mountain	43,600	1984	USFS	Angeles, San Bernardino NFs	45
Siskiyou	153,000	1984	USFS	Six Rivers, Klamath, Siskiyou NFs	2
Snow Mountain	37,000	1984	USFS	Mendocino NF	18
South Sierra	63,000	1984	USFS	Sequoia, Inyo NFs	38
South Warner	70,385	1964, 1984	USFS	Modoc NF	9
Thousand Lakes	16,335	1964	USFS	Lassen NF	11
Trinity Alps (A)	495,377	1984	USFS	Klamath, Six Rivers, Shasta-Trinity NFs	7
Trinity Alps (B)	4,623	1984	BLM	Ukiah District	7
Ventana	164,144	1969, 1978, 1984	USFS	Los Padres NF	34
Yolla Bolly-Middle Eel (A)	145,404	1964, 1984	USFS	Mendocino, Six Rivers, Shasta-Trinity NFs	13
Yolla Bolly-Middle Eel (B)	8,500	1984	BLM	Ukiah District	13
Yosemite	677,600	1984	NPS	Yosemite NP	27

Acres Total: 5,926,158

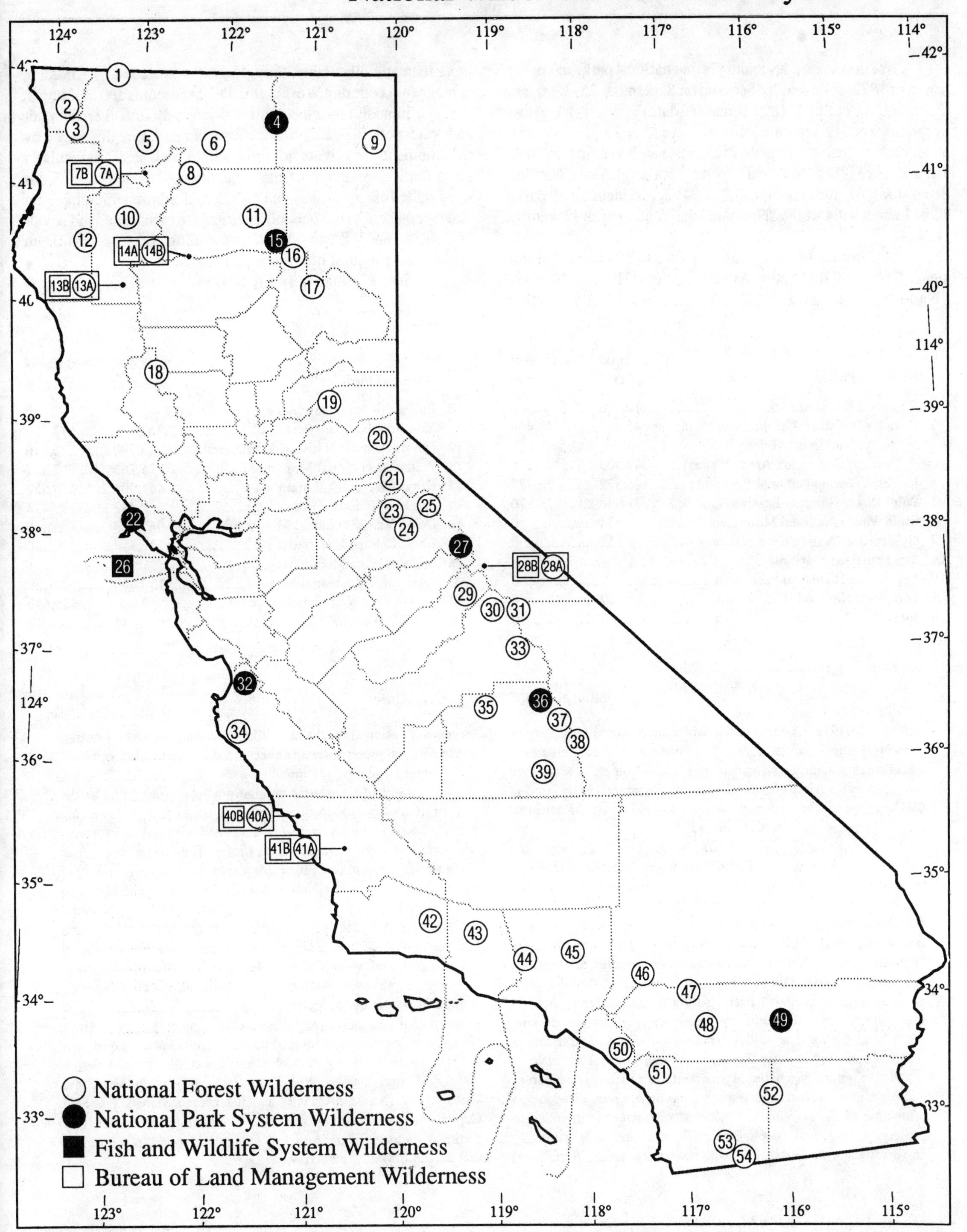

124° 123° 122° 121° 120° 119° 118° 117° 116° 115° 114°
42° 41° 40° 39° 38° 37° 36° 35° 34° 33°
1 2 3 4 5 6 7A 7B 8 9 10 11 12 13A 13B 14A 14B 15 16 17 18 19 20 21 22 23 24 25 26 27 28A 28B 29 30 31 32 33 34 35 36 37 38 39 40A 40B 41A 41B 42 43 44 45 46 47 48 49 50 51 52 53 54
National Forest Wilderness
National Park System Wilderness
Fish and Wildlife System Wilderness
Bureau of Land Management Wilderness

Yellowstone, the country's first national park, was established in 1872, followed by Sequoia on September 25, 1890, and Yosemite on October 1, 1890. Since that date the system has grown to encompass 342 separate units of more than 79,711,334 acres including National Monuments, Preserves, Seashores, Scenic Trails, and a score of other "National" open space designations. National Parks alone, 61 sites, contain 48,055,941 acres (includes 10 parks, 32,041 acres without the "National" designation as of November 1988).

California holds 6 National Parks, 7 National Monuments, 5 National Recreation Areas, 4 National Historic Sites, and 1 National Seashore for a total of 23. This state has more distinct units than any other state, though greatly exceeded by Alaska in acreage (not counting Washington DC's museums and buildings).

In addition to the continual jeopardy of soil erosion, air and water pollution, and visitor abuse, our parks are threatened by developments on private holdings within park areas, particularly since funds earmarked for the acquisition of these private lands were stalled during the Reagan presidential administration. Immediately adjacent to the park boundaries, the menace of land development, oil and gas exploration, and mining also pose significant threats to the natural life of our parks.

See: Bibliography 91, 109, 162.

	National Park	Total Acres	Wilderness Acres
1	Redwood National Park	110,178	0
2	Lava Beds National Monument	45,560	28,460
3	Whiskeytown-Shasta-Trinity National Recreation Area (3 sites)	42,503	0
4	Lassen Volcanic National Park	106,372	78,982
5	Point Reyes National Seashore	71,046	25,370
6	Muir Woods National Monument	554	0
7	Golden Gate National Recreation Area	73,117	0
8A	San Francisco Maritime National Historical Park	12	0
8B	Fort Point National Historic Site	29	0
9	John Muir National Historic Site	9	0
10	Eugene O'Neill National Historic Site	13	0
11	Yosemite National Park	761,170	677,600
12	Devils Postpile National Monument	798	0
13	Pinnacles National Monument	16,265	12,952
14	Kings Canyon National Park	461,901	456,552
15	Sequoia National Park	402,482	280,428
16	Death Valley National Monument	2,067,628	0
17	Channel Islands National Park	249,354	0
18	Santa Monica Mountains National Recreation Area	150,000	0
19	Joshua Tree National Monument	559,960	429,690
20	Cabrillo National Monument	144	0

National Park Service Nomenclature

In 1916, Congress established the National Park Service in the Department of the Interior to provide unified administration of the parks, monuments, historic sites, and other natural areas then managed by a variety of federal agencies. Seventeen years later, in 1933, an executive order transferred an additional 63 national monuments and military sites to the N.P.S.

In the earlier years of park development, titles were used loosely and the term "National Monument," for instance, was applied to great national reservations, historic military fortifications, prehistoric ruins, fossil sites, and the Statue of Liberty.To further confuse the issue, not all areas having a "national" designation are part of the N.P.S. and several in the N.P.S. do not bear the "national" label. These earlier designations continue to cloud the nomenclature system, though for the last few decades there has been some attempt to establish basic criteria for the different official titles, which then indicate either protection or recreation for the site. Many titles such as National Seashore and National Grassland are self explanatory, but other titles are not readily apprehensible.

Generally, a national park contains a variety of resources and encompasses large land or water areas to help provide adequate protection of the resources. A national monument is intended to preserve at least one nationally significant resource. It is usually smaller than a national park, with fewer resources. A national preserve is designed for the protection of certain resources. Activities in national preserves may be restricted if they would jeopardize the resources under protection.

National lakeshores and national seashores focus on the preservation of natural resources while at the same time providing water related recreation. National rivers and wild and scenic rivers preserve ribbons of land bordering on free flowing streams. They also provide opportunities for rafting, canoeing, hiking, and other similar outdoor activity. National scenic trails are generally long distance paths winding through areas of natural beauty. National military park, national battlefield park, national battlefield site, and national battlefield have all been used to designate areas of military history. National memorial is a designation primarily for commemorative purposes. National recreation areas were originally limited in scope, but today include practically all lands set aside primarily for recreation purposes, and include major urban sites. National parkways are ribbons of land flanking roadways designed to offer leisurely driving through scenic areas.

Other titles employed are National Mall, International Historic Site (between Canada and the United States), National Capital Park in Washington, D.C., National Historic Site, and National Historical Park, which is generally applied to complexes that extend beyond single sites.

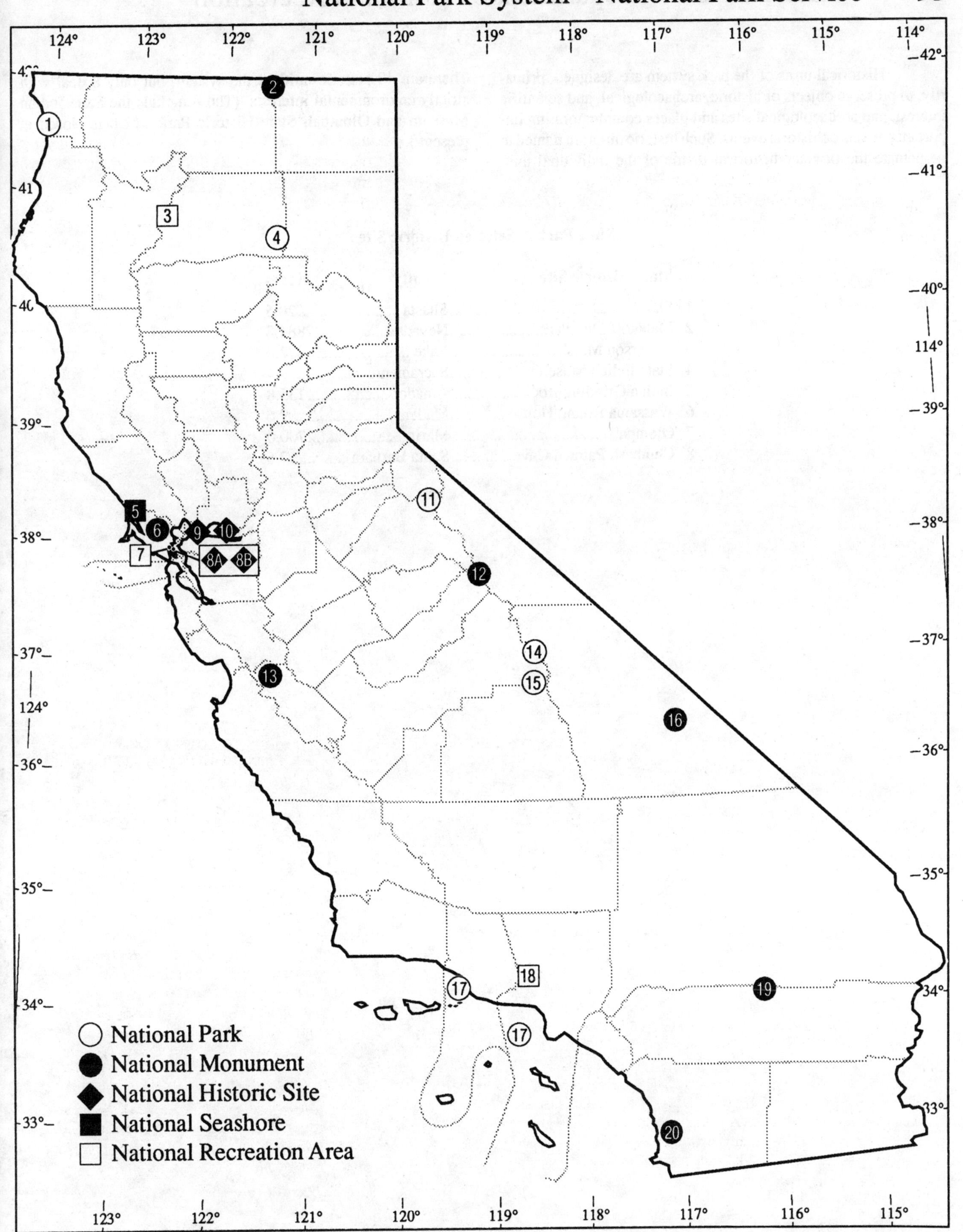
124°
123°
122°
121°
120°
119°
118°
117°
116°
115°
114°
42°
41°
40°
39°
38°
37°
36°
35°
34°
33°
1
2
3
4
5
6
7
8A
8B
9
10
11
12
13
14
15
16
17
18
19
20
National Park
National Monument
National Historic Site
National Seashore
National Recreation Area

Historical units of the park system are designed, primarily, to preserve objects of historic, archaeological, and scientific interest, and archaeological sites and places commemorating important persons or historic events. Such historic units are named to perpetuate the primary historical theme of the individual unit. There are 48 historic units in the system, but only 8 deal with strictly environmental interests. (The 8 include the State Indian Museum and Olompali State Historic Park which is closed at present.)

State Parks - Selected Historic Sites

	State Historic Site	County	Acres
1	Shasta	Shasta	20.5
2	Malakoff Diggins	Nevada	2884.8
3	Anderson Marsh	Lake	872.0
4	State Indian Museum	Sacramento	- -
5	Indian Grinding Rock	Amador	135.8
6	Wassama Round House	Madera	9.6
7	Olompali	Marin	700.0
8	Chumash Painted Cave	Santa Barbara	7.5

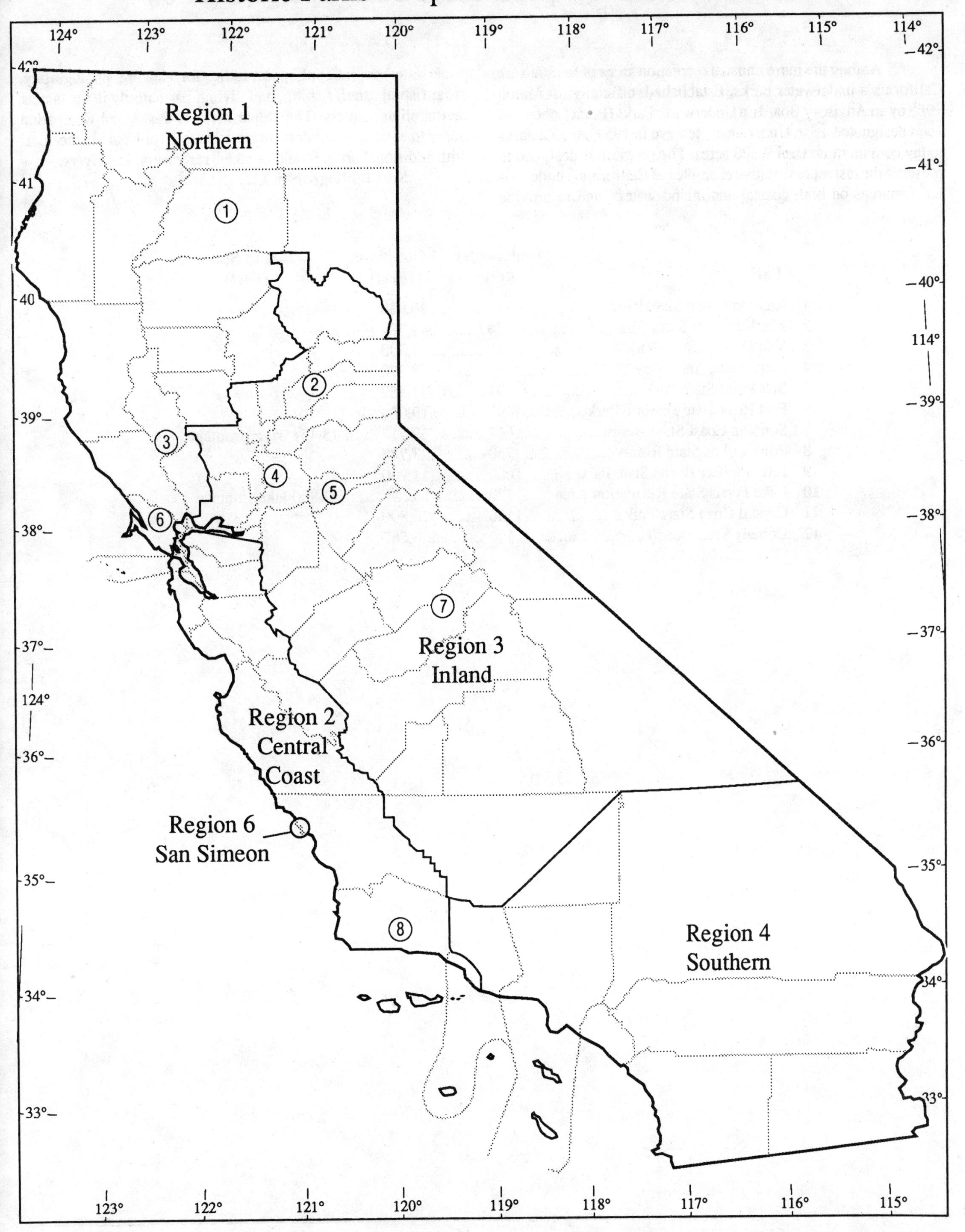
Region 1
Northern
Region 2
Central
Coast
Region 3
Inland
Region 4
Southern
Region 6
San Simeon
1
2
3
4
5
6
7
8
124°
123°
122°
121°
120°
119°
118°
117°
116°
115°
114°
42°
41°
40°
39°
38°
37°
36°
35°
34°
33°

Among the more unusual recreation areas of the state are California's underwater parks. Established, officially, in March 1968 by an Advisory Board on Underwater Parks (Point Lobos had been designated as an Underwater Reserve in 1959), the 12 parks today contain more than 9,800 acres. The program is designed to preserve the last representative examples of California's underwater resources on both coastal and inland waters, and to provide underwater recreational opportunities for viewing, photography, spear fishing, shell fishing, and diving, particularly in areas near metropolitan centers. The Department of Parks and Recreation hopes to enlarge the underwater system to 30 units along the coast with additional areas inland at lakes, reservoirs, and rivers.

See: Bibliography 123.

	Park	Underwater Acres	Coastline (Feet)	Inland Shore (Feet)
1	Russian Gulch State Park	47	10,930	
2	MacKerricher State Park	454	41,787	
3	Van Damme State Park	20	1,700	
4	Manchester State Beach	3651	18,570	
5	Salt Point State Park	94	31,362	
6	Fort Ross State Historic Park	90	19,000	
7	Sonoma Coast State Beach	667	69,512	15,451 river frontage
8	Point Lobos State Reserve	750	23,777	
9	Julia Pfeiffer Burns State Park	1000	11,900	
10	Lake Perris State Recreation Area	120	- -	52,800 lake frontage
11	Crystal Cove State Park	1150	16,800	
12	Doheny State Beach	192	6,567	

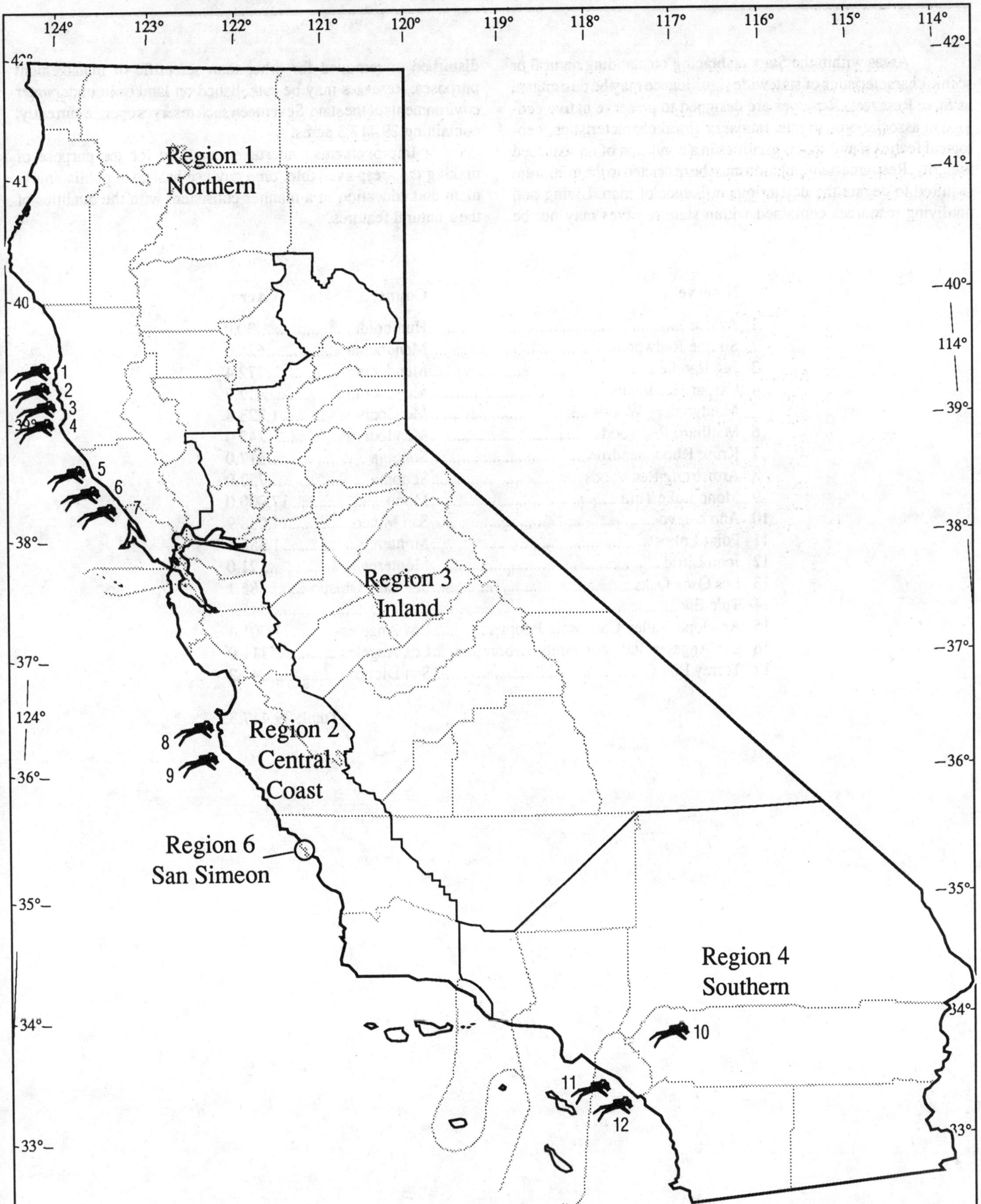

Region 1
Northern

Region 2
Central
Coast

Region 3
Inland

Region 4
Southern

Region 6
San Simeon

1 2 3 4 5 6 7 8 9 10 11 12

124° 123° 122° 121° 120° 119° 118° 117° 116° 115° 114°

42° 41° 40° 39° 38° 37° 36° 35° 34° 33°

Areas within the State embracing outstanding natural or scenic characteristics of statewide significance may be designated as State Reserves. Reserves are designed to preserve native ecological associations, unique faunal or floral characteristics, geological features, and scenic qualities in a condition of undisturbed integrity. Resource manipulation must be restricted to the minimum required to negate the deleterious influence of man. Living and nonliving resources contained within state reserves may not be disturbed or removed for other than scientific or management purposes. Reserves may be established on land or in underwater environments of the state. Seventeen such reserves operate currently, containing 29,417.3 acres.

Improvements undertaken shall be for the purpose of making the areas available, on a day use basis, for public enjoyment and education in a manner consistent with the qualities of their natural features.

	Reserve	County	Acres
1	Azalea	Humboldt	30.0
2	Smithe Redwoods	Mendocino	628.0
3	Jug Handle	Mendocino	772.4
4	Caspar Headlands	Mendocino	2.7
5	Montgomery Woods	Mendocino	1,323.8
6	Mailliard Redwoods	Mendocino	242.0
7	Kruse Rhododendron	Sonoma	317.0
8	Armstrong Redwoods	Sonoma	752.0
9	Mono Lake Tufa	Mono	17,000.0
10	Año Nuevo	San Mateo	4.088.9
11	Point Lobos	Monterey	1,355.1
12	John Little	Monterey	21.0
13	Los Osos Oaks	San Luis Obispo	85.1
14	Tule Elk	Kern	945.7
15	Antelope Valley California Poppy	Los Angeles	1,701.4
16	Los Angeles State & County Arboretum	Los Angeles	111.0
17	Torrey Pines	San Diego	41.2

Total: 29,417.3

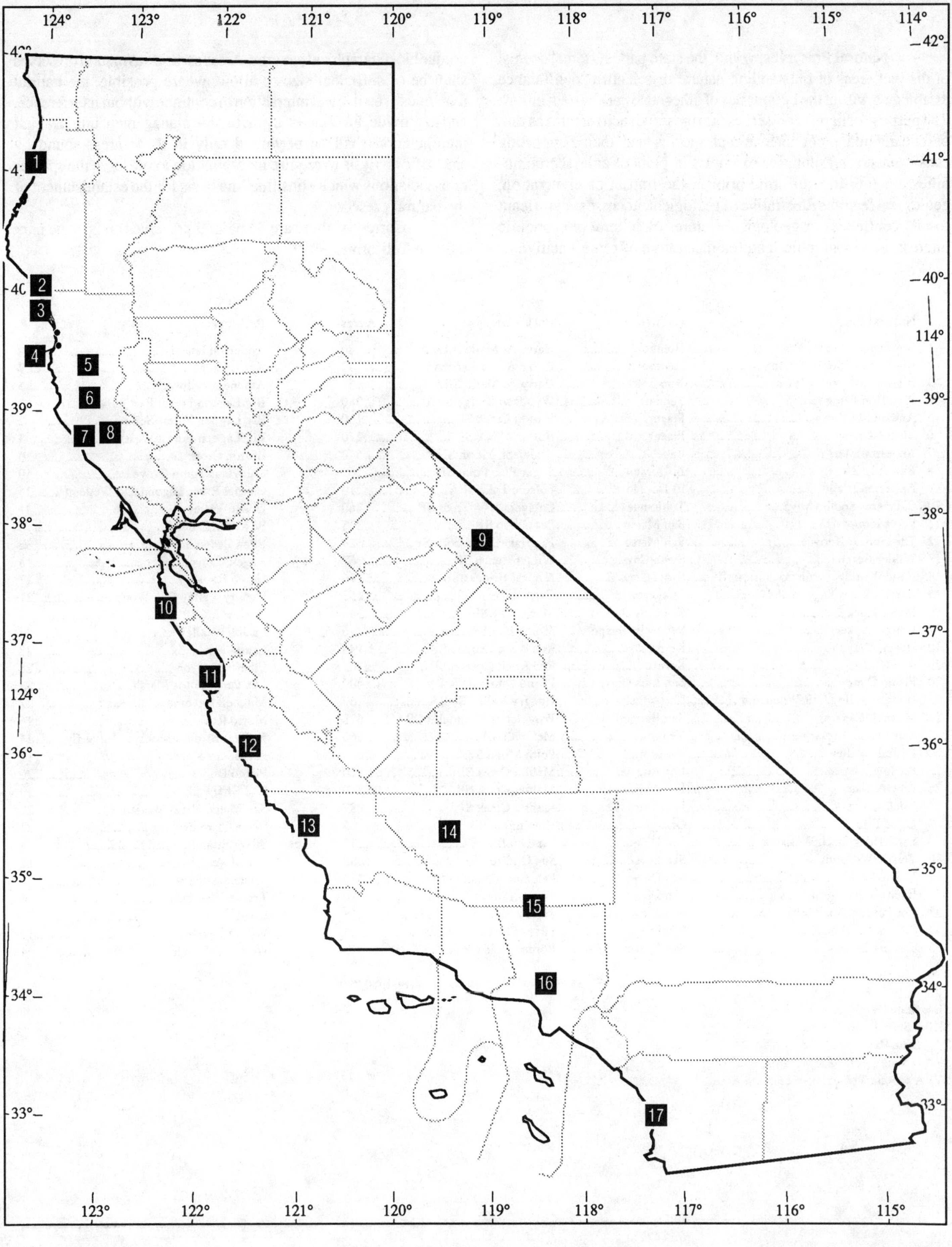
124°
123°
122°
121°
120°
119°
118°
117°
116°
115°
114°
42°
41°
40°
39°
38°
37°
36°
35°
34°
33°
1
2
3
4
5
6
7
8
9
10
11
12
13
14
15
16
17

Natural Preserves, within the state park system, "consist of distinct areas of outstanding natural or scientific significance established within the boundaries of other state park system units". The purpose of natural preserves is to preserve such features as rare or endangered plant and animal species and their supporting ecosystems, representative examples of plant or animal communities existing in California prior to the impact of civilization, geological features illustrative of geological processes, significant fossil occurrences or geological features of cultural or economic interest, or topographic features illustrative of representative or unique biogeographical patterns. Areas set aside as natural preserves shall be of sufficient size to allow, where possible, the natural dynamics of ecological interaction to continue without interference, and to provide, in all cases, a practicable management unit. Habitat manipulation shall be permitted only in those areas found by scientific analysis to require manipulation to preserve the species or associations which constitute the basis for the establishment of the natural preserve.

Currently, there are 35 natural preserves on 26 state park units in California.

	Natural Preserve	County	Park Unit	Acres
1	Big Lagoon Forest East	Humboldt	Harry A. Merlo SRA	80
2	Big Lagoon Forest South	Humboldt	Harry A. Merlo SRA	62
3	Big Lagoon Forest West	Humboldt	Harry A. Merlo SRA	51
4	Woodson Bridge	Tehama	Woodson Bridge SRA	280
5	Antone Meadows	Placer	Burton Creek SP	160
6	Burton Creek	Placer	Burton Creek SP	170
7	Anderson Marsh	Lake	Anderson Marsh SHP	540
8	Edwin L. Z'berg	El Dorado	Sugar Pine Point SP	160
9	Anderson Island	El Dorado	Folsom Lake SRA	10
10	Calaveras South Grove	Tuolumne	Calaveras Big Trees SP	1,260
11	Pescadero Marsh	San Mateo	Pescadero SB	235
12	Theodore J. Hoover	San Mateo	Big Basin Redwoods SP	23
13	Wilder Beach	Santa Cruz	Wilder Ranch SP	67
14	Natural Bridges Monarch Butterfly	Santa Cruz	Natural Bridges SB	16
15	Carmel River Lagoon and Wetland	Monterey	Carmel River SB	53
16	Morro Rock	San Luis Obispo	Morro Bay SP	45
17	Heron Rookery	San Luis Obispo	Morro Bay SP	6
18	Hagen Canyon	Kern	Red Rock Canyon SP	1,145
19	Red Cliffs	Kern	Red Rock Canyon SP	365
20	Pismo Dunes	San Luis Obispo	Pismo Dunes SVRA	400
21	Hungry Valley Oak Woodland	Los Angeles	Hungry Valley SVRA	60
22	Mitchell Caverns	San Bernardino	Providence Mountains SRA	628
23	Santa Clara Estuary	Ventura	McGrath SB	160
24	La Jolla Valley	Ventura	Point Mugu SP	600
25	Kaslow	Los Angeles	Malibu Creek SP	1,920
26	Liberty Canyon	Los Angeles	Malibu Creek SP	810
27	Udell Gorge	Los Angeles	Malibu Creek SP	287
28	Least Tern	Orange	Huntington SB	5
29	San Mateo Creek Wetlands	San Diego	San Onofre SB	37
30	Trestle Wetlands	San Diego	San Onofre SB	82
31	Doane Valley	San Diego	Palomar Mountain SP	450
32	Ellen Browning Scripps	San Diego	Torrey Pines SB	72
33	Los Penasquitos Marsh	San Diego	Torrey Pines SB	218
34	Silver Strand	San Diego	Silver Strand SB	26
35	Tijuana Estuary	San Diego	Border Field SP	327

Total: 10,790

Preserve	#
Anderson Island	9
Anderson Marsh	7
Antone Meadows	5
Big Lagoon Forest East	1
Big Lagoon Forest South	2
Big Lagoon Forest West	3
Burton Creek	6
Calaveras South Grove	10
Carmel River Lagoon and Wetland	15
Doane Valley	31
Edwin L. Z'berg	8
Ellen Browning Scripps	32
Hagen Canyon	18
Heron Rookery	17
Hungry Valley Oak Woodland	21
Kaslow	25
La Jolla Valley	24
Least Tern	28
Liberty Canyon	26
Los Penasquitos Marsh	33
Mitchell Caverns	22
Morro Rock	16
Natural Bridges Monarch Butterfly	14
Pescadero Marsh	11
Pismo Dunes	20
Red Cliffs	19
San Mateo Creek Wetlands	29
Santa Clara Estuary	23
Silver Strand	34
Theodore J. Hoover	12
Tijuana Estuary	35
Trestle Wetlands	30
Udell Gorge	27
Wilder Beach	13
Woodson Bridge	4

SB = State Beach
SHP = State Historic Park
SP = State Park
SRA = State Recreation Area
SVRA = State Vehicular Recreation Area

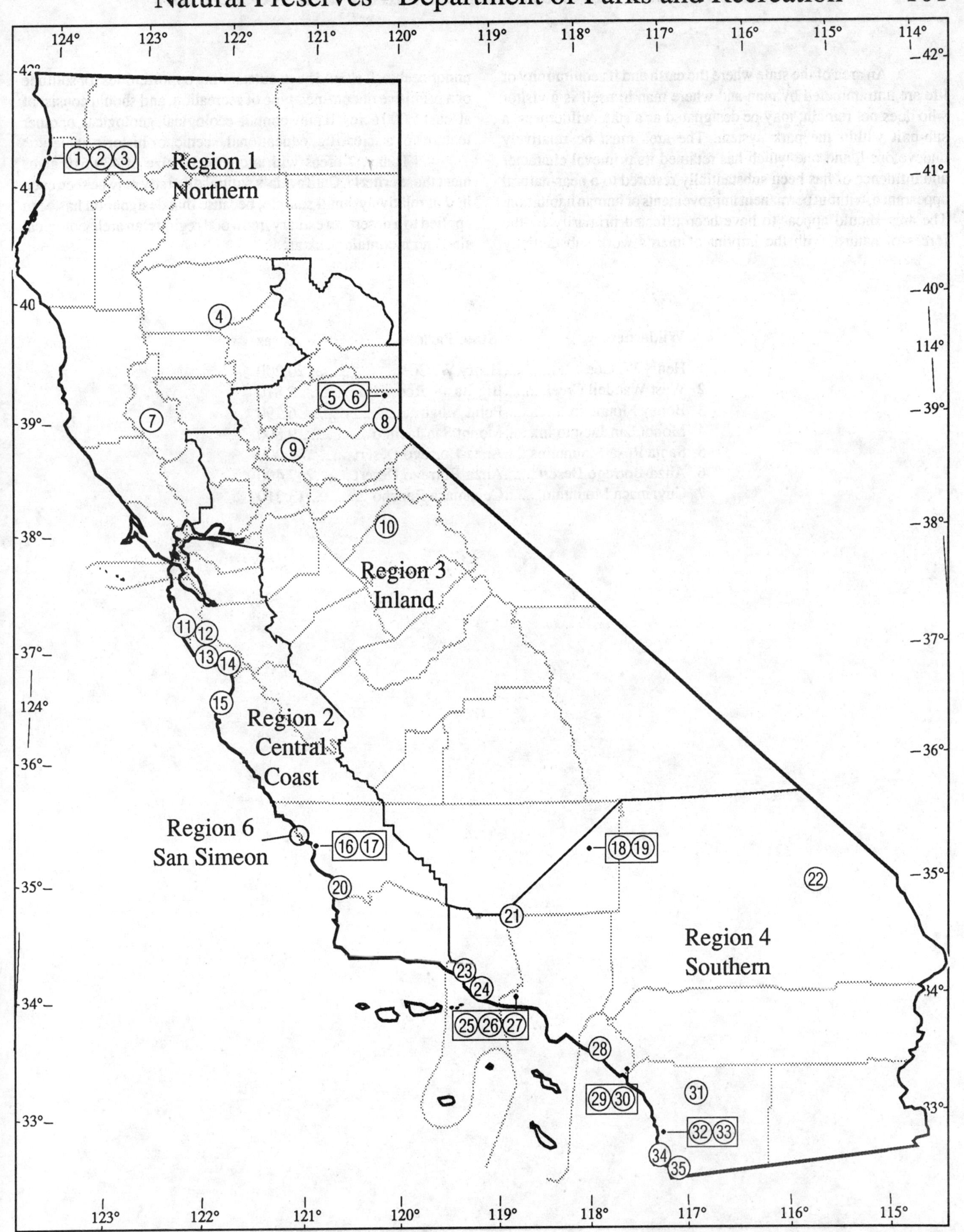
Region 1
Northern
Region 3
Inland
Region 2
Central
Coast
Region 6
San Simeon
Region 4
Southern

An area of the state where the earth and its community of life are untrammeled by man and where man himself is a visitor who does not remain, may be designated as a state wilderness, a sub-unit within the park system. The area must be relatively undeveloped, and one which has retained its primeval character and influence or has been substantially restored to a near-natural appearance, without permanent improvements or human habitation. The area should appear to have been affected primarily by the forces of nature, with the imprint of man's work substantially unnoticeable. It should have outstanding opportunities for solitude or a primitive unconfined type of recreation, and should consist of at least 5,000 acres. It may contain ecological, geological, or other features of a scientific, educational, scenic, or historic value.

Today, 7 areas with a combined size of 439,610 acres meet those criteria. California's natural diversity may be seen even in that relatively small sample, because that designation has been applied to a desert, an estuary, redwood regions, an archaeological site, and mountain peak areas.

	Wilderness	State Park	Acres
1	Henry W. Coe	Henry W. Coe	20,200
2	West Waddell Creek	Big Basin Redwoods	5,810
3	Boney Mountain	Point Mugu	6,190
4	Mount San Jacinto	Mount San Jacinto	9,800
5	Santa Rosa Mountains	Anza-Borrego Desert	87,000
6	Anza-Borrego Desert	Anza-Borrego Desert	297,400
7	Cuyamaca Mountain	Cuyamaca Rancho	13,210

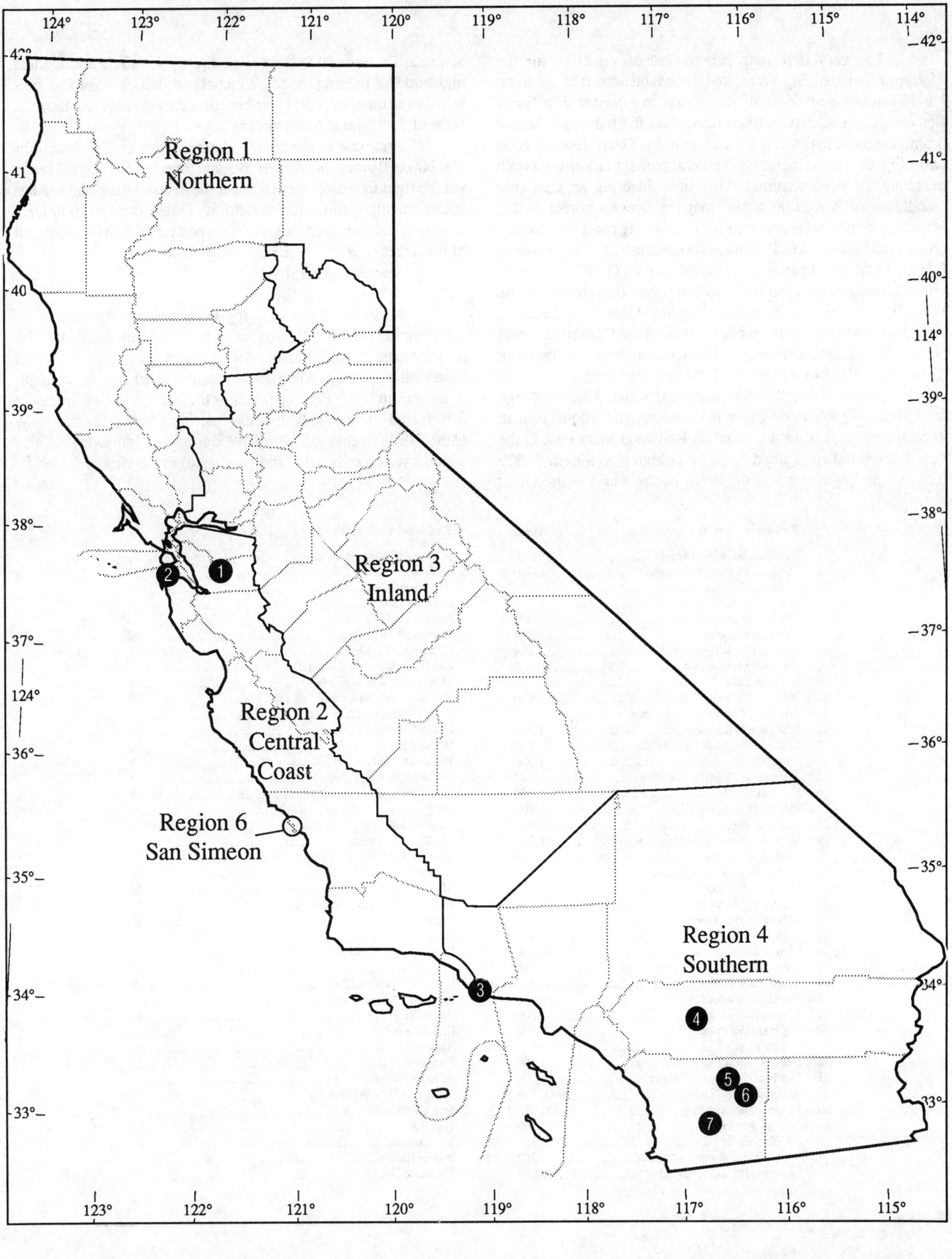

Region 1
Northern

Region 2
Central
Coast

Region 3
Inland

Region 4
Southern

Region 6
San Simeon

1
2
3
4
5
6
7

124° 123° 122° 121° 120° 119° 118° 117° 116° 115° 114°

42° 41° 40° 39° 38° 37° 36° 35° 34° 33°

The very first state park included protection for the Mariposa Grove of Big Trees, and the establishment of the State Park System stemmed from the efforts to save the Big Basin Redwoods. Thus, it is not surprising that the redwoods hold a prominent place in the state park's hierarchy. Thirty-five redwood parks line the coast (including 2 federal parks) in a narrow swath matching the coastal summer fog line, covering an area (not woodland) of 2,300,000 acres from the Oregon border to the Monterey-San Luis Obispo county line. (The big coast trees extend only 14 miles north of California along the southern Oregon coast.) Inland, 4 additional parks bring the total to 39. Only 3 regions of the world naturally harbor the great trees; the Coast Redwood, the Dawn Redwood in a remote area of western China, and the Giant Sequoia found only on the western flank of the California Sierra Nevada in 75 isolated groves (35,600 acres) at elevations between 5,000 and 7,500 feet mostly south of the Kings River.

The world's tallest tree, unnamed, stands 367.8 feet high on the banks of Redwood Creek in the Redwood National Forest. The Dyerville Giant in the Humbolt Redwood State Park is the largest coast redwood standing 362 feet high with a girth of 629 inches. The largest natural organism in the world is the Giant Sequoia, The General Sherman, in Sequoia National Park, 275 feet high and 998 inches (38 feet, 2 inches) around. Several Sequoias have been found over 300 feet high, but they do not have the total mass of the General Sherman tree.

Ancestor redwood trees have been in California for 160,000,000 years and the current giant trees for 20,000,000 years, yet the fight to protect them against the axe and the chainsaw has been particularly bitter and unyielding. Today, only 300,000 acres of an original 2,000,000 acres of redwood forest remain and much of that acreage is still under the threat of the axe.

See: Bibliography 197.

Note: Champion trees registered with the American Forest Association National Register of Big Trees are determined by the total mass of the tree calculated on the circumference of the tree measured in inches, 4.5 feet above the ground, plus the total height of the tree in feet, plus 25 percent of the average crown spread in feet. Based on those calculations, California holds 72 champions of the 748 different species registered as natives (plus a few exotics which have reproduced in the wild state) in the United States.

	Redwood Park or Preserve	Acres
1	Jedediah Smith Redwoods	9,500
2	Del Norte Coast Redwoods	6,400
3	Redwood National Park	110,131
4	Prairie Creek Redwoods	12,500
5	Redwood National Park	(see #3)
6	Humboldt Lagoons	1,504
7	Grizzly Creek Redwoods	394
8	Humboldt Redwoods	50,522
9	Benbow Lake	780
10	Richardson Grove	900
11	Reynolds Wayside Campground	66
12	Sinkyone Wilderness	3,600
13	Smithe Redwoods	600
14	Standish-Hickey	1,000
15	Admiral William H. Standley	45
16	Russian Gulch	1,300
17	Pygmy Forest (Jug Handle State Reserve)	700
18	Van Damme	2,100
19	Montgomery Woods State Reserve	1,142
20	Paul M. Dimmick Wayside Campground	12
21	Hendy Woods	690
22	Mailliard Redwoods	240
23	Armstrong Redwoods	750
24	Calaveras Big Trees	5,900
25	Samuel P. Taylor	2,700
26	Mount Tamalpais	6,200
26	Muir Woods National Monument	510
27	Yosemite National Park	760,917
28	Portola Redwoods	2,000
29	Butano Redwoods	2,200
30	Big Basin Redwoods	16,000
31	Wilder Ranch	4,511
32	Henry Cowell Redwoods	4,100
33	The Forest of Nisene Marks	10,000
34	Kings Canyon National Park	460,136
35	Sequoia National Park	403,023
36	Andrew Molera	4,700
37	Pfeiffer Big Sur	800
38	Julia Pfeiffer Burns	1,863
39	Landels-Hill Big Creek Reserve	3,858

Redwood Park or Preserve	#
Admiral William H. Standley	15
Andrew Molera	36
Armstrong Redwoods	23
Benbow Lake	9
Big Basin Redwoods	30
Butano Redwoods	29
Calaveras Big Trees	24
Del Norte Coast Redwoods	2
The Forest of Nisene Marks	33
Grizzly Creek Redwoods	7
Hendy Woods	21
Henry Cowell Redwoods	32
Humboldt Lagoons	6
Humboldt Redwoods	8
Jedediah Smith Redwoods	1
Julia Pfeiffer Burns	38
Kings Canyon National Park	34
Landels-Hill Big Creek Reserve	39
Mailliard Redwoods	22
Montgomery Woods State Reserve	19
Mount Tamalpais	26
Muir Woods National Monument	26
Paul M. Dimmick Wayside Campground	20
Pfeiffer Big Sur	37
Portola Redwoods	28
Prairie Creek Redwoods	4
Pygmy Forest	17
Redwood National Park	3
Redwood National Park	5
Reynolds Wayside Campground	11
Richardson Grove	10
Russian Gulch	16
Samuel P. Taylor	25
Sequoia National Park	35
Sinkyone State Wilderness	12
Smith Redwoods	13
Standish - Hickey	14
Van Damme	18
Wilder Ranch	31
Yosemite National Park	27

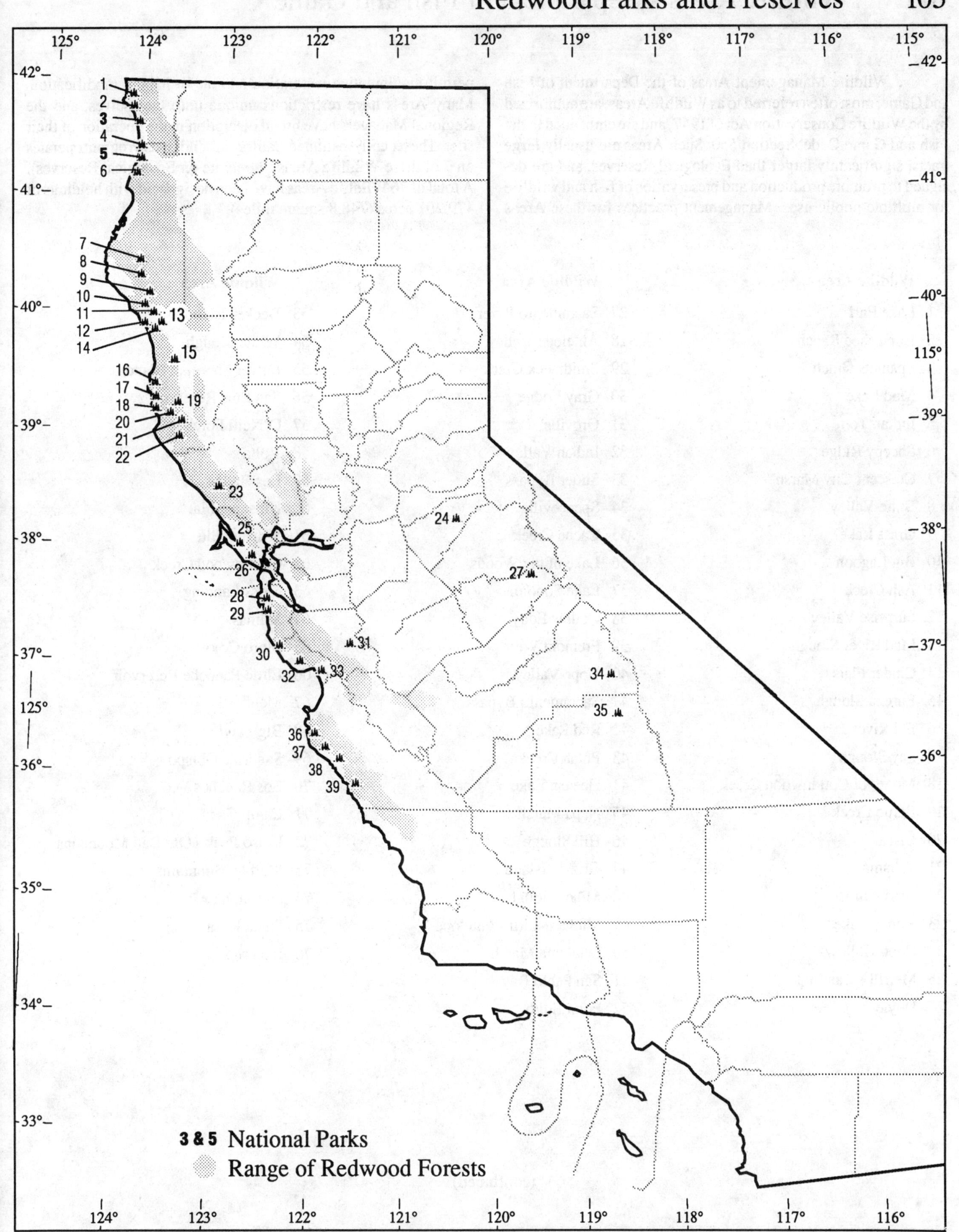
125°
124°
123°
122°
121°
120°
119°
118°
117°
116°
115°
42°
41°
40°
39°
38°
37°
36°
35°
34°
33°
1
2
3
4
5
6
7
8
9
10
11
12
13
14
15
16
17
18
19
20
21
22
23
24
25
26
27
28
29
30
31
32
33
34
35
36
37
38
39
3 & 5 National Parks
Range of Redwood Forests

Wildlife Management Areas of the Department of Fish and Game, most often referred to as Wildlife Areas, are authorized by the Wildlife Conservation Act of 1947, and are embedded in the Fish and Game Code, Section 550. Such Areas are usually large tracts, significantly larger than Ecological Reserves, and are designed to promote production and preservation of fish and wildlife for multiple public uses. Management practices for these Areas permit manipulative restoration and major habitat modification. Many Areas have restrictions unique unto themselves, and the Regional Managers have broad discretion in the operation of their sites. The recently instituted California Wildlands Program operates on 7 of these Wildlife Areas (2 operate on Ecological Reserves). A total of 76 Wildlife Areas have been designated with holdings of 479,201 acres (748.8 square miles).

	Wildlife Area
1	Lake Earl
2	Horseshoe Ranch
3	Spannus Gulch
4	Mud Lake
5	Indian Tom
6	Sheepy Ridge
7	Crescent City Marsh
8	Butte Valley
9	Grass Lake
10	Big Lagoon
11	Ash Creek
12	Surprise Valley
13	Mad River Slough
14	Cinder Flats
15	Eureka Slough
16	Eel River
17	Fay Slough
18	Mouth of Cottonwood Creek
19	Battle Creek
20	Biscar
21	Tehama
22	Bass Hill
23	Honey Lake
24	Coon Hollow
25	Merrill's Landing
26	Doyle

	Wildlife Area
27	Sacramento River
28	Antelope Valley
29	Smithneck Creek
30	Gray Lodge
31	Oroville
32	Indian Valley
33	Sutter Bypass
34	Spenceville
35	Cache Creek
36	Lake of the Woods
37	Lake Sonoma
38	Collins Eddy
39	Fremont Weir
40	Hope Valley
41	Sacramento Bypass
42	Red Lake
43	Putah Creek
44	Heenan Lake
45	Napa Marshes
46	Hill Slough
47	Grizzly Island
48	Miner Slough
49	Slinkard-Little Antelope
50	Petaluma Marsh
51	San Pablo Bay
52	Point Edith

	Wildlife Area
53	Decker Island
54	White Slough
55	Lower Sherman Island
56	San Luis Reservoir
57	O'Neill Forebay
58	Volta
59	Los Banos
60	Kinsman Flat
61	Watsonville
62	Cottonwood Creek
63	Moss Landing
64	Salinas
65	Moro Cojo
66	Little Panoche Reservoir
67	Mendota
68	Big Sandy
69	San Luis Obispo
70	Los Ranchos
71	Camp Cady
72	Kelso Peak - Old Dad Mountains
73	Marble Mountains
74	San Jacinto
75	Santa Rosa
76	Imperial

(continued)

Region 1

Region 2

Region 3

Region 3

Region 4

Region 5

Areas - Region 1	County	#	Founded	Acres
Ash Creek	Lassen-Modoc	11	8-1-86	13,897
Bass Hill	Lassen	22	12-9-77	3,515
Battle Creek	Shasta-Tehama	19	8-5-83	106
Big Lagoon	Humboldt	10	8-13-76	1,600
Biscar	Lassen	20	6-29-73	468
Butte Valley	Siskiyou	8	8-28-81	13,323
Cinder Flats	Shasta	14	12-9-77	720
Crescent City Marsh	Del Norte	7	8-29-80	338
Doyle	Lassen	26	8-23-68	14,066
Eel River	Humboldt	16	8-23-68	1,404
Eureka Slough	Humboldt	15	8-29-80	3
Fay Slough	Humboldt	17	8-5-88	483
Grass Lake	Siskiyou	9	8-6-82	30
Honey Lake	Lassen	23	1953	5,669
Horseshoe Ranch	Siskiyou	2	12-9-77	5,017
Indian Tom	Siskiyou	5	8-23-68	59
Lake Earl	Del Norte	1	8-29-80	2,612
Mad River Slough	Humboldt	13	8-4-89	510
Merrill's Landing	Tehama	25	8-29-80	312
Cottonwood Creek	Shasta	18	8-6-82	317
Mud Lake	Siskiyou	4	8-6-82	39
Sheepy Ridge	Siskiyou	6	8-23-68	320
Spannus Gulch	Siskiyou	3	8-6-82	132
Surprise Valley	Modoc	12	8-31-79	460
Tehama	Tehama	21	8-23-68	49,639
			Total:	115,039

Areas - Region 2	County	#	Founded	Acres
Antelope Valley	Sierra	28	8-29-80	4,480
Collins Eddy	Sutter-Yolo	38	8-05-88	32
Coon Hollow	Buttle	24	8-29-80	729
Fremont Weir	Sutter	39	8-28-81	215
Gray Lodge	Butte-Sutter	30	1953	12,177
Heenan Lake	Alpine	44	8-5-83	1,662
Hope Valley	Alpine	40	4-6-89	2,759
Lake of the Woods	Yuba	36	8-25-78	623
Lower Sherman Isl.	Sacramento	55	8-13-76	3,115
Miner Slough	Solano	48	8-28-81	37
Oroville	Butte	31	8-23-68	11,870
Putah Creek	Solano	43	8-28-81	669
Red Lake	Alpine	42	8-5-88	760
Sacramento Bypass	Yolo	41	8-5-88	345
Sacramento River	Butte-Glenn	27	8-29-80	1,222
Smithneck Creek	Sierra	29	4-8-88	1,385
Spenceville	Nevada-Yuba	34	8-23-68	11,233
Sutter Bypass	Coloma-Sutter	33	8-23-68	5,271
White Slough	San Joaquin	54	8-29-80	800
			Total:	59,384

Areas - Region 3	County	#	Founded	Acres
Big Sandy	S.L.O./Mont	68	8-13-76	754
Cache Creek	Lake	35	8-5-88	1,519
Decker Island	Solano	53	8-5-88	33
Grizzly Island	Solano	47	1954	11,225
Hill Slough	Solano	46	8-31-79	1,795
Indian Valley	Lake	32	8-4-89	1,305
Lake Sonoma	Sonoma	37	8-1-86	8,000
Los Ranchos	San Luis Obispo	70	8-1-86	2
Moro Cojo	Monterey	65	8-1-86	26
Moss Landing	Monterey	63	8-30-85	665
Napa Marches	Napa-Solano	45	1953	1,650
Petaluma Marsh	MarinSonoma	50	8-31-79	2,249
Point Edith	Contra Costa	52	8-30-85	760
Salinas	Monterey	64	1975	518
San Luis Obispo	San Luis Obispo	69	8-31-84	453
San Pablo Bay	Marin	51	8-13-76	10,637
Watsonville	Santa Cruz	61	8-30-85	109
			Total:	41,700

Areas - Region 4	County	#	Founded	Acres
Cottonwood Creek	Santa Cl-Mrcd	62	8-25-78	6,136
Kinsman Flat	Madera	60	8-7-87	450
Little Panoche Resrv.	Fresno	66	6-29-73	780
Los Banos	Merced	59	1954	5,586
Mendota	Fresno	67	1956	12,429
O'Neill Forebay	Merced	57	8-16-74	700
San Luis Reservoir	Merced	56	8-16-74	900
Volta	Merced	58	6-29-73	2,700
			Total:	29,681

Areas - Region 5	County	#	Founded	Acres
Camp Cady	San Bernardino	71	8-29-80	1,546
Imperial	Imperial	76	1951	7,924
Kelso Peak and Old Dad Mountains	San Bernardino	72	8-4-89	124,240 *
Marble Mountains	San Bernardino	73	8-4-89	56,860 *
San Jacinto	Riverside	74	8-6-82	5,731
Santa Rosa	Riverside	75	8-5-83	25,476
Slinkard and Little Antelope	Monterey	49	6-29-79	11,620
			Total:	233,397

Grand Total: 479,201

Prior to 1959 Wildlife Areas were known as Waterfowl Management Areas under Section 271 of Title 14

* Cooperatively managed with BLM for bighorn hunting boundaries

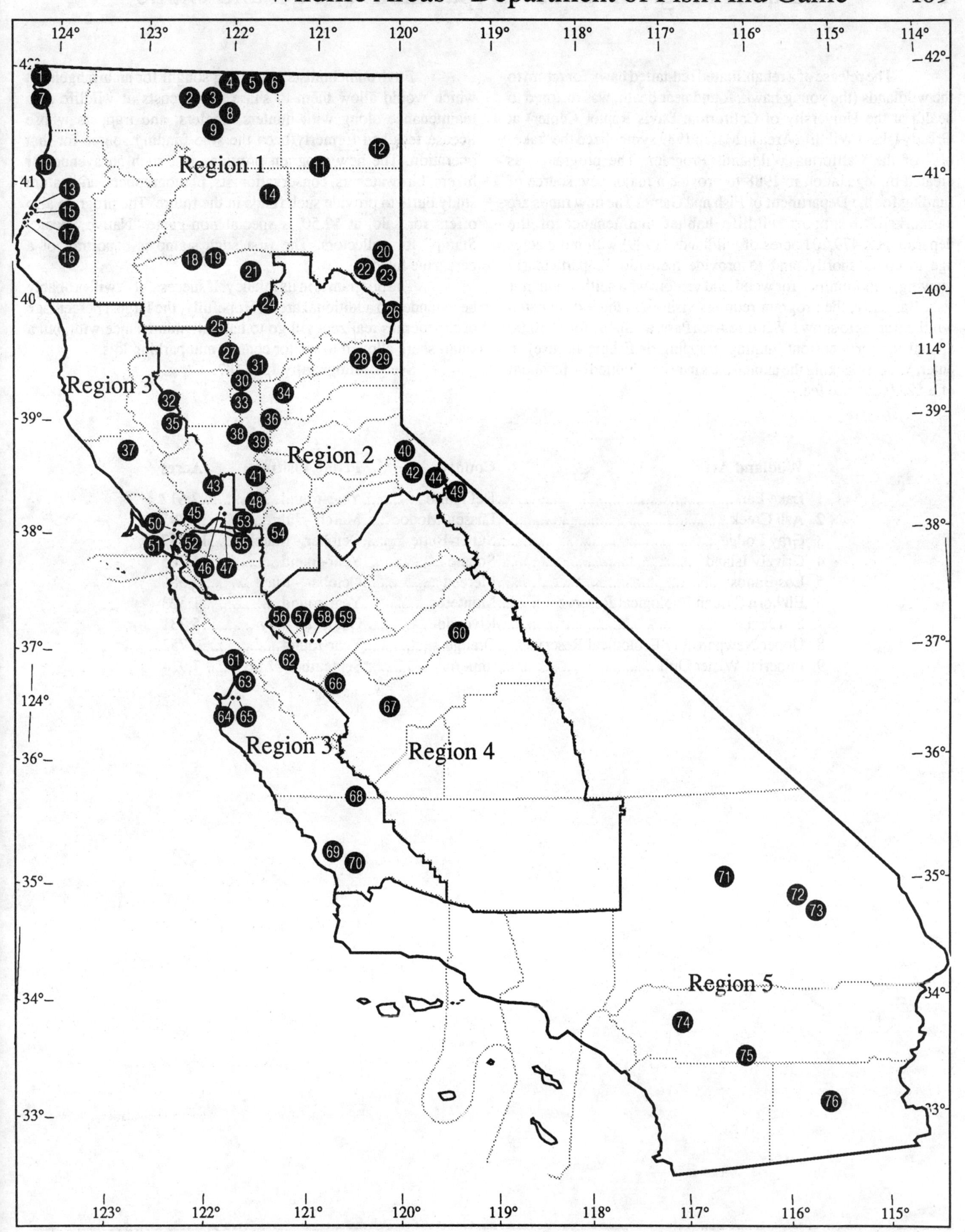
Region 1
Region 2
Region 3
Region 3
Region 4
Region 5

The release of a rehabilitated red-tailed hawk for return to the wildlands (the young hawk, found near death, was restored to health at the University of California, Davis Raptor Center) at Grizzly Island Wildlife Area in March 1989 symbolized the "take-off" of the California Wildlands Program. The program was created by legislation in 1988 to provide a major new source of funding for the Department of Fish and Game. The new funds are earmarked to support wildlife habitat maintenance of the department's 479,201 acres of wildlands (1989) with more acreage to come shortly, and to provide recreational, particularly viewing, opportunities for woodland visitors who neither hunt nor fish. Basically, the program requires visitors to the 9 designated wildland areas to show a Wildlife Area Pass, available for a $10.00 annual fee (or a current hunting, trapping or fishing license) to enter. Visitors lacking the annual pass may be admitted on payment of a $2.00 day use fee.

Environmentalists have long sought for an arrangement which would allow them to share in the costs of wildlife area maintenance along with hunters, anglers, and trappers whose license fees had formerly been the sole funding source for that operation. The new program provides just such an avenue for hikers, birdwatchers, conservationists, photographers, and nature study buffs to provide such funds in the future. The program also offers for sale, at $7.50, a special non-game "Native Species Stamp" for collectors. The first such stamp is a portrait of a peregrine falcon.

The program is in its infancy. If successful, it will probably be extended to additional areas. Hopefully, the largest percentage of the monies realized will go to habitat maintenance with, but a small share allotted to visitor centers and parking lots.

See: Bibliography 131, 187.

	Wildland Area	**County**	**Prime Visiting**	**Acres**
1	Lake Earl	Del Norte	Year-round	2,612
2	Ash Creek	Lassen-Modoc	March - July	13,897
3	Gray Lodge	Sutter-Butte	October - March	12,177
4	Grizzly Island	Solano	Year-round	11,225
5	Los Banos	Merced	October - May	5,586
6	Elkhorn Slough Ecological Reserve	Monterey	Year-round	1,323
7	San Jacinto	Riverside	October - May	5,731
8	Upper Newport Bay Ecological Reserve	Orange	Year-round	752
9	Imperial-Wister Unit	Imperial	September - May	7,924

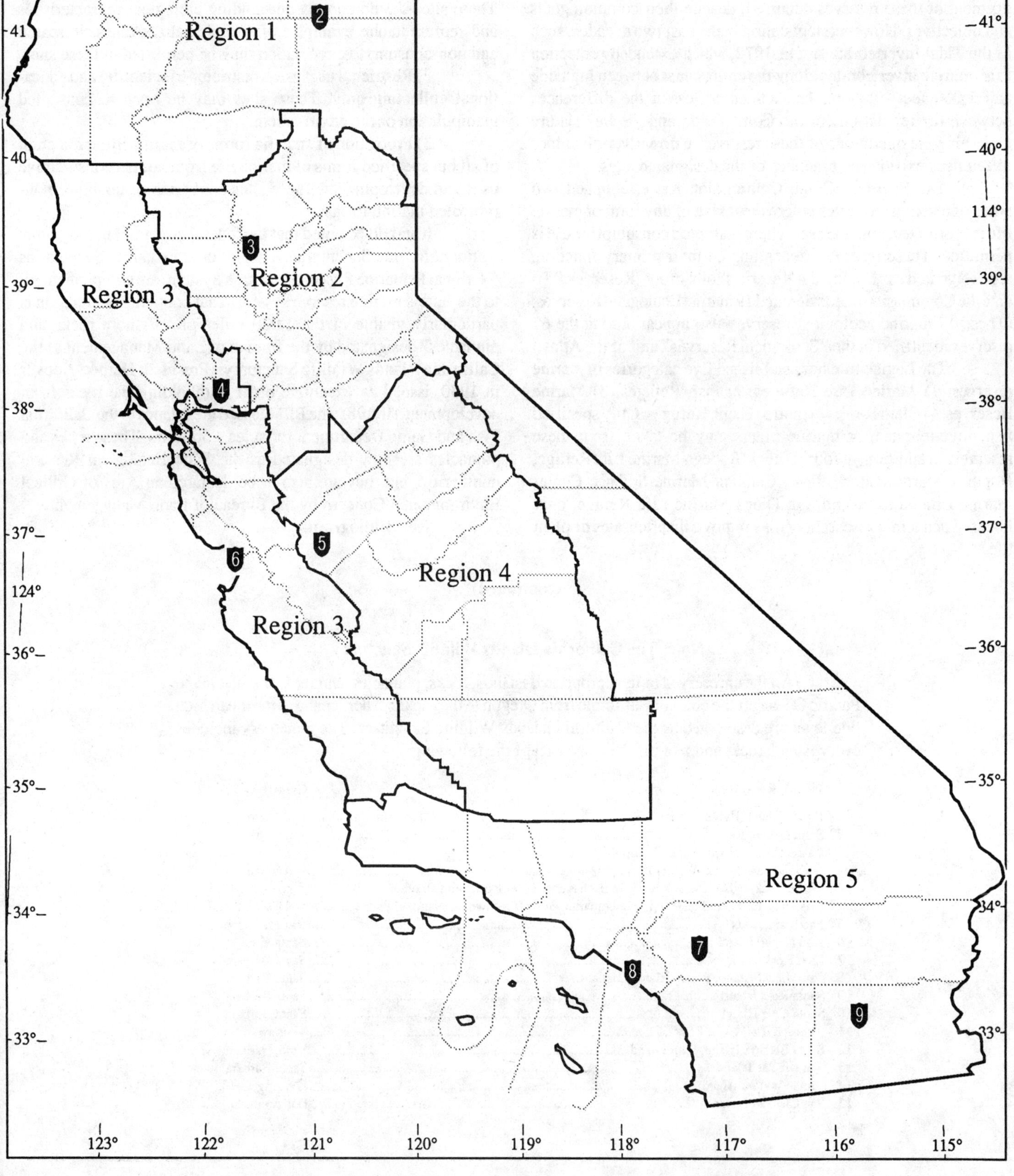
124°
123°
122°
121°
120°
119°
118°
117°
116°
115°
114°
42°
41°
40°
39°
38°
37°
36°
35°
34°
33°
Region 1
Region 2
Region 3
Region 3
Region 4
Region 5
1
2
3
4
5
6
7
8
9

Both the Legislature and the Fish and Game Commission designate state marine reserves. Originally, their independent actions lacked common guidelines for terminology, description, or use restrictions. Consequently, significant differences in the management of these reserves occurred, despite their common goals and objectives. However, legislation in the past two decades, such as the Tidal Invertebrate Act of 1972, which extended protection to all marine invertebrates along the entire coast between high tide and 1,000 feet offshore, has tended to lessen the differences between the regulations of the Commission and the Legislature governing the operations of these reserves. Currently, only minor differences exist in the practices of the designated sites.

The Fish and Game Commission has established two types of reserves: reserves where the taking of any form of marine life is prohibited, and reserves where restricted consumptive use is permitted. These reserves, depending on their primary function, are designated as "Ecological Reserves" or simply "Reserves." To date the Commission has designated 19 marine Ecological Reserves. (These 19 marine ecological reserves also appear among the 67 reserves identified in the "Ecological Reserves" unit of the Atlas.)

The Legislature has established five categories of marine reserves: 1) Marine Life Refuges, 2) Fish Refuges, 3) Marine Reserves, 4) Clam Reserves, and 5) Game Refuges. Only specified fish, invertebrates, or marine plants may be taken from these reserves. In addition, at four refuges: Bodega Marine Life Refuge, Hopkins Marine Life Refuge, Catalina Marine Science Center Marine Life Refuge, and San Diego Marine Life Refuge, only licensed academic researchers may remove invertebrates or plant life.

Several studies on the value of marine reserves have led to agreement on the need for 3 forms of protection:

1) Total protection: protection of marine ecosystems from all forms of human consumptive uses and developments. These sites should contain outstanding biological characteristics and representative examples of various habitat types. Research and non-consumptive recreation may be permitted in these sites.

2) Restricted use: use is restricted to scientific and educational collecting only. These sites may be open to controlled manipulation of the environment.

3) Protection of specific forms of marine life: protection of all but specified forms of marine life from human consumptive uses and developments. These sites could permit taking of non-protected marine forms.

It should be noted that the federal government programs: National Marine Sanctuaries and National Estuarine Sanctuaries (National Estuarine Reserve Research System) are complementary to the state's marine reserve system. That cooperative situation is particularly visible in the 1,000 miles of "offshore rocks and pinnacles" designated by the Bureau of Land Management as the California Islands Wildlife Sanctuary. (President Herbert Hoover in 1930, issued an executive order withholding the rocks from development.) In 1983, the BLM leased these lands to the California Fish and Game Department for at least 50 years. These rocks and pinnacles are now designated as the State Ecological Reserve number 67, but they are also to be declared an Area of Critical Environmental Concern by the Bureau of Land Management.

See: Bibliography 104.

(continued)

Note: The California Islands Wildlife Sanctuary

All the unreserved or unappropriated islands, rocks, pinnacles, and reefs situated in the Pacific Ocean off the coast of California from Oregon to the Mexican border above the mean high tide level are designated as the California Islands Wildlife Sanctuary. The Sanctuary includes every island, rock, and reef off the coast **except** the following:

	Island, Rock, Reef	County
1	Prince Island, Hunter Rocks	Del Norte
2	Saint George Reef	Del Norte
3	Castle Rock, Pelican Rock, Round Rock, Lighthouse Island, Whaler Island, Prestons Island	Del Norte
4	White Rock, False Klamath Rock, Wilson Rock, Sister Rocks and all other islands, rocks, and reefs, lying one-quarter mile off Redwoods National Park	Del Norte, Humboldt
5	Farallon Islands	San Francisco
6	Año Nuevo Island	Santa Cruz
7	Lion Rock	Santa Barbara
8	San Miguel Island, Prince Island, Castle Rock	Santa Barbara
9	Santa Rosa Island	Santa Barbara
10	Santa Cruz Island	Santa Barbara
11	Anacapa Island, Cat Rock	Ventura
12	Santa Barbara Island, Sutil Island	Santa Barbara
13	San Nicolas Island	Santa Barbara
14	Santa Catalina Island	Los Angeles
15	San Clemente Island	Los Angeles

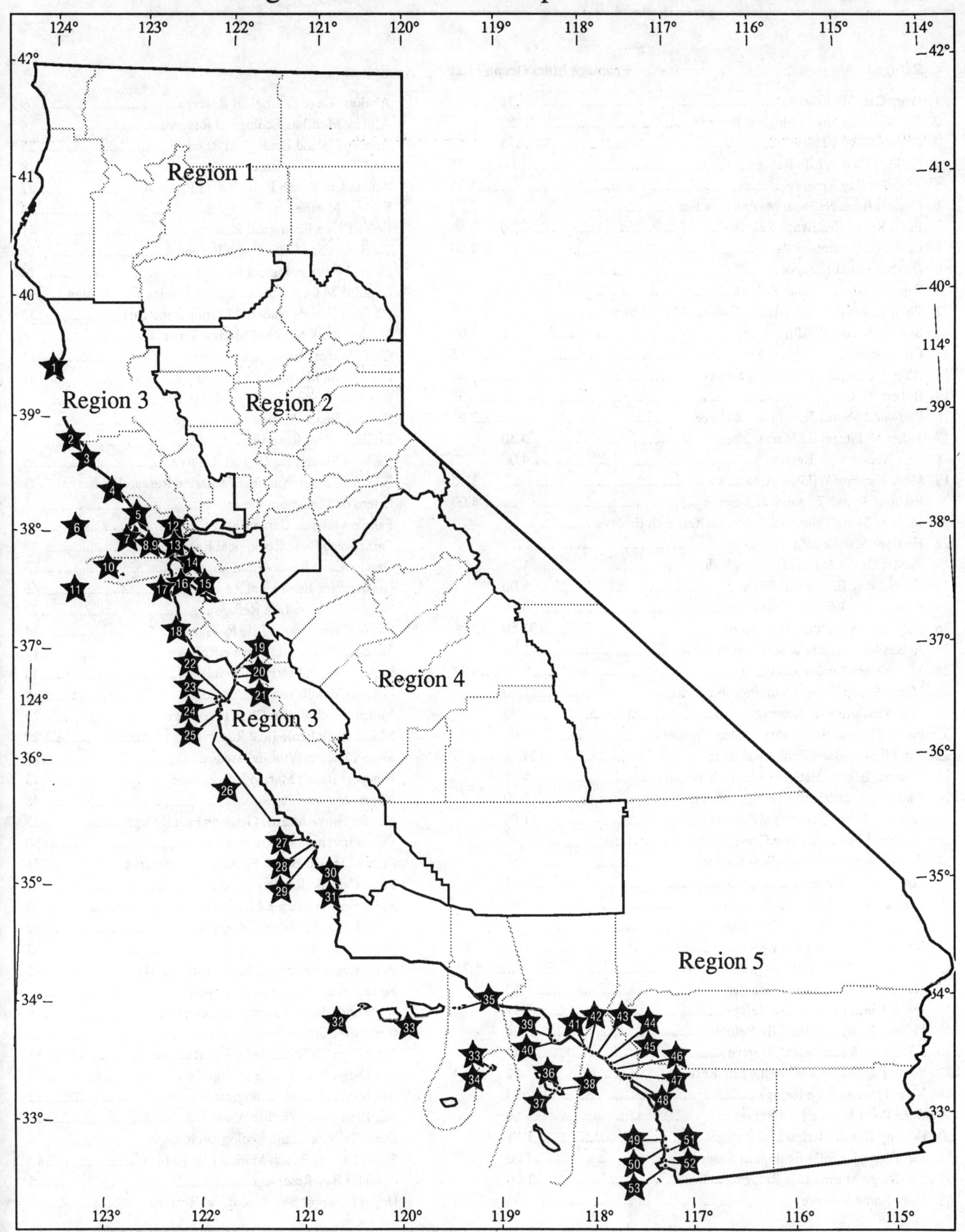
Region 1
Region 2
Region 3
Region 3
Region 4
Region 5
124°
123°
122°
121°
120°
119°
118°
117°
116°
115°
114°
42°
41°
40°
39°
38°
37°
36°
35°
34°
33°

Marine Life Refuges and Reserves - Department of Fish and Game

	Refuge	Frontage Miles Ocean	Bay
1	Point Cabrillo Reserve	.38	
2	Del Mar Landing Ecological Reserve	.59	
3	Gerstle Cove Reserve	.25	
4	Bodega Marine Life Refuge	1.00	.75
5	Tomales Bay Reserve	- -	5.25
6	Cordell Bank National Marine Sanctuary	- -	
7	Point Reyes Headlands Reserve	3.50	
8	Estero de Limantour Reserve	- -	7.90
9	Duxbury Reef Reserve	2.10	
10	Farallon Islands Game Refuge	- -	
11	Point Reyes Farallon Islands National Marine Sanctuary	- -	
12	San Pablo Bay Wildlife Area	- -	5.00
13	Corte Madera Ecological Reserve	- -	2.11
14	Albany Mudflats Ecological Reserve	- -	.60
15	Robert W. Crown Reserve	- -	.53
16	Redwood Shores Ecological Reserve	- -	3.21
17	James V. Fitzgerald Marine Reserve	3.30	
18	Año Nuevo State Reserve	4.00	
19	Moss Landing Wildlife Area	- -	3.30
20	Elkhorn Slough Ecological Reserve	- -	4.00
21	Elkhorn Slough National Estuarine Research Reserve	- -	
22	Hopkins Marine Life Refuge	.25	
23	Pacific Grove Marine Gardens Fish Refuge	1.75	
24	Carmel Bay Ecological Reserve	5.00	
25	Point Lobos Ecological Reserve	4.50	
26	California Sea Otter Game Refuge	101.50	
27	Atascadero Beach Pismo Clam Preserve	1.50	
28	Morro Rock Ecological Reserve	9.37	9.37
29	Morro Beach Pismo Clam Preserve	1.90	
30	Pismo Invertebrate Reserve	.30	
31	Pismo-Oceano Beach Pismo Clam Preserve	4.60	
32	San Miguel Island Ecological Reserve	24.00	
33	Channel Islands National Marine Sanctuary	- -	
34	Santa Barbara Island Ecological Reserve	5.00	
35	Anacapa Island Ecological Reserve	11.00	
36	Catalina Marine Science Center Marine Life Refuge	1.00	
37	Farnsworth Bank Ecological Reserve	- -	
38	Lovers Cove Reserve	.31	
39	Abalone Cove Ecological Reserve	1.00	
40	Point Fermin Marine Life Refuge	1.00	
41	Bolsa Chica Ecological Reserve	- -	7.60
42	Upper Newport Bay Ecological Reserve	- -	5.75
43	Newport Beach Marine Life Refuge	.53	
44	Irvine Coast Marine Life Refuge	3.41	
45	Laguna Beach Marine Life Refuge	.96	
46	Heisler Park Ecological Reserve	.49	
47	South Laguna Beach Marine Life Refuge	.52	
48	Niguel Marine Life Refuge	2.14	
49	Dana Point Marine Life Refuge	.66	
50	Doheny Beach Marine Life Refuge	1.11	
51	San Diego-La Jolla Ecological Reserve	1.00	
52	San Diego Marine Life Refuge	1.00	
53	Point Loma Reserve	.66	

Refuge	#
Abalone Cave Ecological Reserve	39
Albany Mudflats Ecological Reserve	14
Anacapa Island Ecological Reserve	35
Año Nuevo State Reserve	18
Atascadero Beach Pismo Clam Preserve	27
Bodega Marine Life Refuge	4
Bolsa Chica Ecological Reserve	41
California Sea Otter Game Refuge	26
Carmel Bay Ecological Reserve	24
Catalina Marine Science Center Marine Life Refuge	36
Channel Islands National Marine Sanctuary	33
Cordell Bank National Marine Sanctuary	6
Corte Madera Ecological Reserve	13
Dana Point Marine Life Refuge	49
Del Mar Landing Ecological Reserve	2
Doheny Beach Marine Life Refuge	50
Duxbury Reef Reserve	9
Elkhorn Slough Ecological Reserve	20
Elkhorn Slough National Estuarine Research Reserve	21
Estero de Limantour Reserve	8
Farallon Islands Game Refuge	10
Farnsworth Bank Ecological Reserve	37
Gerstle Cove Reserve	3
Heisler Park Ecological Reserve	46
Hopkins Marine Life Refuge	22
Irvine Coast Marine Life Refuge	44
James V. Fitzgerald Marine Reserve	17
Laguna Beach Marine Life Refuge	45
Lovers Cove Reserve	38
Morro Beach Pismo Clam Preserve	29
Morro Rock Ecological Reserve	28
Moss Landing Wildlife Area	19
Newport Beach Marine Life Refuge	43
Niguel Marine Life Refuge	48
Pacific Grove Marine Gardens Fish Refuge	23
Pismo Invertebrate Reserve	30
Pismo-Oceano Beach Pismo Clam Preserve	31
Point Cabrillo Reserve	1
Point Fermin Marine Life Refuge	40
Point Lobos Ecological Reserve	25
Point Loma Reserve	53
Point Reyes Farallon Islands National Marine Sanctuary	11
Point Reyes Headlands Reserve	7
Redwood Shores Ecological Reserve	16
Robert W. Crown Reserve	15
San Diego-La Jolla Ecological Reserve	51
San Diego Marine Life Refuge	52
San Miguel Island Ecological Reserve	32
San Pablo Bay Wildlife Area	12
Santa Barbara Island Ecological Reserve	34
South Laguna Beach Marine Life Refuge	47
Tomales Bay Reserve	5
Upper Newport Bay Ecological Reserve	42

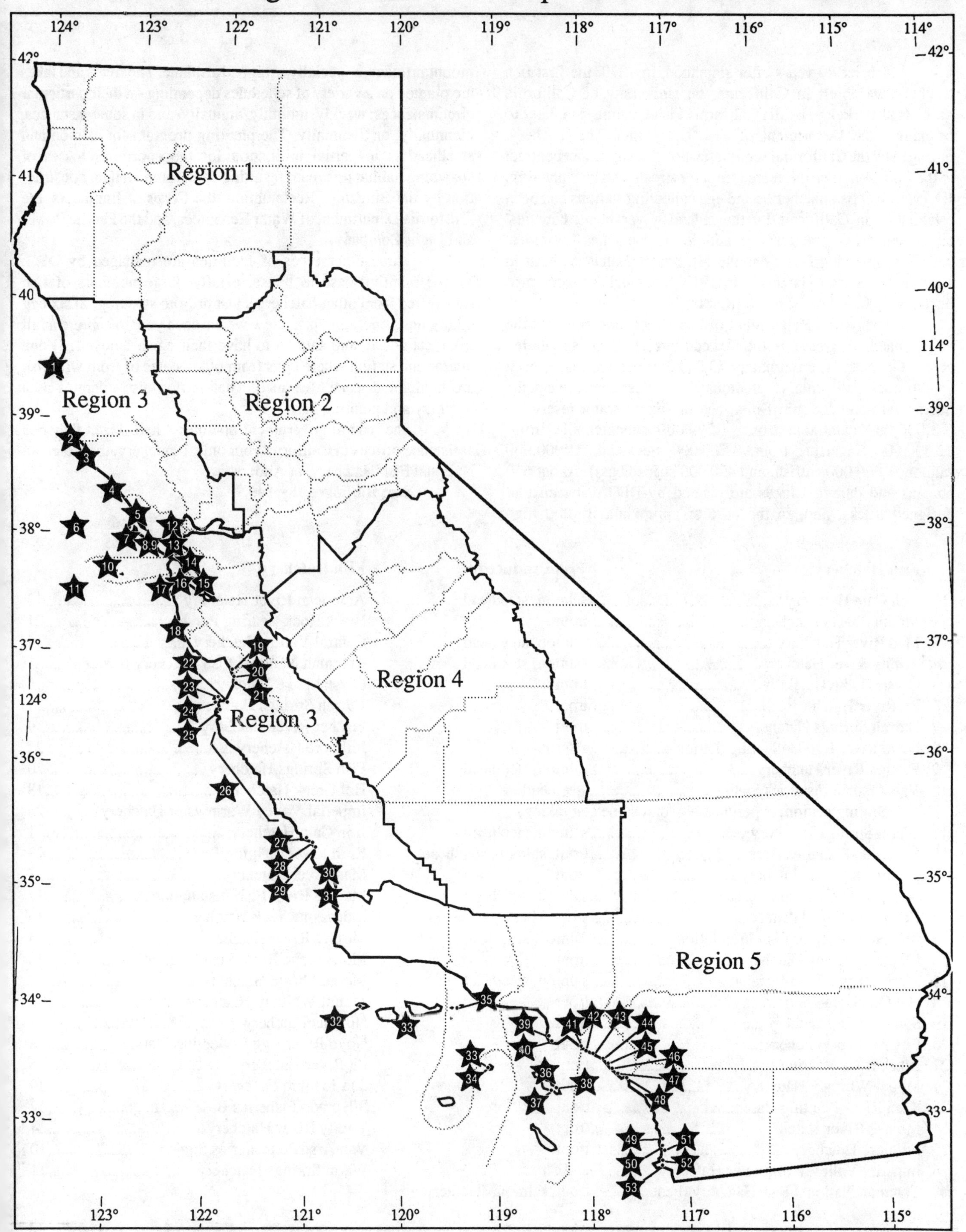

Region 1
Region 2
Region 3
Region 3
Region 4
Region 5
124°
123°
122°
121°
120°
119°
118°
117°
116°
115°
114°
42°
41°
40°
39°
38°
37°
36°
35°
34°
33°
1
2
3
4
5
6
7
8,9
10
11
12
13
14
15
16
17
18
19
20
21
22
23
24
25
26
27
28
29
30
31
32
33
33
34
35
36
37
38
39
40
41
42
43
44
45
46
47
48
49
50
51
52
53

A brief 20 years after statehood, in 1870, the first fish hatchery was built in California, on University of California grounds at Berkeley, by the California Fish Commission, later to be named the Department of Fish and Game. The hatchery, managed by the California Acclimatization Society, under contract to the Fish Commission, reared trout for stream and lake planting. By 1960, 169 fish hatcheries and egg collecting stations had been established in California by the federal government, counties, cities, and private interests, in addition to the state. The oldest currently operating facility is the Mt. Shasta Hatchery, built in 1888 as the Sisson Hatchery. In 1910, that hatchery alone produced 66,000,000 salmon and trout fry.

Today, the Department of Fish and Game manages the largest hatchery system in the United States at 26 sites scattered across the state. Approximately 53,000,000 fish are reared each year to stock California's high mountain lakes, streams, rivers, the Sacramento-San Joaquin Delta, and the warm water reservoirs (13,000,000 catchable trout, 1,200,000 subcatchable trout, 12,300,000 fingerling trout, 3,500,000 steelhead, 21,000,000 salmon, 1,600,000 catfish, and 400,000 striped bass). About 650 streams and dammed lakes are planted by DFG using custom designed trucks, and, in the case of approximately 900 high mountain lakes, a specially equipped airplane. The rivers and lakes are planted on a variety of schedules depending on their particular circumstances: weekly, monthly, annually, and in some instances, biennually, or triennially. The planting program for salmon and steelhead is in partial mitigation for the enormous losses of spawning habitat upstream resulting from rampant dam construction by the Bureau of Reclamation, the Corps of Engineers, the California Department of Water Resources, and the Pacific Power and Light Company.

Three other types of facilities are managed by DFG: 1) one planting base to house, briefly, large numbers of fish transferred from other hatcheries just prior to stocking, 2) six egg taking units operating just a few weeks each year to capture adult trout, steelhead, and salmon to have their eggs removed, 3) one quarantine station where eggs from other states or from wild fish are held for disease clearance. That station also operates as a hatchery and planting base.

The federal government operates a half-dozen fisheries assistance offices in California, but only 1 hatchery, the Coleman National Fish Hatchery, at Anderson.

See: Bibliography 146.

	Fish Hatchery	**Fish Produced**
1	Iron Gate Hatchery	salmon, steelhead
2	Mount Shasta Hatchery	trout
3	Mad River Hatchery	salmon, steelhead
4	Trinity River Hatchery	salmon, steelhead
5	Crystal Lake Hatchery	trout
6	Pit River Hatchery	trout
7	Darrah Springs Hatchery	trout
8	Noyo River Egg Collecting Station	salmon
9	Feather River Hatchery	salmon, steelhead
10	Van Arsdale Fisheries Station (counting station, experimental work, some egg taking)	steelhead
11	Warm Springs Hatchery	salmon, steelhead
12	Silverado Fisheries Base	trout, salmon, steelhead
13	American River Hatchery	trout
13	Nimbus Hatchery	salmon, steelhead
14	Central Valley Hatchery	striped bass
15	Mokelumne River Fish Installation	salmon, steelhead
16	Moccasin Creek Hatchery	trout
17	Merced River Fish Installation	salmon, steelhead
18	Hot Creek Hatchery	trout
19	San Joaquin Hatchery	trout
20	Fish Springs Hatchery	trout
21	Black Rock Rearing Ponds	trout
22	Mount Whitney Hatchery	trout
23	Kern River Planting Base	trout
24	Mojave River Hatchery	trout
25	Fillmore Hatchery	trout
26	Imperial Valley Warmwater Hatchery	catfish
A	Coleman National Fish Hatchery (Federal)	trout, salmon, steelhead

Fish Hatchery	**#**
American River Hatchery	13
Black Rock Rearing Ponds	21
Central Valley Hatchery	14
Coleman National Fish Hatchery (Federal)	**A**
Crystal Lake Hatchery	5
Darrah Springs Hatchery	7
Feather River Hatchery	9
Fillmore Hatchery	25
Fish Springs Hatchery	20
Hot Creek Hatchery	18
Imperial Valley Warmwater Hatchery	26
Iron Gate Hatchery	1
Kern River Planting Base	23
Mad River Hatchery	3
Merced River Fish Installation	17
Moccasin Creek Hatchery	16
Mojave River Hatchery	24
Mokelumne River Fish Installation	15
Mount Shasta Hatchery	2
Mount Whitney Hatchery	22
Nimbus Hatchery	13
Noyo River Egg Collecting Station	8
Pit River Hatchery	6
San Joaquin Hatchery	19
Silverado Fisheries Base	12
Trinity River Hatchery	4
Van Arsdale Fisheries Station	10
Warm Springs Hatchery	11

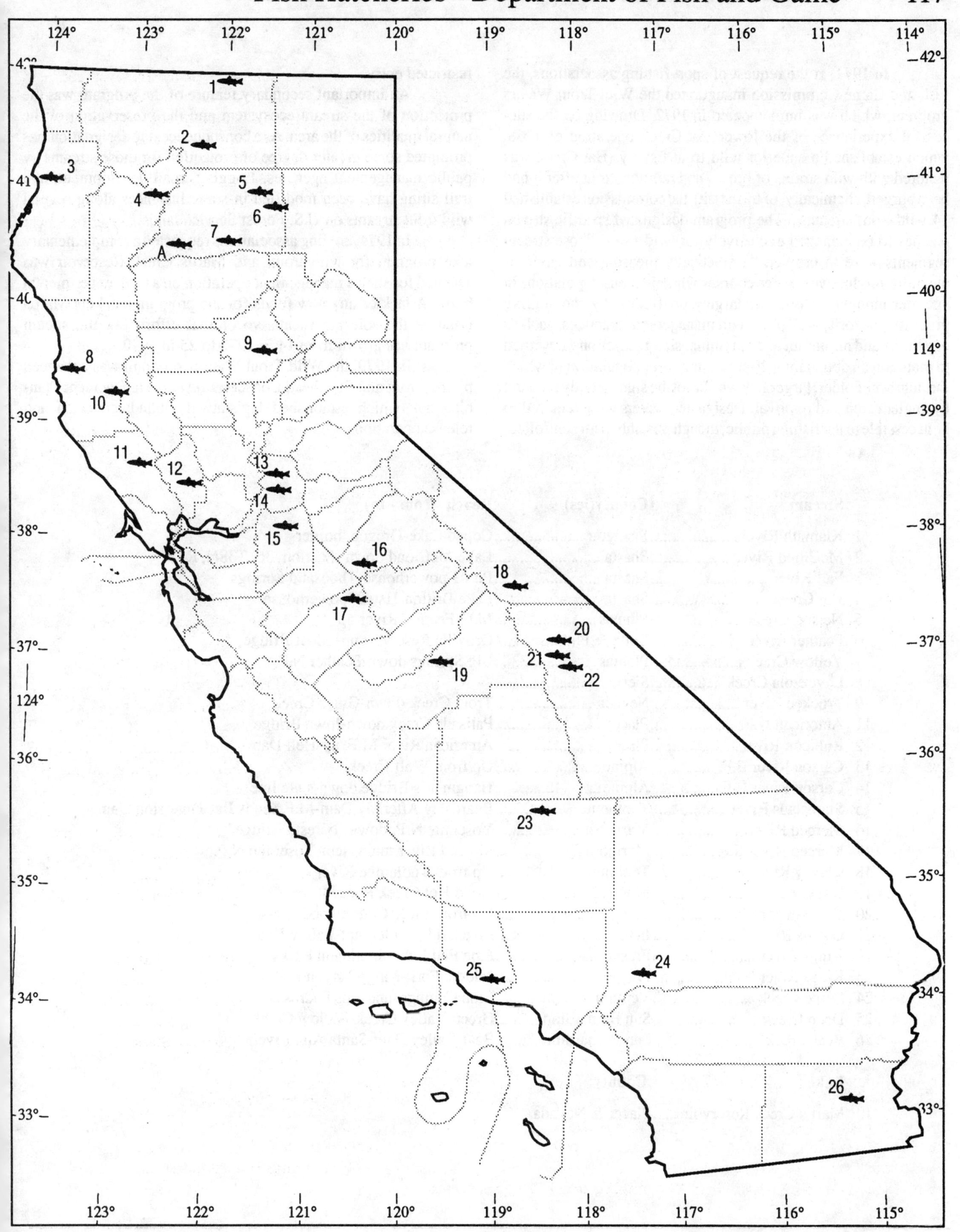
124°
123°
122°
121°
120°
119°
118°
117°
116°
115°
114°
42°
41°
40°
39°
38°
37°
36°
35°
34°
33°
1
2
3
4
5
6
7
A
8
9
10
11
12
13
14
15
16
17
18
19
20
21
22
23
24
25
26

In 1971, at the request of sport fishing associations, the Fish and Game Commission inaugurated the Wild Trout Waters program, which was implemented in 1972. Drawing on the successful experience of the lower Hat Creek operation of 1968, which established a superior wild trout fishery (Hat Creek was restored with wild strains of brown and rainbow trout after it had been purged, chemically, of trash fish), the commission established 14 wild trout streams. The program designated specific stream reaches to be managed exclusively for wild trout. Those stream segments were to provide "aesthetically pleasing and environmentally productive" water courses wherein trout population, in age and numbers, would be largely unaffected by the angling process. A priority was placed on management practices, such as bag limits and minimum and maximum size restrictions, designed to maintain an abundant self-sustaining trout population in which the number of older, larger fish would not be significantly reduced by angler catch and removal. Designated waters were required to be accessible to the fishing public, though possibly with controlled, restricted entry.

An important secondary feature of the program was the protection of the stream ecosystem and the preservation of the natural qualities of the area. As a consequence, the designation has prompted some greater degree of protection for those streams by public management agencies. Logging, road development, and trail siting have been modified in some instances along several wild trout streams on U.S. Forest Service lands.

In 1974, angling associations requested a complementary lake program for wild trout, and Martis Creek Reservoir was selected for such a management operation on a trial, experimental basis. A lack of any new funds for the program has kept Martis Lake as the sole representative of that effort, but the stream program has grown from 14 in 1972 to 25 in 1990.

In 1979, the Wild Trout Waters program was enhanced by the Trout and Steelhead Conservation and Management Planning act, which established legislatively mandated catch and release operations.

	Stream	County(ies)	Reach (from - to)
1	Klamath River	Siskiyou	Copco Lake-Oregon border
2	McCloud River	Shasta	Lake McCloud down -Section 36, T38N, R3W.
3	Fall River	Shasta	Pit #1 powerhouse-Thousand Springs
4	Hat Creek	Shasta	Lake Britton-Hat#2 powerhouse
5	Nelson Creek	Plumas	M.F. Feather River up
6	Feather River M.F.	Butte & Plumas	Oroville Reservoir up- Sloat Bridge
7	Yellow Creek	Plumas	Big Springs down-Feather N.F.
8	Lavezzola Creek	Sierra	- -
9	Truckee River	Nevada	Trout Creek down-Grays Creek
11	American N.F.	Placer	Palisade Creek down-Iowa Bridge
12	Rubicon River	Placer	American River M.F. up-Hell Dam
13	Carson River E.F.	Alpine	Up from Wolf Creek
14	Carson River E.F.	Alpine	Hangman's Bridge down-state line
15	Stanislaus River	Tuolumne	Beardsley Afterbay Dam-M.F.Sands Bar Diversion Dam
16	Merced River	Mariposa	Yosemite N.P. down- Foresta Bridge
17	Merced River S.F.	Mariposa	Merced River main stem-Yosemite N.P.
18	Clavey River	Tuolumne	Up from Tuolumne River
19	Hot Creek	Mono	Up to Hot Creek Ranch
20	Cottonwood Creek	Inyo	Up from Little Cottonwood Creek
21	Owens River	Inyo	Five Bridges-Pleasant Valley Dam
22	Kings River	Fresno	Pine Flat Lake-South and Forks
23	Kings River S.F.	Fresno	Middle Fork-KingsCanyon N.P.
24	Sespe Creek	Ventura	Lion Campground-Los Padres N.F.
25	Deep Creek	San Bernardino	Green Valley Creek-Willow Creek
26	Bear Creek	San Bernardino	Bear Valley Dam-Santa Ana River

	Lake	County
10	Martis Creek Reservoir	Placer & Nevada

1 North Coast Region

2 San Francisco Bay Region

3 Central Coast Region

4 Los Angeles Region

5 Central Valley Region

6 Lahontan Region

7 Colorado River Basin Region

8 Santa Ana Region

9 San Diego Region

The Trout and Steelhead Conservation and Management Planning Act of 1979, familiarly known as the Catch and Release Act, was designed as an enhancement of the Fish and Game Commission's Wild Trout Waters program. The Act notes that, "In order to provide for a diversity of available angling experiences throughout the state, it is the intent of the Legislature that the commission maintain the existing wild trout program and as part of such program develop catch and release fisheries in the more than 20,000 miles of trout streams and approximately 5,000 lakes containing trout in California."

The act provided for bag limits of zero, one, or two fish only, and recommended "imposing minimum and maximum size limits." Further restrictions have been imposed by the department such as fishing with artificial lures and/or barbless hooks only.

The regulations also required that the department recommend additional waters annually, "no less than 25 miles of stream or stream segments and at least one lake that it deems suitable for consideration as catch and release trout fisheries." However, since the act declared that the program was, a continuation and perpetuation of the department's existing wild trout program and other programs," no funds were appropriated for the required review and analysis of state waters. Consequently, those provisions have been extremely difficult to maintain. As of 1990, 28 segments of 24 rivers, and 7 lakes have been designated as catch and release waters.

#	Stream	Length (miles)	County(ies)	Designated	Species	Limits: Bag	Size
1	Sacramento River	14.00	Shasta	1981	Rainbow	2	
2	McCloud River	4.00	Shasta	1980	Rainbow, Brown	2	
2	McCloud River	1.00	Shasta	1980	Rainbow, Brown	0	
3	Fall River (Pit)	23.00	Shasta	1980	Rainbow	2	14"max.
4	Hat Creek	3.50	Shasta	1980	Rainbow, Brown	2	18"min.
6	Eel River M.F.	28.00	Mendocino, Lake	1986	Steelhead, Salmon	2	
7	Stony Creek M.F.	10.00	Colusa, Glenn, and Mendocino	1989	Rainbow	2	8"max.
8	Yellow Creek	2.00	Plumas	1983	Rainbow, Brown	2	10"max.
9	Yuba River	5.00	Sierra	1986	Rainbow, Brown	2	10"min.
10	Truckee River	4.00	Nevada	1983	Rainbow, Brown	2	15"min.
10	Truckee River	4.00	Nevada	1986	Rainbow, Brown	2	
14	Stanislaus River M.F.	2.00	Tuolumne	1986	Rainbow, Brown	2	14"min.
14	Stanislaus River M.F.	17.00	Tuolumne	1987	Rainbow, Brown	2	
16	East Walker River	8.00	Mono	1981	Brown	2	14"min.
17	Merced River	4.00	Mariposa	1986	Rainbow, Brown	2	12"min.
19	Hot Creek	1.00	Mono	1980	Rainbow, Brown	0	
21	Owens River	16.00	Inyo	1980	Brown	2	
22	San Lorenzo River	15.00	Santa Cruz	1984	Steelhead	2	
23	Kings River S.F.	18.00	Fresno	1985	Rainbow	2	
24	Kings River	20.00	Fresno	1987	Rainbow, Brown	2	
25	Carmel River	12.00	Monterey	1984	Steelhead	2	
26	Kaweah River, Marble F.	8.00	Tulare	1988	Rainbow, Brown	0/5 *	
27	Tule River N.F.	6.00	Tulare	1981	Rainbow, Brown	2	
28	Kern River	4.00	Tulare	1989	Rainbow, Brown	2	14"min.
29	Piru Creek	1.25	Los Angeles	1989	Rainbow	0	
30	San Gabriel River W.F.	5.50	Los Angeles	1985	Rainbow	0	
31	Deep Creek	16.00	San Bernardino	1983	Rainbow, Brown	2	8"min.
32	Bear Creek	9.00	San Bernardino	1986	Rainbow, Brown	2	8"min.
	Cottonwood Creek	4.00	Inyo	Proposed	Golden	0	
	Pit River	5.00	Shasta	Proposed	Rainbow	2	18"min.
	Stanislaus River, lower	4.50	(several)	Proposed	Rainbow	0	
	Owens River, upper	15.00	Mono	Proposed	Rainbow, Brown	2	

#	Lakes	Surface Area (acres)	County	Designated
5	Manzanita Lake	35.00	Shasta	1985
11	Milton Lake	70.00	Nevada	1984
12	Martis Lake	70.00	Nevada, Placer	1980
13	Heenan Lake	135.00	Alpine	1983
15	Kirman Lake	45.00	Mono	1982
18	McCloud Lake	10.00	Mono	1981
20	Crowley Lake	5,000.00	Mono	1986
	Sotcher Lake	- -	Mono	Designation Delayed
	Twin Lakes	30.00	Mono	Designation Delayed

Streams	#
Bear Creek	32
Carmel River	25
Deep Creek	31
East Walker River	16
Eel River M.F	6
Fall River	3
Hat Creek	4
Hot Creek	19
Kaweah River Marble Fork	26
Kern River	28
Kings River	23
Kings River S.F.	24
McCloud River (4 mile reach)	2
McCloud River (1 mile reach)	2
Merced River	17
Owens River	21
Piru Creek	29
Sacramento River	1
San Gabriel River	30
San Lorenzo River	22
Stanislaus River (2 mile reach)	14
Stanislaus River (17 mile reach)	14
Stony Creek M.F.	7
Truckee River	10
Truckee River	10
Tule River N.F.	27
Yellow Creek	8
Yuba River N.F.	9

Lakes	#
Crowley Lake	20
Heenan Lake	13
Kirman Lake	15
Manzanita Lake	5
Martis Lake	12
McCloud Lake	18
Milton Lake	11

* 0-Rainbow, 5-Brown (pending Sequoia National Park approval).

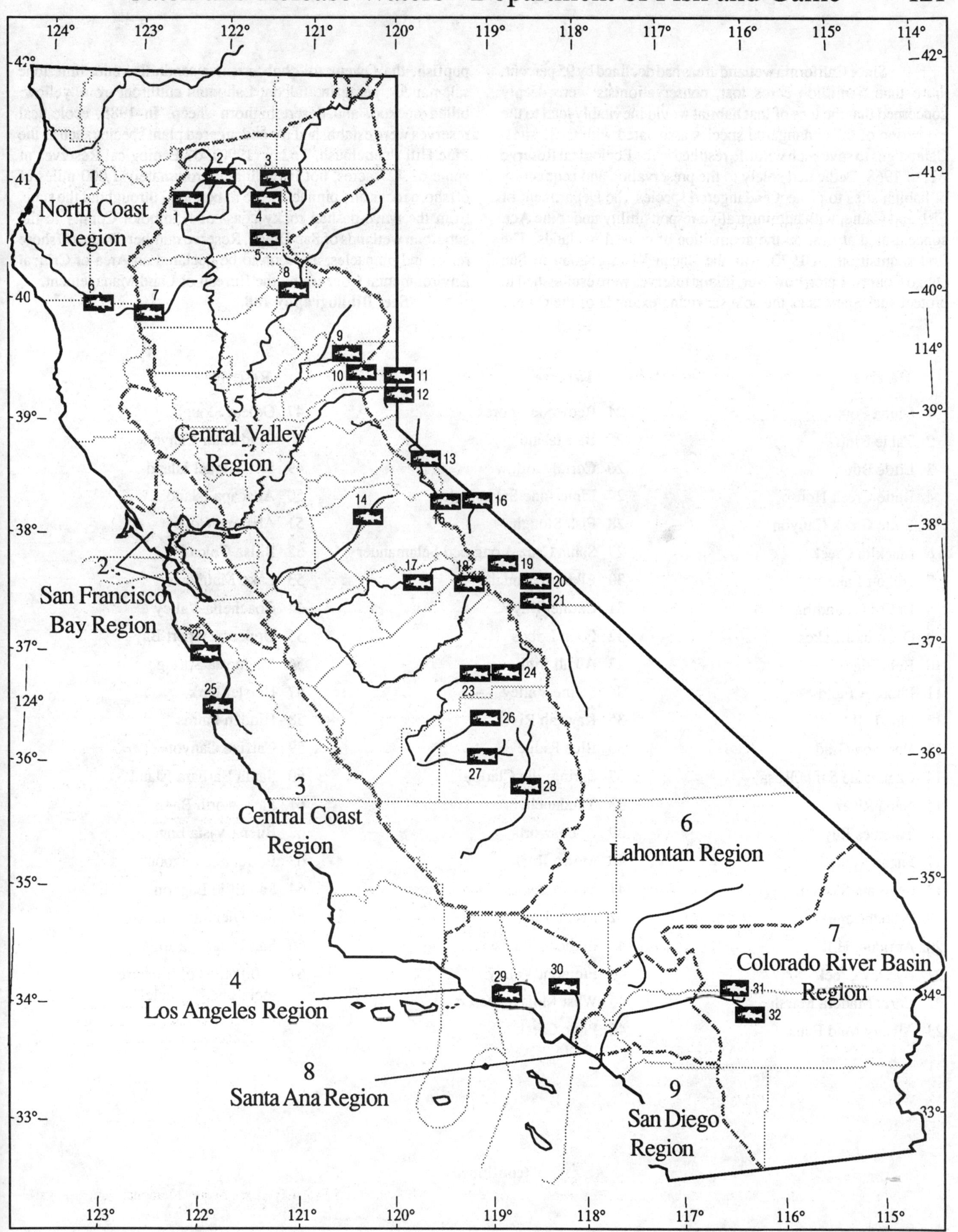
1
North Coast
Region
2
San Francisco
Bay Region
3
Central Coast
Region
4
Los Angeles Region
5
Central Valley
Region
6
Lahontan Region
7
Colorado River Basin
Region
8
Santa Ana Region
9
San Diego
Region

Since California wetland areas had declined by 95 percent, more than 5 million acres lost, conservationists were deeply concerned that the loss of that habitat would inevitably lead to the extinction of the endangered species associated with those sites. Campaigns to save such wildlife resulted in the Ecological Reserve Act of 1968. Dedicated solely to the preservation and acquisition of habitat sites to protect endangered species, the Department of Fish and Game, with administrative responsibility under the Act, concentrated at first on the acquisition of coastal wetlands. The first acquisition, in 1970, was the Buena Vista Lagoon in San Diego County. Later, however, inland reserves were established to protect such species as the sole surviving example of the Owens pupfish, the Owens toi-chub, greater sandhill crane, limestone salamander, peregrine falcon, Lahontan cutthroat trout, yellow-billed cuckoo, and desert bighorn sheep. In 1985, ecological reserves were established for endangered plant species such as the Pine Hill flannelbush. Today (1990), 67 Ecological Reserves of some 61,427 acres, not including approximately 1,000 miles of offshore rocks and pinnacles, are to be found throughout the state, from the wave-dashed rocky coast of Humboldt County to the suburban wetlands of San Diego. Reserve number 67, the offshore rocks and pinnacles, are soon to be declared an Area of Critical Environmental Concern by the Bureau of Land Management.

See: Bibliography 188.

	Reserve
1	China Point
2	Table Bluff
3	Little Butte
4	Butte Creek House
5	Butte Creek Canyon
6	Macklin Creek
7	Abbott Lake
8	Del Mar Landing
9	O'Connor Lakes
10	Bobelaine
11	Phoenix Field
12	Pine Hill
13	Harrison Grade
14	Laguna de Santa Rosa
15	Napa River
16	Tomales Bay
17	Fagan Slough
18	Peytonia Slough
19	Woodbridge
20	Apricum Hill
21	By Day Creek
22	Corte Madera Marsh
23	Albany Mud Flats
24	Redwood Shores
25	Bair Island
26	Corral Hollow
27	Limestone Salamander
28	Fish Slough
29	Santa Cruz Longtoed Salamander
30	Elkhorn Slough
31	Carmel Bay
32	Point Lobos
33	Alkali Sink
34	Saline Valley
35	Kaweah River
36	Blue Ridge
37	Springville Clarkia
38	Yaudanchi
39	Allensworth
40	Morro Rock
41	Morro Dunes
42	Pismo Lake
43	Elkhorn Plains
44	Fremont Valley
45	West Mojave Desert
46	Piute Creek
47	Goleta Slough
48	Coldwater Canyon
49	San Miguel Island
50	Anacapa Island
51	Abalone Cove
52	Bolsa Chica
53	Lake Matthews
54	Coachella Valley
55	Upper Newport Bay
56	Magnesia Spring
57	Heisler Park
58	Hidden Palms
59	Carrizo Canyon
60	Santa Barbara Island
61	Farnsworth Bank
62	Buena Vista Lagoon
63	Batiquitos Lagoon
64	San Elijo Lagoon
65	San Dieguito Lagoon
66	San Diego-La Jolla
67	1000 miles of offshore rocks and pinnacles

(continued)

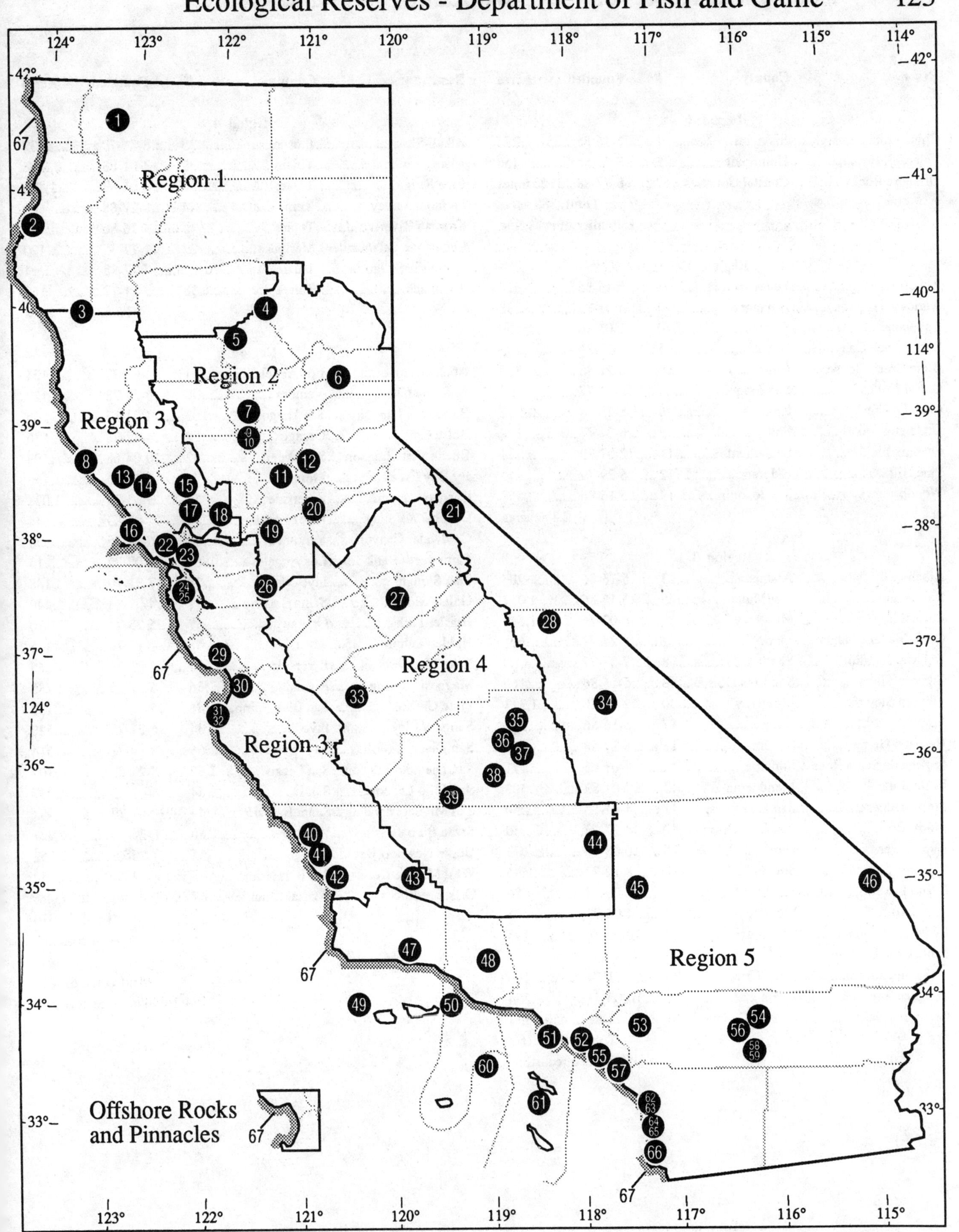

Region 1
Region 2
Region 3
Region 3
Region 4
Region 5
Offshore Rocks and Pinnacles
67

Reserve	County	#	Founded	Acres
Region 1				
China Point	Siskiyou	1	5-16-86	238
Table Bluff	Humboldt	2	8-27-88	150
Offshore Rocks	Coastal Counties	67	8-27-88	150 miles
				Total: 388 acres
				+ 150 miles of coastline
Region 2				
Abbott lake	Sutter	7	5-16-86	624
Apricum Hill	Amador	20	4-01-83	38
Bobelaine	Sutter	10	1-03-86	430
Butte Creek Canyon	Butte	5	8-27-88	285
Butte Creek House	Butte	4	8-27-88	320
Corral Hollow	San Joaquin	26	6-24-77	92
Macklin Creek	Nevada	6	8-25-78	32
O'Connor Lakes	Sutter	9	6-24-77	178
Phoenix Field	Sacramento	11	12-07-79	8
Pine Hill	Eldorado	12	6-29-79	240
Woodbridge	San Joaquin	19	5-16-86	559
				Total: 2,806 acres
Region 3				
Albany Mudflats	Alameda	23	5-16-86	206
Bair Island	San Mateo	25	5-16-86	1,017
Carmel Bay	Monterey	31	4-02-76	1,642
Corte Madera Marsh	Marin	22	6-24-77	621
Del Mar Landing	Sonoma	8	7-27-72	60
Elkhorn Plains	San Luis Obispo	43	5-16-86	160
Elkhorn Slough	Monterey	30	12-05-80	1,323
Fagan Slough	Napa	17	5-16-86	292
Harrison Grade	Sonoma	13	8-27-88	32
Laguna de Santa Rosa	Sonoma	14	4-01-83	71
Little Butte	Mendocino	3	4-01-83	403
Morro Dunes	San Luis Obispo	41	4-01-83	50
Morro Rock	San Luis Obispo	40	4-27-23	30
Napa River	Napa	15	10-08-76	73
Peytonia Slough	Solano	18	8-12-77	595
Pismo Lake	San Luis Obispo	42	12-09-77	70
Pt. Lobos	Monterey	32	4-06-73	775
Redwood Shores	San Mateo	24	10-08-76	175
Santa Cruz Long-toed Salamander	Santa Cruz	29	1-11-74	196
Tomales Bay	Marin	16	10-05-73	554
Offshore Rocks	Coastal Counties	67	8-27-88	450 miles
				Total: 8,345 acres
				+ 450 miles of coastline

Reserve	County	#	Founded	Acres
Region 4				
Alkali Sink	Fresno	33	8-31-79	932
Allensworth	Tulare	39	4-01-83	486
Blue Ridge	Tulare	36	4-01-83	1,476
Fremont Valley	Kern	44	8-27-88	640
Kaweah River	Tulare	35	5-16-86	98
Limestone Salamander	Mariposa	27	1-17-75	120
Springville-Clarkia	Tulare	37	8-27-88	4
Yaudanchi	Tulare	38	3-11-77	162
				Total: 3,918 acres
Region 5				
Abalone Cove	Los Angeles	51	4-01-77	124
Anacapa Island	Ventura	50	6-23-78	9,113
Batiquitos Lagoon	San Diego	63	4-01-83	135
Bolsa Chica	Orange	52	4-27-73	536
Buena Vista Lagoon	San Diego	62	10-04-68	199
By Day Creek	Mono	21	5-16-86	160
Carrizo Canyon	Riverside	59	3-03-78	1,021
Coachella Valley	Riverside	54	4-01-83	644
Coldwater Canyon	Ventura	48	6-27-75	58
Farnsworth Bank	Los Angeles	61	12-08-72	14
Fish Slough	Inyo-Mono	28	7-31-70	188
Goleta Slough	Santa Barbara	47	4-01-83	440
Heisler Park	Orange	57	5-25-73	30
Hidden Palms	Riverside	58	1-11-74	135
Lake Matthews	Riverside	53	4-01-83	2,565
Magnesia Spring	Riverside	56	6-27-75	1,285
Piute Creek	San Bernardino	46	5-16-86	139
Saline Valley	Inyo	34	6-27-75	520
San Diego-La Jolla	San Diego	66	8-19-71	514
San Dieguito Lagoon	San Diego	65	8-27-88	107
San Elijo Lagoon	San Diego	64	4-01-83	737
San Miguel Island	Santa Barbara	49	10-07-77	17,993
Santa Barbara Island	Ventura	60	6-23-78	7,921
Upper Newport Bay	Orange	55	7-25-75	752
West Mojave Desert	San Bernardino	45	8-27-88	640
Offshore Rocks	Coastal Counties	67	8-27-88	400 miles
				Total: 45,970
				+ 400 miles of coastline

Grand Total: 61,472
+ 1,000 miles of coastline

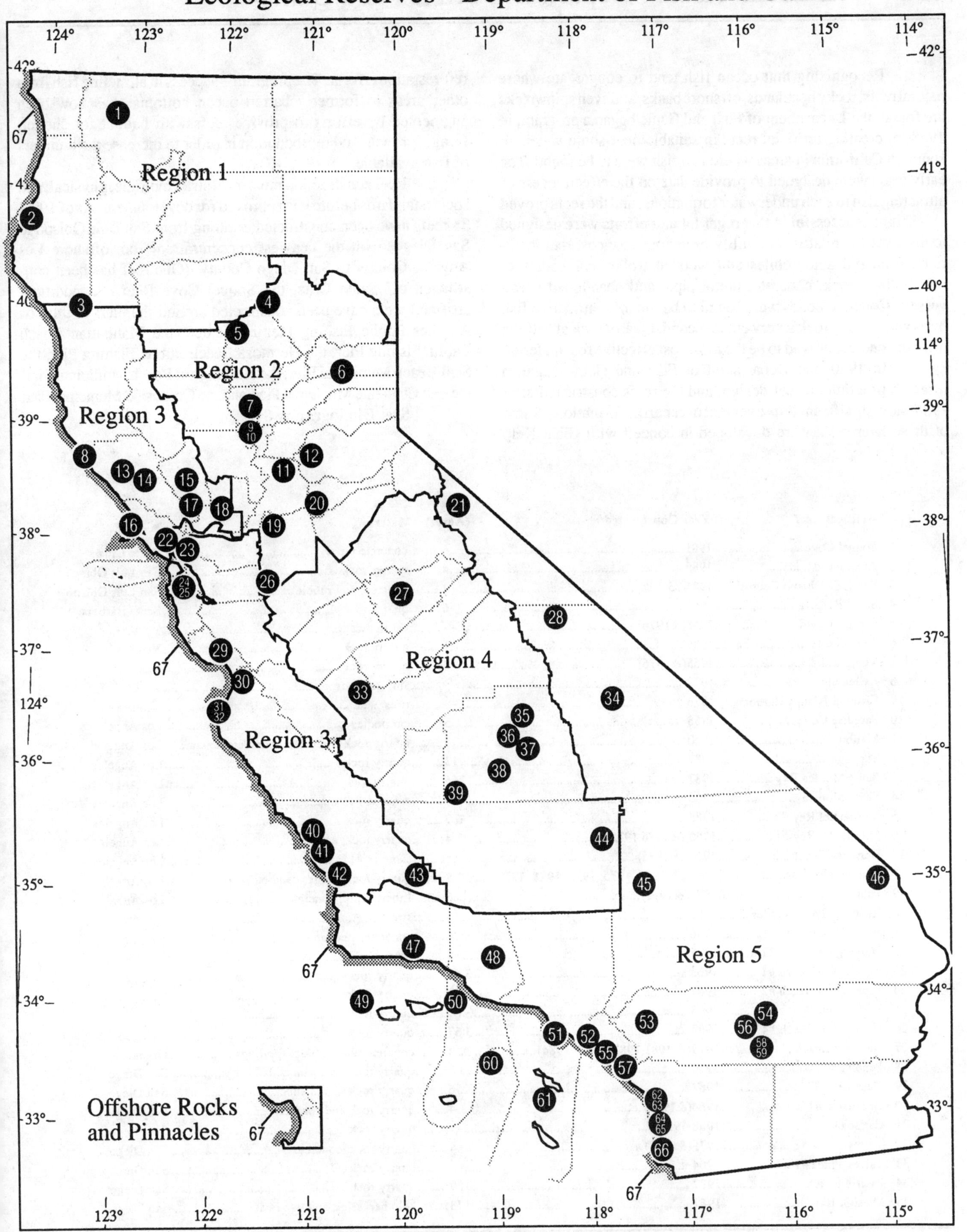
Region 1
Region 2
Region 3
Region 4
Region 3
Region 5
Offshore Rocks
and Pinnacles
67
1
2
3
4
5
6
7
9
10
8
11
12
13
14
15
16
17
18
19
20
21
22
23
24
25
26
27
28
29
30
31
32
33
34
35
36
37
38
39
40
41
42
43
44
45
46
47
48
49
50
51
52
53
54
55
56
57
58
59
60
61
62
63
64
65
66
124°
123°
122°
121°
120°
119°
118°
117°
116°
115°
114°
42°
41°
40°
39°
38°
37°
36°
35°
34°
33°

Recognizing that ocean fish tend to congregate where natural reefs, rocky headlands, offshore banks, and even shipwrecks are found, the Department of Fish and Game began a program, in 1958, of creating artificial reefs in suitable near-shore waters in southern California in areas where few fish were to be found. The early trials were designed to provide data on the effectiveness of attracting fish to such underwater formations, and the reefs proved to be highly successful in that regard. Later efforts were designed to discover the relative durability of various low-cost reef materials, from old automobiles and wooden trolley car bodies to rubber tires, rock, concrete, metal pipe, and abandoned ocean vessels. Concrete boxes were found to be the most attractive fish lures with quarry rock a very close second. Quarry rock at half the price of concrete proved to be the most cost effective reef material.

In 1980, the Department of Fish and Game began a research program on reef design, and the reefs constructed after 1984 show significant improvement over earlier formations. Some of these later reefs were developed in concert with Giant Kelp reforestation efforts. The program's success in attracting fish from other areas to formerly barren ocean bottoms has now been augmented by efforts to provide reefs with habitat for shelter, forage, growth, and reproduction in order to increase the numbers of fish available.

Reef construction must meet strict biologic, physical, and social standards before it is approved for development. As of 1990, 35 reefs have been constructed, running from San Luis Obispo to San Diego, with the heaviest concentration from offshore Los Angeles County to San Diego County. (One reef has been constructed off Santa Cruz, the Soquel Cove Reef.) In addition, artificial reefs have been constructed around the piers at the Los Angeles Public Fishing Pier in Venice, the Manhattan Beach Public Fishing Pier, the Hermosa Beach Public Fishing Pier, the Seal Beach Municipal Pier, the Huntington Beach Municipal Pier, the San Clemente Municipal Pier, and the Oceanside Municipal Pier.

See: Bibliography 81.

	Artifical Reef	Year Constructed	Area (Acres)	Material	County
1	Soquel Cove	1981	.6	concrete pipes	Santa Cruz
2	Atascadero	1985	.4	quarry rock	San Luis Obispo
3	San Luis Obispo County	1984/85	13	concrete and rubble	San Luis Obispo
4	Santa Barbara	1988	256	quarry rock	Santa Barbara
5	Rincon Island	1958(&1976)	- -	car bodies	Ventura
6	Pitas Point	1984	1.1	quarry rock	Ventura
7	Ventura	1965(&1976)	8.8	quarry rock	Ventura
8	La Jenelle	1978	- -	ship sections	Ventura
9	Channel Islands Harbor	1976-1979	- -	tires	Ventura
10	Paradise Cove	1958	.5	car bodies	Los Angeles
11	Malibu	1960	.5	quarry rock, streetcars, concrete, cars	Los Angeles
12	Topanga	1987	13	quarry rock	Los Angeles
13	Santa Monica Bay	1987	256	quarry rock	Los Angeles
14	Santa Monica	1960	.5	same as #11	Los Angeles
15	Marina del Rey #2	1985	6.9	quarry rock	Los Angeles
16	Marina del Rey #1	1965 (&1976,1978)	3.4	quarry rock, concrete	Los Angeles
17	Hermosa Beach	1960 (&1975)	.5	same as #11	Los Angeles
18	Redondo Beach	1962 (& 1974, 1975, 1976, 1978, 1979)	1.8	quarry rock, barge, concrete, pipe,	Los Angeles
19	Palawan	1977(&1978)	.6	Liberty ship, concrete	Los Angeles
20	Redondo-Palos Verdes	1958	.1	streetcars	Los Angeles
21	Bolsa Chica	1986	220	concrete, steel and concrete barges	Orange
22	Huntington Beach Tire	1975	35	tires	Orange
23	Huntington Beach #1	1963	3.67	quarry rock	Orange
24	Huntington Beach #2	1963	3.67	quarry rock	Orange
25	Huntington Beach #3	1963	3.67	quarry rock	Orange
26	Huntington Beach #4	1963	3.67	quarry rock	Orange
27	Newport Beach	1979(& 1981,1982, 1983, 1984)	4	concrete rubble, blocks, pilings	Orange
28	Pendleton	1980	3.5	quarry rock	San Diego
29	Oceanside #2	1987	256	quarry rock	San Diego
30	Oceanside #1	1964(&1987)	4	quarry rock and concrete	San Diego
31	Carlsbad	1988-1989	- -	quarry rock	San Diego
32	Torrey Pines #2	1975(&1978)	.4	quarry rock, concrete	San Diego
33	Torrey Pines #1	1964	- -	quarry rock	San Diego
34	Pacific Beach	1987	109	quarry rock	San Diego
35	Mission Bay	1987	173	kelp harvester, and two boats	San Diego

Reef	#
Atascadero	2
Bolsa Chica	21
Carlsbad	31
Channel Islands Harbor	9
Hermosa Beach	17
Huntington Beach Tire	22
Huntington Beach #1	23
Huntington Beach #2	24
Huntington Beach #3	25
Huntington Beach #4	26
La Jenelle	8
Malibu	11

Reef	#
Marina del Rey #2	15
Marina del Rey #1	16
Russian Bay	31
Newport Beach	27
Oceanside #2	29
Oceanside #1	30
Pacific Beach	34
Palawan	19
Paradise Cove	10
Pendleton	28
Pitas Point	6
Redondo Beach	18

Reef	#
Redondo-Palos Verdes	20
Rincon Island	5
San Luis Obispo County	3
Santa Barbara	4
Santa Monica	14
Santa Monica Bay	13
Soquel Cove	1
Topanga	12
Torrey Pines #2	32
Torrey Pines #1	33
Ventura	7

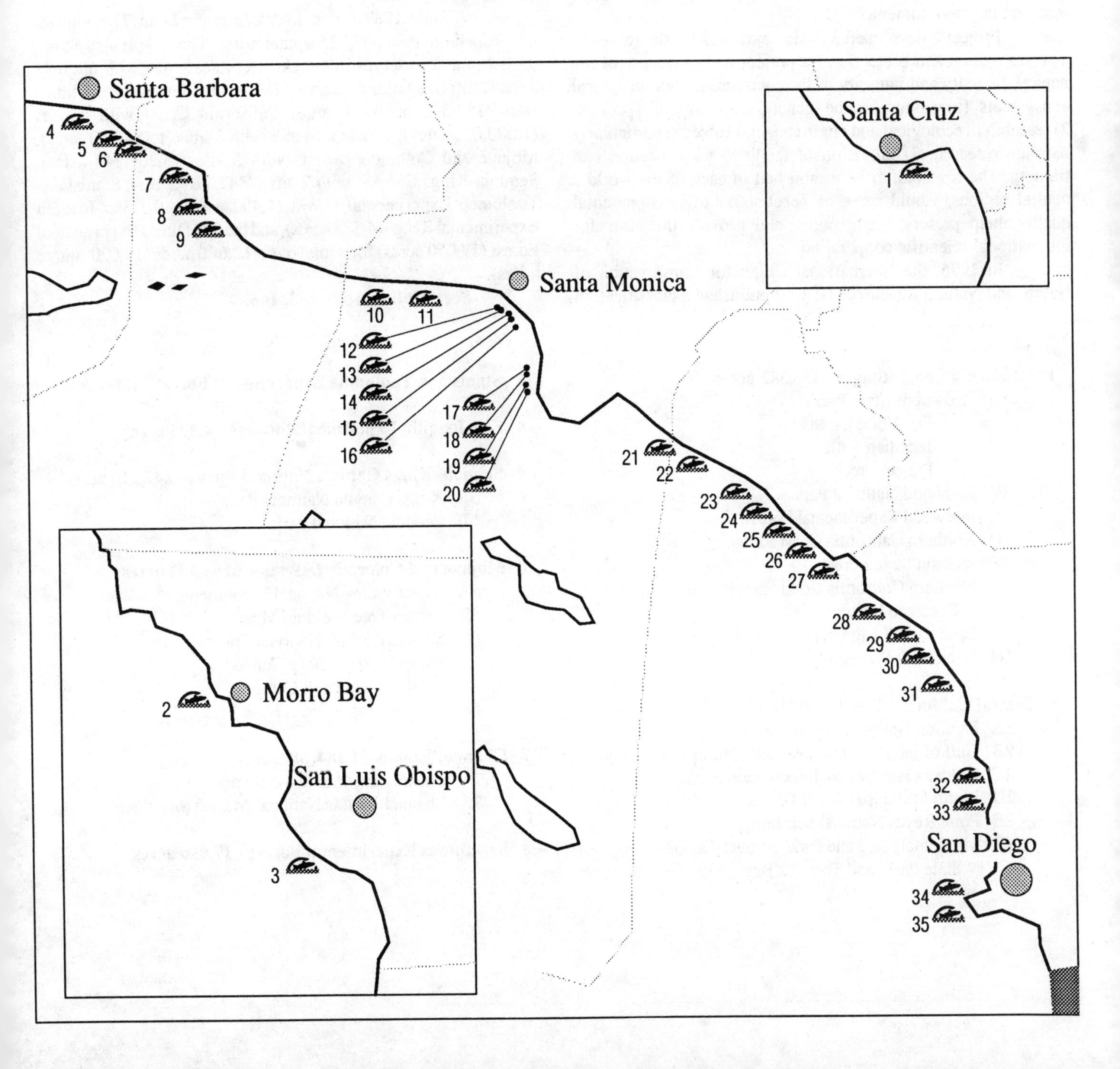

The world's most ambitious enterprise for the protection of natural diversity is the Biosphere Reserve undertaking, sometimes called Project 8, of UNESCO's Man and the Biosphere Programme (MAB). Organized in 1971, on the work of its predecessor body, the International Biological Programme, the MAB Council first proposed 13, later 14, project areas for research and education: 1) Tropical Forests, 2) Temperate Forests, 3) Grazing Land, 4) Arid Lands, 5) Fresh Water and Coasts, 6) Mountain and Tundra, 7) Islands, 8) Biosphere Reserves, 9) Cultivation Systems, 10) Major Engineering Works, 11) Urban Ecosystems, 12) Demographic Changes, 13) Perception of Environmental Quality, and 14) Environmental Pollution. Projects 1-7 relate to major national regions and 9-14 are concerned with the interaction of man and the environment.

Project 8 developed 3 basic objectives for the reserves: 1) conservation and protection for present and future use of the natural diversity and integrity of the communities within natural ecosystems (to safeguard the genetic diversity of species); 2) research in ecological and environmental subjects, particularly baseline research; 3) provision of facilities for education and training. The reserves, to be established in each of the world's natural regions, would serve as benchmarks of environmental quality, help preserve gene pools, and provide the basis for international scientific cooperation.

In 1975, the International Union for Conservation of Nature and Natural Resources (IUCN) published a classification scheme and a map of 193 natural regions of the earth, named "Biogeographical Provinces." Criteria adopted in 1974 (MAB Report 22) and a refined version of the IUCN scheme along with more detailed national classification systems govern the selection of sites within the provinces. Four such provinces lie wholly or partially within California: 1) Oregonian, 2) Californian, 3) Sierra Cascade, and 4) Sonoran. In the early 1980s, the United States MAB Programme began to establish multiple site reserves to create large ecologically linked conservation areas and to encourage cooperation among complementary and/or contiguous protected sites. That concept continues with the most recent designation in February, 1989, of the Central California Coast Biosphere Reserve with 8 discrete sites.

A June 1989 report lists 276 reserves in 71 countries covering more than 557,725 square miles. That report also shows 46 Biosphere Reserves in the United States, most with multiple sites. California holds 8 reserves: California Coast Ranges with 10 sites (153,382 acres), Central California Coast with 8 sites (1,000,012 acres), Channel Islands with 2 sites (1,184,740 acres), Mojave and Colorado Deserts with 5 sites (3,204,242 acres), Sequoia-Kings Canyon with 2 sites (847,210 acres), Stanislaus-Tuolumne Experimental Forest (1,499 acres), the San Joaquin Experimental Range (4,525 acres), and the San Dimas Experimental Forest (17,050 acres) for a total of 6,412,660 acres (10,020 square miles).

See: Bibliography 143, 166.

1 California Coast Ranges - 153,382 acres
- 1A Redwoods State Parks
  - Del Norte Coast
  - Jedediah Smith
  - Prairie Creek
- 1B Redwood National Park
- 1C Redwood Experimental Forest
- 1D Northern California Coast Range
- 1E Jackson State Forest
- 1F Northern California Coast Range Preserve Research Natural Area
- 1G Landels Hill Big Creek Reserve
- 1H Cone Peak Western Slope

2 Central California Coast - 1,000,012 acres
- 2A Farallon National Wildlife Refuge
- 2B Gulf of the Farallones National Marine Sanctuary
- 2C Golden Gate National Recreation Area
- 2D Marin Municipal Water District
- 2E Point Reyes National Seashore
- 2F Mt. Tamalpais State Park, Samuel Taylor State Park, and Tomales Bay

3 Stanislaus-Tuolumne Experimental Forest - 1,499 acres

4 San Joaquin Experimental Range - 4,525 acres

5 Sequoia-Kings Canyon National Parks - 847,210 acres
- 5A Kings Canyon National Park
- 5B Sequoia National Park

6 Mojave and Colorado Deserts - 3,204,242 acres
- 6A Death Valley National Monument
- 6B Joshua Tree National Monument
- 6C San Bernardino National Forest
- 6D Philip L. Boyd Deep Canyon Desert Research Center
- 6E Anza Borrego Desert State Park

7 Channel Islands - 1,184,740 acres
- 7A Channel Islands National Park
- 7B Channel Islands National Marine Sanctuary

8 San Dimas Experimental Forest - 17,050 acres

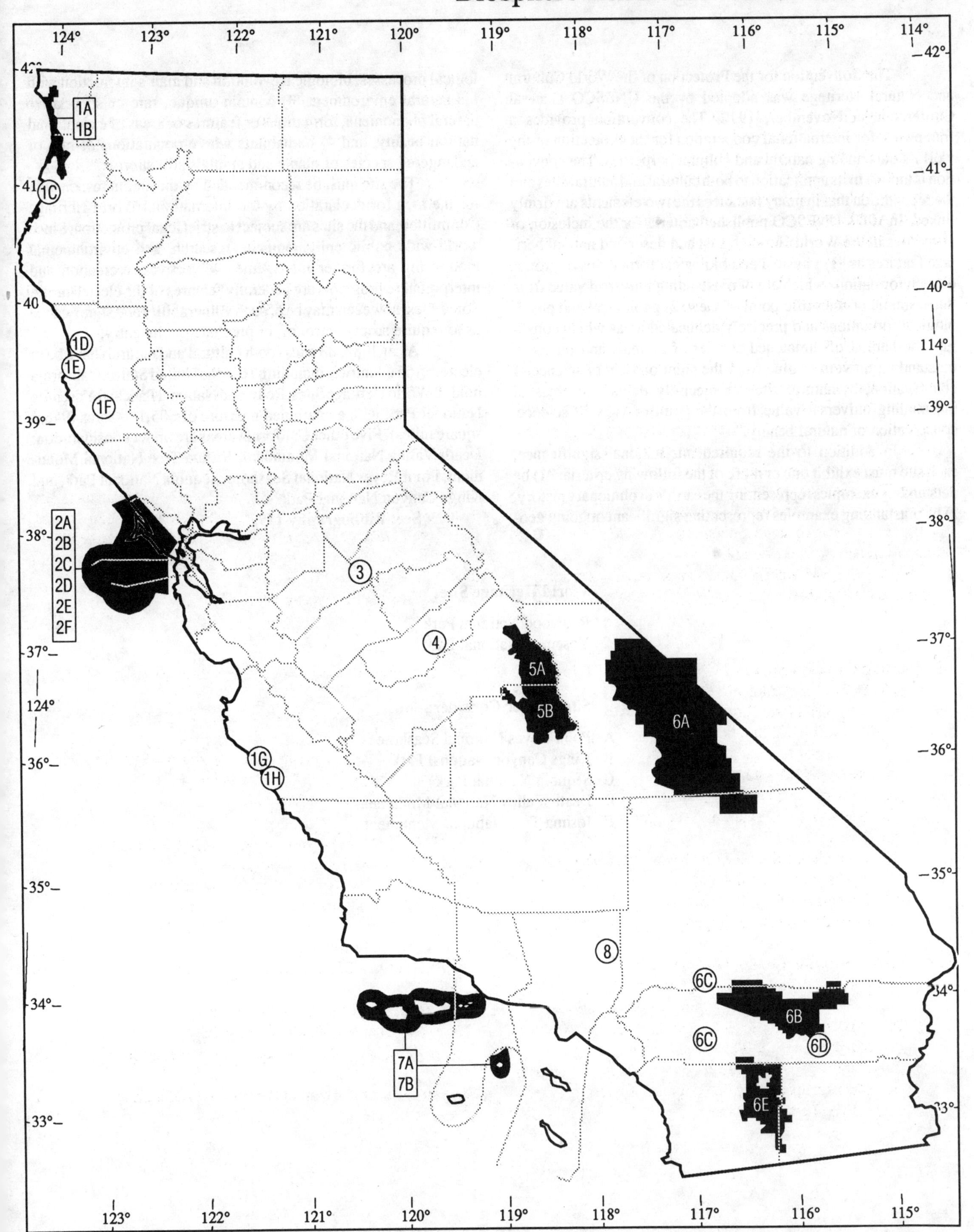
124°
123°
122°
121°
120°
119°
118°
117°
116°
115°
114°
42°
41°
40°
39°
38°
37°
36°
35°
34°
33°
1A
1B
1C
1D
1E
1F
1G
1H
2A
2B
2C
2D
2E
2F
3
4
5A
5B
6A
6B
6C
6D
6E
7A
7B
8

The Convention for the Protection of the World Cultural and Natural Heritage was adopted by the UNESCO General Conference in November, 1972. The convention provides a framework for international cooperation for the protection of the world's outstanding natural and cultural properties. The convention is unique in its application to both cultural and natural sites and the recognition that in many instances the two elements are firmly linked. In 1980, UNESCO published criteria for the inclusion of properties in the World Heritage List and described natural heritage features as "1) physical and biological formations or groups of such formations which are of outstanding universal value from the aesthetic or scientific point of view, 2) geological and physiological formations and precisely delineated areas which constitute the habitat of threatened species of animals and plants of outstanding universal value from the point of view of science or conservation, 3) natural sites or precisely delineated areas of outstanding universal value from the point of view of science, conservation or natural beauty."

In addition to the required international significance, each site must exhibit one or more of the following criteria: "1) be outstanding examples representing the earth's evolutionary history, 2) be outstanding examples representing significant ongoing geological processes, biological evolution and man's interaction with his natural environment, 3) contain unique, rare or superlative natural phenomena, formations or features or areas of exceptional natural beauty, and 4) be habitats where populations of rare or endangered species of plants and animals still survive."

The site must be recommended by the nation responsible for the area for declaration by the international World Heritage Committee, and the sites are subject to strict legal protection. Since world-wide public enlightenment, research, and environmental monitoring are further management objectives, recreation and interpretative programs are generally features of the elected areas. Some sites, however, may be of such vulnerability and significance as to require strict control of, or preclude public entry.

As of 1988, 88 sites (both cultural and natural) have been elected in 65 countries, including 16 in the United States. California holds 2 World Heritage Sites, Redwood National Park and Yosemite National Park with a combined measure of 605,150 acres (946.6 square miles). Five other California areas are under consideration: Death Valley National Monument, Joshua Tree National Monument, Point Reyes National Seashore, Sequoia National Park, and Kings Canyon National Park.

See: Bibliography 119.

World Heritage Site

1 Redwood National Park
2 Yosemite National Park

Sites Under Consideration

A Point Reyes National Seashore
B Kings Canyon National Park
C Sequoia National Park
D Death Valley National Monument
E Joshua Tree National Monument

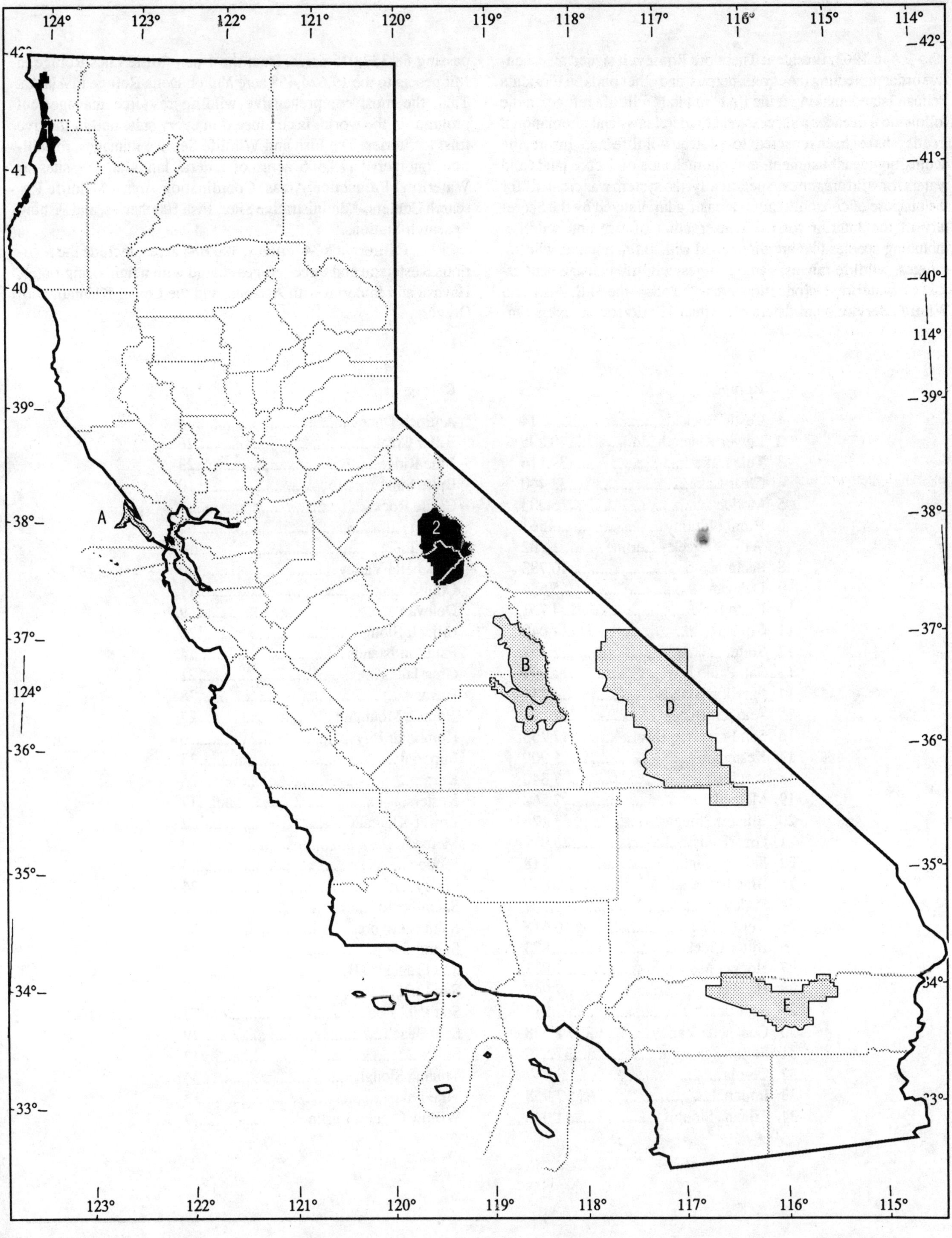
124°
123°
122°
121°
120°
119°
118°
117°
116°
115°
114°
42°
41°
40°
39°
38°
37°
36°
35°
34°
33°
1
2
A
B
C
D
E

In 1903, President Theodore Roosevelt signed an executive order protecting the egrets, herons, and other birds on Florida's Pelican Island, making it the first "natural" wildlife refuge. In the following 8 decades a succession of federal laws and international treaties have been enacted to protect wildlife and insure the acquisition, establishment, and maintenance of secure lands and waters for wild creatures. Specifically, the system was created "for the purpose of consolidating...areas...administered by the Secretary of the Interior for the conservation of fish and wildlife, including species that are threatened with extinction...as wildlife refuges, wildlife ranges, game refuges, wildlife management areas, or waterfowl production areas." Today the U.S. Fish and Wildlife Service administers more than 437 diverse areas encompassing 88,337,015 acres, from the 1 acre Mille Lacs Refuge in Minnesota to the 19,624,458 acre Yukon Delta Refuge in Alaska. This, the most comprehensive wildlife resource management program in the world, is conducted in every state and in the five trust territories. The Fish and Wildlife Service manages, in addition, another 2,178,506 acres of federal lands at 349 sites as Waterfowl Production Areas, Coordination Areas, Wildlife Research Centers, Administrative Sites, Fish Hatcheries, and Fishing Research Stations.

California's 34 refuges, 300,684 acres, include the notorious Kesterson and three refuges shared with neighboring states, Havasu and Imperial with Arizona, and the Lower Klamath with Oregon.

	Refuge	Acres
1	Castle Rock	14
2	Lower Klamath	40,295
3	Tule Lake	39,116
4	Clear Lake	33,440
5	Modoc	6,283
6	Humboldt Bay	561
7	Willow Creek - Lurline	1,202
8	Sacramento	10,783
9	Delevan	5,634
10	Butte Sink	4,761
11	Colusa	4,040
12	Sutter	2,591
13	San Pablo Bay	11,697
14	Farallon Islands	211
15	Antioch Dunes	55
16	San Francisco Bay	17,255
17	Kesterson	5,900
18	San Luis	7,340
19	Merced	2,562
20	Ellicott Slough	126
21	Grasslands	26,079
22	Salinas Lagoon	518
23	Blue Ridge	897
24	Pixley	5,187
25	Kern	10,618
26	Bitter Creek	873
27	Hopper Mountain	1,871
28	Havasu	7,747
29	Seal Beach	911
30	Coachella Valley	2,338
31	Salton Sea	37,579
32	Cibola	3,647
33	Imperial	7,958
34	Tijuana Slough	1,023

Refuge	#
Antioch Dunes	15
Bitter Creek	26
Blue Ridge	23
Butte Sink	10
Castle Rock	1
Cibola	32
Clear Lake	4
Coachella Valley	30
Colusa	11
Delevan	9
Ellicott Slough	20
Farallon Islands	14
Grasslands	21
Havasu	28
Hopper Mountain	27
Humboldt Bay	6
Imperial	33
Kern	25
Kesterson	17
Lower Klamath	2
Merced	19
Modoc	5
Pixley	24
Sacramento	8
Salinas Lagoon	22
Salton Sea	31
San Francisco Bay	16
San Luis	18
San Pablo Bay	13
Seal Beach	29
Sutter	12
Tijuana Slough	34
Tule Lake	3
Willow Creek - Lurline	7

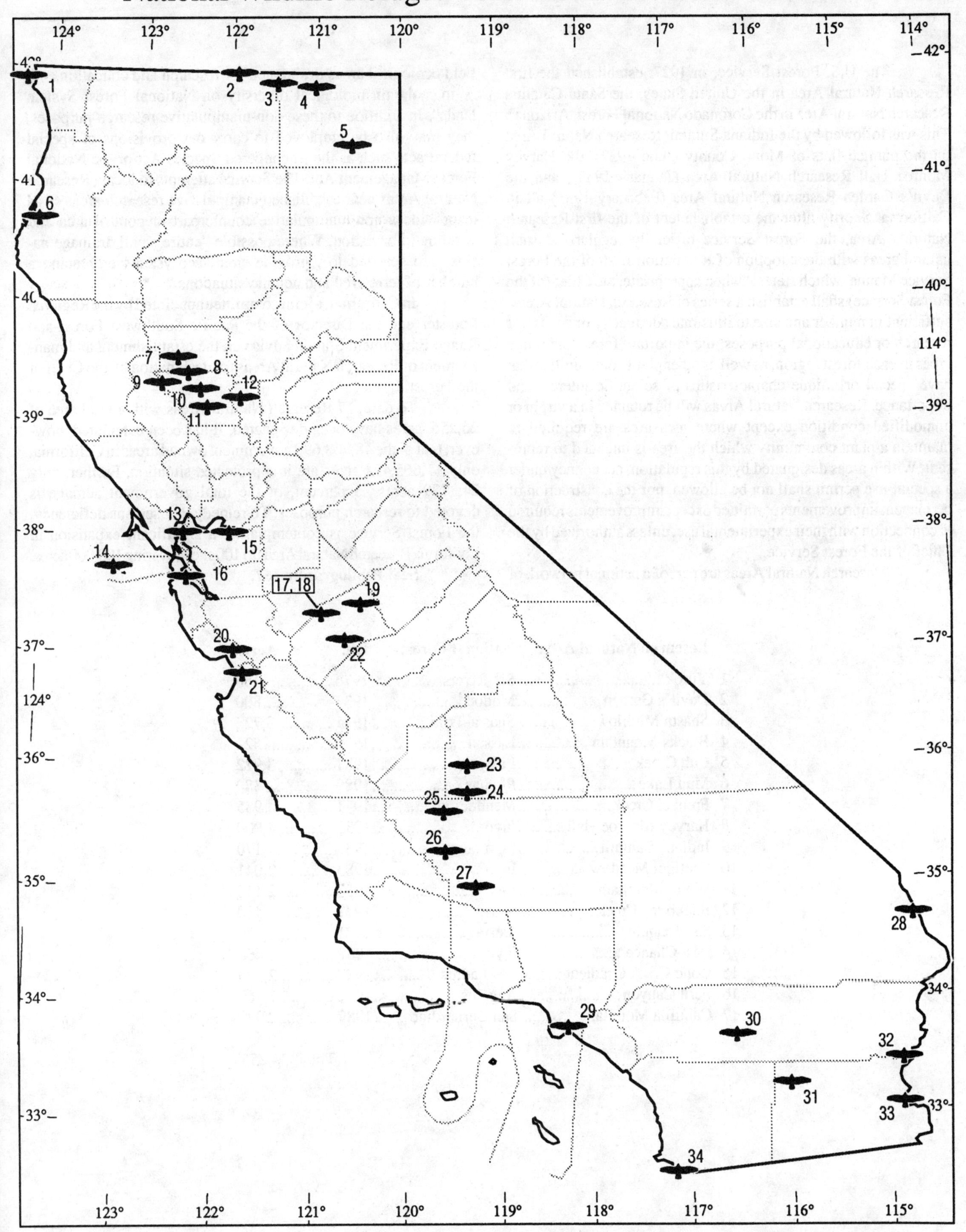
124°
123°
122°
121°
120°
119°
118°
117°
116°
115°
114°
42°
41°
40°
39°
38°
37°
36°
35°
34°
33°
1
2
3
4
5
6
7
8
9
10
11
12
13
14
15
16
17,18
19
20
21
22
23
24
25
26
27
28
29
30
31
32
33
34

The U.S. Forest Service, in 1927, established the first Research Natural Area in the United States, the Santa Catalina Research Natural Area in the Coronado National Forest, Arizona.; This was followed by the Indiana Summit Research Natural Area on the pumice flats of Mono County (June 1932), the Harvey Monroe Hall Research Natural Area (January 1933), and the Devil's Garden Research Natural Area (February 1933) all in California. Shortly after the establishment of the first Research Natural Area, the Forest Service officially regularized such natural areas with the adoption of Regulation L-20 of the Forest Service Manual which states, "when appropriate, the Chief (of the Forest Service) shall establish a series of Research Natural Areas, sufficient in number and size to illustrate adequately or typify for research or educational purposes, the important forest and range types in each forest region, as well as other plant communities that have special or unique characteristics of scientific interest and importance. Research Natural Areas will be retained in a virgin or unmodified condition except where measures are required to maintain a plant community which the area is intended to represent. Within areas designated by this regulation, occupancy under a special-use permit shall not be allowed, nor the construction of permanent improvements permitted except improvements required in connection with their experimental use, unless authorized by the Chief of the Forest Service."

"Research Natural Areas are part of a national network of field ecological areas designated for research and education and/ or to maintain biological diversity on National Forest System lands". In addition to these non-manipulative research purposes, they may also be employed to carry out provisions of special federal acts such as the Endangered Species Act or the National Forest Management Act. The Service attempts to locate Research Natural Areas near established manipulative research areas so as to provide a non-manipulative counterpart or control area and baseline information. Where possible "entire small drainage basins" are selected to ."provide small ecosystems embracing a number of terrestrial and aquatic situations."

In California, a joint committee appointed by the Regional Forester and the Director of the Pacific Southwest Forest and Range Experiment Station advise on the establishment and management of Research Natural Areas for designation by the Chief of the Forest Service.

To date, 17 Research Natural Areas with a total area of 26,255 acres have been designated. It has been calculated, however, that of the 48,968,690 government owned areas in California, only 11,665,474 acres are in a protected situation. Further, only 245,729 acres, .5 percent, of the total government acreage is devoted to research purposes. To remedy that serious deficiency, the Forest Service is contemplating a significant expansion in California Research Natural Areas to 100 sites covering 100,000 acres.

See: Bibliography 157.

	Research Natural Area	National Forest	Est.	Acres
1	Yurok	Six Rivers	1976	150
2	Devil's Garden	Modoc	1933	800
3	Shasta Mudflow	Shasta-Trinity	1971	3,722
4	Blacks Mountain	Lassen	1976	521
5	Cub Creek	Lassen	1981	3,922
6	Mud Lake	Plumas	1989	380
7	Frenzel Creek	Mendocino	1971	935
8	Harvey Monroe Hell	Inyo	1933	4,000
9	Indiana Summit	Inyo	1932	1,170
10	Sentinel Meadow	Inyo	1983	2,041
11	White Mountain	Inyo	1953	2,110
12	Backbone Creek	Sierra	1971	430
13	San Joaquin	Sierra	1971	70
14	Last Chance	Inyo	1982	660
15	Cone Creek Gradient	Los Padres	1987	2,955
16	Fern Canyon	Angeles	1972	1,460
17	Cahuilla Mountain	San Bernardino	1989	929

Total: 26,255

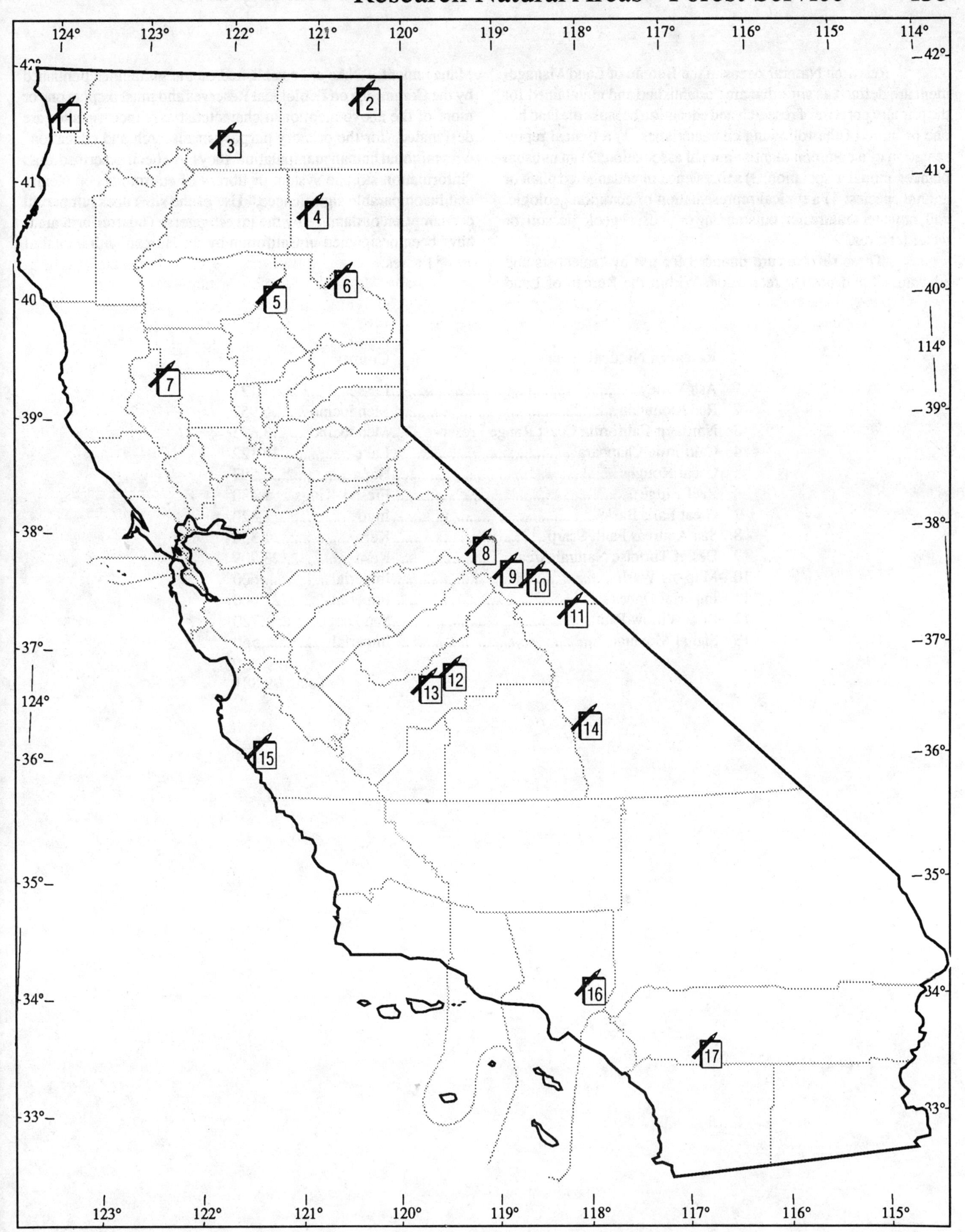
124°
123°
122°
121°
120°
119°
118°
117°
116°
115°
114°
42°
41°
40°
39°
38°
37°
36°
35°
34°
33°
1
2
3
4
5
6
7
8
9
10
11
12
13
14
15
16
17

Research Natural Areas in the Bureau of Land Management are defined as sites that are "established and maintained for the primary purpose of research and education because the land has one or more of the following characteristics: 1) a typical representation of a common plant or animal association; 2) an unusual plant or animal association; 3) a threatened or endangered plant or animal species; 4) a typical representation of common geologic, soil, or water features; 5) outstanding or unusual geologic, soil, or water features."

These reserves are intended for use by "scientists and educators," and not for recreation. Within the Bureau of Land Management, RNAs are established on the guidelines published by the Committee on Ecological Reserves and must display one or more of the above mentioned characteristics. Since the sites are designated "for the primary purpose of research and education" with minimal human manipulation, they have been described as an "information storage system or library of enormous complexity and incomparable significance." Use of the sites does not permit consumption or damage by the investigators. Thirteen such areas have been designated in California by the Bureau with a total of 79,881 acres.

	Research Natural Area	County	Acres
1	Ash Valley	Lassen	1,121
2	Red Mountain	Mendocino	6,957
3	Northern California Coast Range Preserve	Mendocino	3,695
4	California Chapparal	Lake	10,122
5	Cedar Roughs	Napa	5,597
6	Reef Ridge	Fresno-Kings	1,430
7	Great Falls Basin	Inyo	1,820
8	San Andreas Fault Scarp	Kern	1,350
9	Desert Tortoise Natural Area	Kern	23,909
10	Milpitas Wash	Imperial	600
11	Imperial Dunes	Imperial	22,000
12	Bow Willow Palms	San Diego	720
13	Signal Mountain	Imperial	560

Total: 79,881

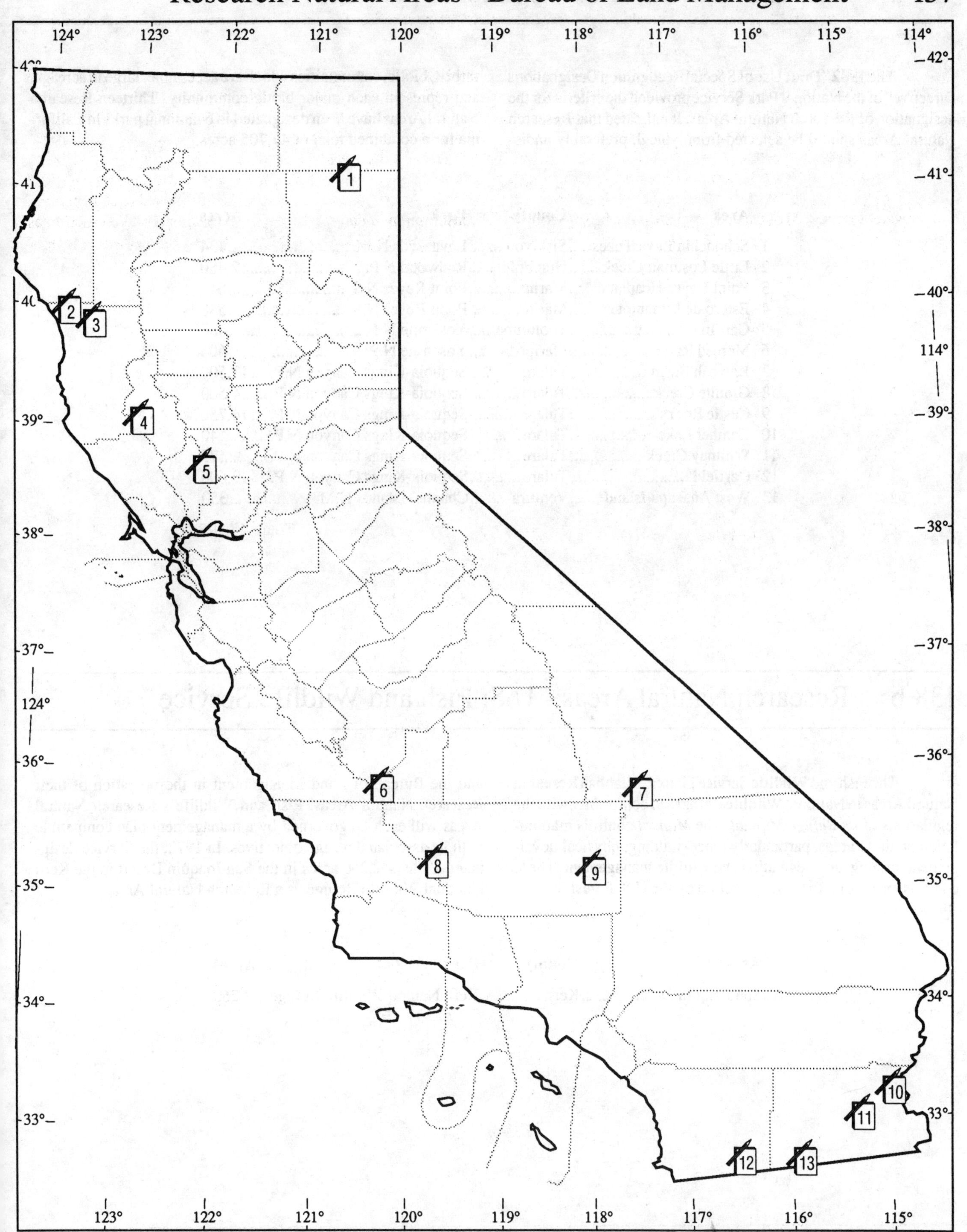
124°
123°
122°
121°
120°
119°
118°
117°
116°
115°
114°
42°
41°
40°
39°
38°
37°
36°
35°
34°
33°
1
2
3
4
5
6
7
8
9
10
11
12
13

138 a Research Natural Areas - National Park Service

The 1982 "Draft Use of Special Recognition Designations Directive" of the National Park Service provided the criteria for the designation of Research Natural Areas. It indicated that Research Natural Areas should be selected from typical, preferably undisturbed, biotic, communities which are accessible, large in acreage, and represent each major biotic community. Thirteen Research Natural Areas have been designated in 6 national parks in California for a combined total of 47,705 acres.

	Area	County	Park	Acres
1	Schonchin Lava Tubes	Siskiyou	Lava Beds N.M.	134
2	Little Lostman Creek	Humboldt	Redwood N.P.	2,480
3	Point Reyes Headland	Marin	Point Reyes N.S.	640
4	Estero de Limantour	Marin	Point Reyes N.S.	548
5	Carl Inn	Tuolumne	Yosemite N.P.	488
6	Merced River	Mariposa	Yosemite N.P.	606
7	Kaweah Basin	Tulare	Sequoia-Kings Canyon N.P.	13,500
8	Granite Creek	Tulare	Sequoia-Kings Canyon N.P.	4,500
9	Castle Rocks	Tulare	Sequoia-Kings Canyon N.P.	14,750
10	Heather Lake	Tulare	Sequoia-Kings Canyon N.P.	40
11	Whitney Creek	Tulare	Sequoia-Kings Canyon N.P.	75
12	Garfield	Tulare	Sequoia-Kings Canyon N.P.	9,600
13	West Anacapa Island	Ventura	Channel Islands N.M.	350
				Total: 47,705

138 b Research Natural Areas - U.S. Fish and Wildlife Service

The Fish and Wildlife Service plans to establish Research Natural Areas in National Wildlife Refuges in conformity with the regulations of its *Refuge Manual*. The *Manual* controls manipulation on these areas, particularly fence building, physical development, mining, and vegetation and wildlife management. These controls are similar to those exercised by the U.S. Forest Service and the Bureau of Land Management in the operation of their Research Natural Areas. Fish and Wildlife's Research Natural Areas will each be governed by a management plan compatible with its associated refuge objectives. In 1972, the Service designated a site of 2,260 acres in the San Joaquin Desert in the Kern National Wildlife Refuge as a Research Natural Area.

	Area	County	Park	Acres
A	San Joaquin Desert	Kern	Kern Natural Wildlife Refuge	2,260

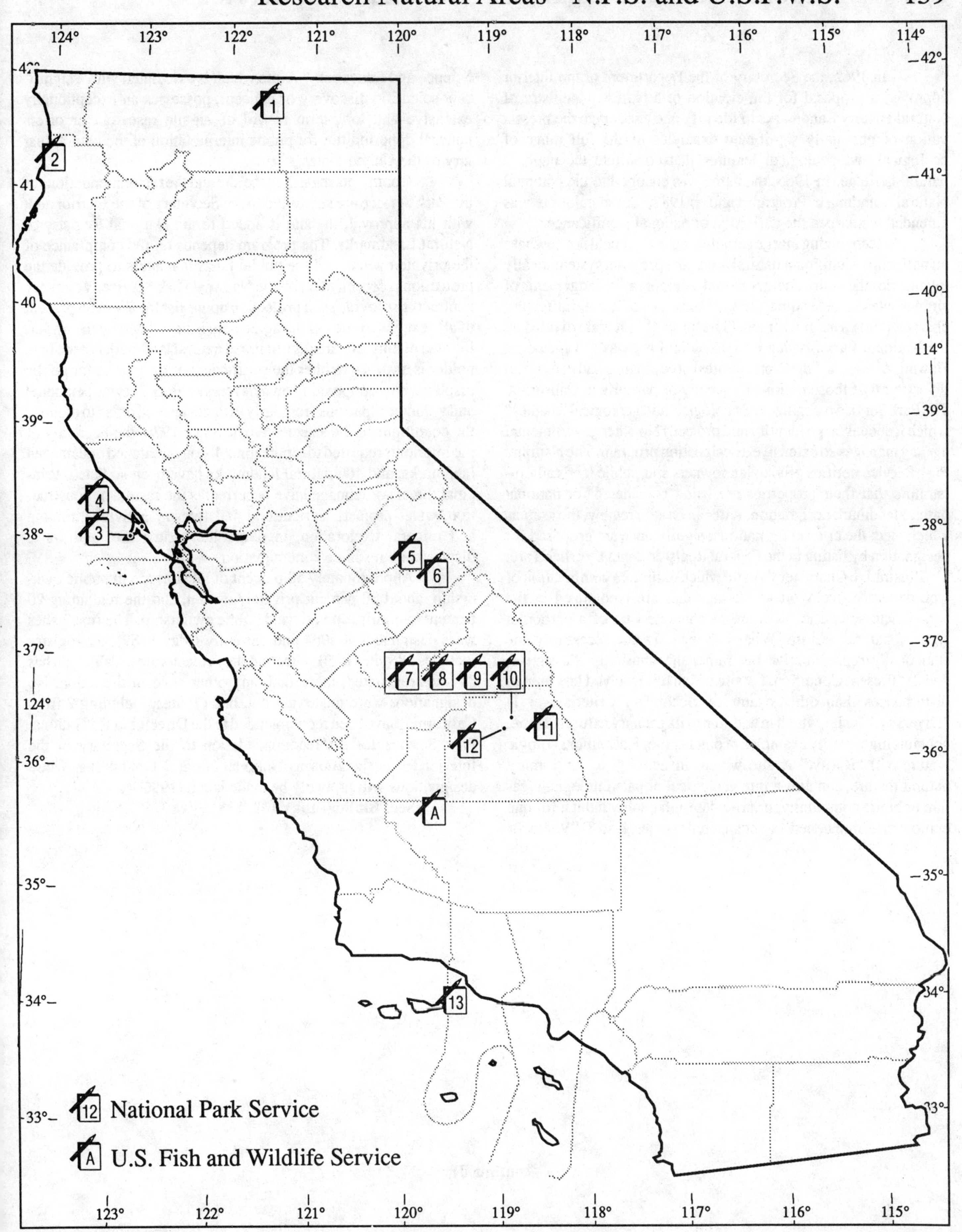
124°
123°
122°
121°
120°
119°
118°
117°
116°
115°
114°
42°
41°
40°
39°
38°
37°
36°
35°
34°
33°
1
2
3
4
5
6
7
8
9
10
11
12
13
A
12 National Park Service
A U.S. Fish and Wildlife Service

In 1962, the Secretary of the Department of the Interior approved a proposal for the creation of a National Registry of Natural History Landmarks to identify and encourage the preservation of nationally significant examples of the full range of ecological and geological features that constitute the nation's natural heritage. In 1965, the name was changed to the National Natural Landmark Program, and in1987, the regulation was amended to sharpen the definition of national significance.

Recognizing that government agencies could not purchase all nationally significant natural areas, the program systematically inventories the entire country to make comparative judgements of similar sites to determine the best example of a natural feature characterizing a natural region. (The lower 48 states are divided by the National Park Service into 26 natural regions [33 including Hawaii, Alaska, and the Protectorates] of ecological and geological diversity; 6 of those regions lie wholly or partially in California. The Park Service recognizes 78 ecological and geological "themes" which it endeavors to identify and protect.) No other governmental agency manages so extensive a designation program. The National Park Service notifies private landowners and public lands administrators that their properties are being considered for national natural landmark designation. After an on-site review, those areas which meet the criteria for national significance are proposed for designation by listing in the Federal Register. Primary criteria are: 1) "Illustrative Character." A site which exhibits a combination of well-developed component features that are recognized in the appropriate scientific literature as characteristic of a particular type of natural feature. What is sought is not necessarily the statistically representative but rather the unusually illustrative, and 2) "Present Condition." A site which has received less human disturbances than other examples. Secondary Criteria are: 1) "Diversity." A site, which in addition to its primary natural feature, contains high quality examples of other ecological and/or geologic features. 2) "Rarity." A site which, in addition to its primary natural feature, contains a rare geological or paleontological feature or biotic community, or provides high quality habitat for one or more rare, threatened, or endangered species. And 3) "Value for Science and Education." A site which is associated with a significant scientific discovery or concept, possesses an exceptionally extensive and long-term record of on-site research, or offers unusual opportunities for public interpretation of the natural history of the United States.

Comments regarding the site and the recommendations of the Park Service are submitted to the Secretary of the Interior and, with his approval, the site is added to the National Registry of Natural Landmarks. The program depends on the compliance of the private owners and the public lands managers to provide the protection recommended by the National Park Service. Voluntary contracts to provide such protection now exist for about 65 percent of all designated sites. If the agreement is not signed by the owner, the area retains its status as a national natural landmark but the land holder is awarded neither the certificate nor the plaque for public display which cooperating owners receive. Park Service personnel and volunteer "patrons" regularly visit designated areas to monitor the conditions of the selected areas. Since 1976, the Secretary of the Interior is required to report annually on threatened or damaged landmarks, and 100 natural landmarks have been so listed. Principal threats or damage have occurred as the results of construction, water projects, agricultural and forestry activities, mining and mineral exploration, incompatible visitor use, and waste disposal and resource contamination.

Approximately 50 percent of all sites are in public ownership, about 30 percent privately owned, and the remaining 20 percent in multiple private and public ownership. The first 7 sites were designated in 1964 and currently (May 1989), the registry includes 586 sites in 50 states and the Protectorates. California has 34 sites registered, more than any other state in the union. No designations were made in 1989, but 17 sites, including 2 from California, have been recommended to the Director of the National Park Service for his recommendation to the Secretary of the Interior for designation as National Natural Landmarks. These designations will probably be made late in 1990.

See: Bibliography 132, 178.

(continued)

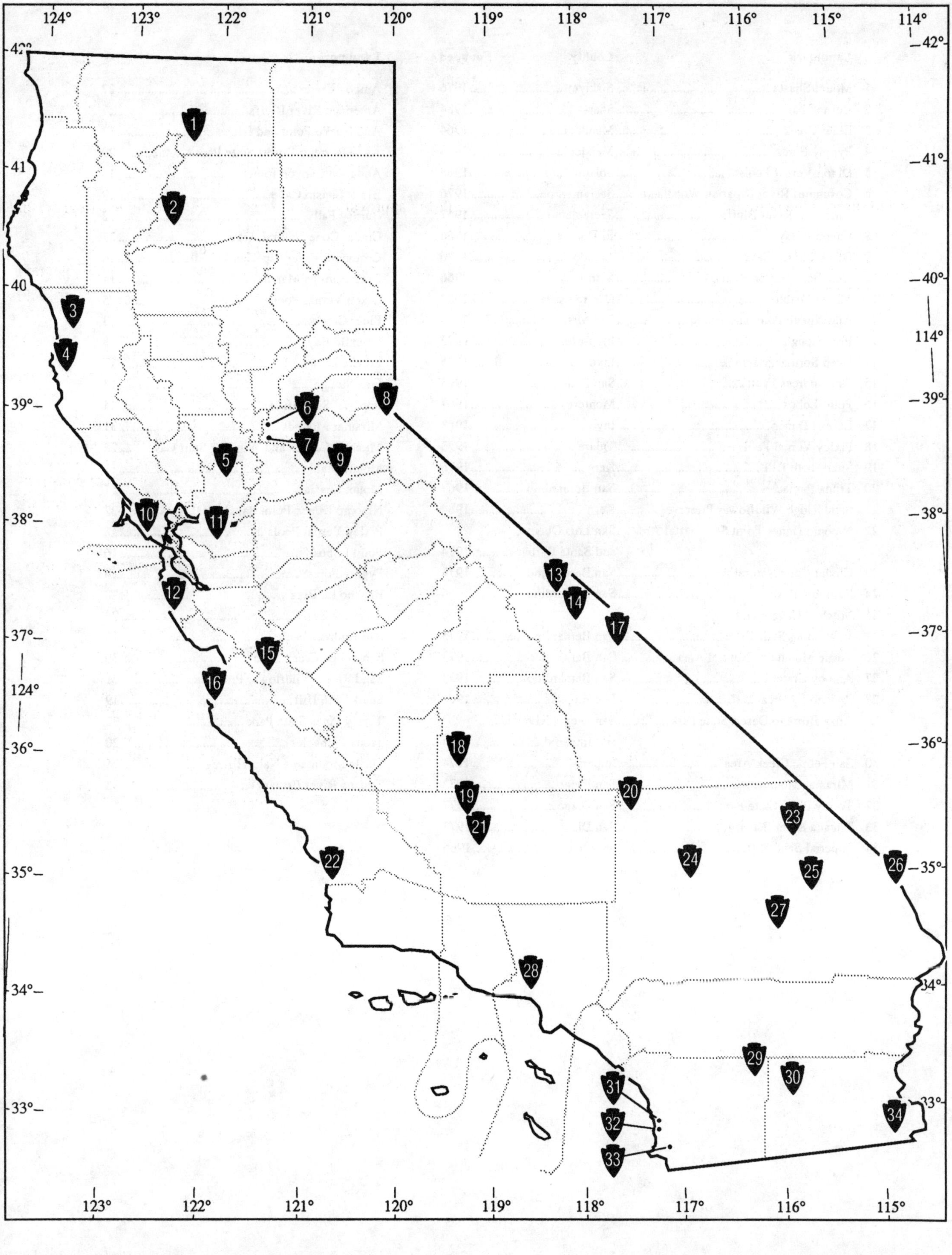
124°
123°
122°
121°
120°
119°
118°
117°
116°
115°
114°
42°
41°
40°
39°
38°
37°
36°
35°
34°
33°
1
2
3
4
5
6
7
8
9
10
11
12
13
14
15
16
17
18
19
20
21
22
23
24
25
26
27
28
29
30
31
32
33
34

	Landmark	County	Founded
1	Mount Shasta	Siskiyou	1976
2	Burney Falls	Shasta	1984
3	Elder Creek	Mendocino	1964
4	Pygmy Forest	Mendocino	1973
5	Dixon Vernal Pools	Solano	1988
6	Cosumnes River Riparian Woodlands	Sacramento	1976
7	American River Bluffs	Sacramento	1967
8	Emerald Bay	El Dorado	1966
9	Black Chasm Cave	Amador	1976
10	Audubon Canyon Ranch	Marin	1966
11	Mount Diablo	Contra Costa	1982
12	Año Nuevo Point and Island	San Mateo	1980
13	Fish Slough	Inyo-Mono	1975
14	Deep Springs Marsh	Inyo	1975
15	San Andreas Fault	San Benito	1965
16	Point Lobos	Monterey	1967
17	Eureka Dunes	Inyo	1983
18	Pixley Vernal Pools	Tulare	1973
19	Sharktooth Hill	Kern	1976
20	Trona Pinnacles	San Bernardino	1967
21	Sand Ridge Wildflower Preserve	Kern	1984
22	Nipomo Dunes-Point Sal Coastal Area	San Luis Obispo, and Santa Barbara	1974
23	Cinder Cone Natural Area	San Bernardino	1973
24	Rainbow Basin	San Bernardino	1966
25	Mitchell Cavern and Winding Stair Cave	San Bernardino	1975
26	Turtle Mountains Natural Area	San Bernardino	1973
27	Amboy Crater	San Bernardino	1973
28	Rancho La Brea	Los Angeles	1964
29	Anza-Borrego Desert State Park	San Diego, Riverside, and Imperial	1974
30	San Felipe Creek Area	Imperial	1974
31	Miramar Mounds	San Diego	1972
32	Torrey Pines State Park	San Diego	1977
33	Tijuana River Estuary	San Diego	1973
34	Imperial Sand Hills	Imperial	1966

Landmarks	#
Amboy Crater	27
American River Bluffs	7
Año Nuevo Point and Island	12
Anza-Borrego Desert State Park	29
Audubon Canyon Ranch	10
Black Chasm Cave	9
Burney Falls	2
Cinder Cone Natural Area	23
Cosumnes River Riparian Woodlands	6
Deep Springs Marsh	14
Dixon Vernal Pools	5
Elder Creek	3
Emerald Bay	8
Eureka Dunes	17
Fish Slough	13
Imperial Sand Hills	34
Miramar Mounds	31
Mitchell Cavern and Winding Stair Cave	25
Mount Diablo	11
Mount Shasta	1
Nipomo Dunes-Point Sal Coastal Area	22
Pixley Vernal Pools	18
Point Lobos	16
Pygmy Forest	4
Rancho La Brea	28
Rainbow Basin	24
San Andreas Fault	15
San Felipe Creek Area	30
Sand Ridge Wildflower Preserve	21
Sharktooth Hill	19
Torrey Pines State Park	32
Trona Pinnacles	20
Turtle Mountains Natural Area	26
Tijuana River Estuary	33

The general provisions of the Organic Administration Act of 1897, and the Forest and Range Renewable Resource Act of 1978, authorize the Secretary of Agriculture to designate experimental forests. The Secretary has delegated this authority to the Chief of the Forest Service, who now has sole responsibility for that designation. The local station directors recommend sites for designation as experimental forests, and if concurrence is given by the Regional Forester, the proposal is forwarded to the Chief.

The forests are designed to provide for the research necessary for the efficient management of the land, "to provide outdoor laboratories, and to serve as sites for pilot testing and demonstrating integrated management techniques." Experimental forests are useful for short-term studies, but are particularly suitable "for long term studies where close control of land management activities and experimental conditions are needed."

Ten experimental forests have been established in California under the administration of the Pacific Southwest Forest and Range Experiment Station, with total holdings of 62,239.7 acres.

	Forest	Founded	County	Acres
1	Yurok Redwood	1940	Del Norte	936.5
2	Blacks Mountain	1934	Lassen	10,259.6
3	Swain Mountain	1932	Lassen	5,999.6
4	Challenge	1942	Yuba	3,573.1
5	Onion Creek	1958	Placer	3,249.4
6	Stanislaus-Tuolomne (2 sites)	1943	Tuolumne	1,499.9
7	San Joaquin Experimental Range	1934	Madera	4,497.2
8	Teakettle Creek	1958	Fresno	2,495.7
9	San Dimas	1933	Los Angeles	17,050.0
10	North Mountain Experimental Area	1963	Riverside	12,678.7

Total: 62,239.7

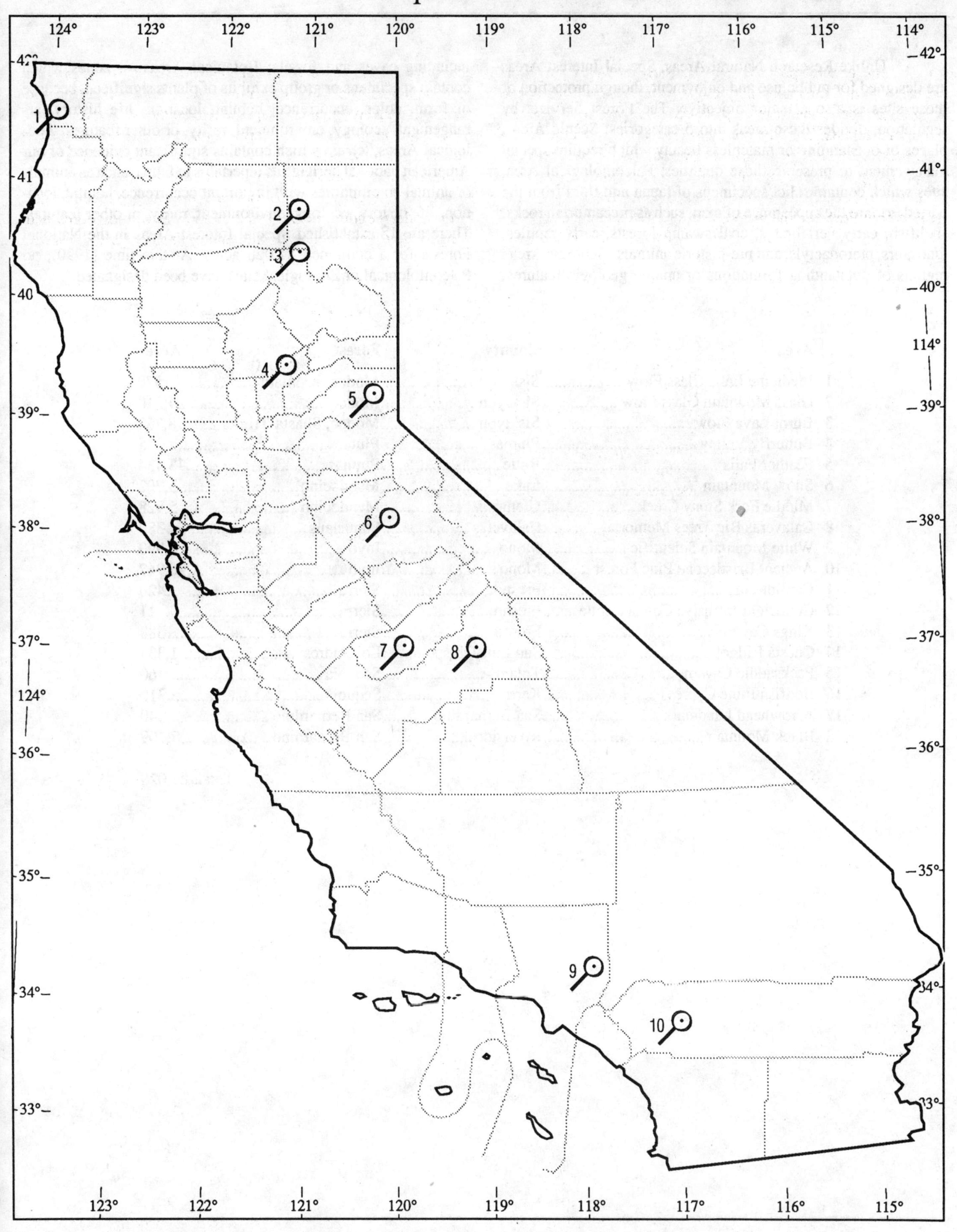
124°
123°
122°
121°
120°
119°
118°
117°
116°
115°
114°
42°
41°
40°
39°
38°
37°
36°
35°
34°
33°
1
2
3
4
5
6
7
8
9
10

Unlike Research Natural Areas, Special Interest Areas are designed for public use and enjoyment, though protection of those sites is also a major objective. The Forest Service, by regulation, divides these areas into 5 categories: Scenic Areas, places of outstanding or matchless beauty which require special management to preserve those qualities; Paleontological Area, sites which contain relict specimens of fauna and flora from the period prior to the appearance of man, such as precambrian rocks, shellfish, early vertebrates, coal swamp forests, early reptiles, dinosaurs, pterodactyls, and pre-historic animals; Geologic Area, regions of outstanding formations or unique geologic features, including caves and fossils; Botanical Areas — zones which contain specimens or group exhibits of plants significant because of form, color, occurrence, habitat, location, life history, arrangement, ecology, environment, rarity, or other features; Zoological Areas, terrain which contains significant evidence of our American national heritage as it pertains to fauna, such as animals or animal communities with important occurrence, habitat, location, life history, ecology, environment, rarity, or other features. There are 18 established Special Interest Areas in the National Forests for a combined 81,028 acres. At this time (1990), no Paleontological or Zoological sites have been designated.

	Area	County	Forest	Acres
1	Medicine Lake Glass Flow	Siskiyou	Modoc, Klamath	570
2	Glass Mountain Glass Flow	Siskiyou	Modoc	4,210
3	Burnt Lava Flow	Siskiyou	Modoc, Shasta-Trinity	8,760
4	Butterfly Valley	Plumas	Plumas	335
5	Feather Falls	Butte	Plumas	13,434
6	Snow Mountain	Lake	Mendocino	5,702
7	Middle Fork Stony Creek	Glenn	Mendocino	5,928
8	Calaveras Big Trees Memorial	Calaveras	Stanislaus	380
9	White Mountain Scientific	Mono	Inyo	4,990
10	Ancient Bristlecone Pine Forest	Mono	Inyo	26,947
11	Carpinteria	Fresno	Sierra	420
12	Courtright Intrusive Contact Zone	Fresno	Sierra	11
13	Kings Cavern	Fresno	Sierra	388
14	Cuesta Ridge	San Luis Obispo	Los Padres	1,334
15	Packsaddle Cavern	Tulare	Sequoia	160
16	Bodfish Piute Cypress	Kern	Sequoia	310
17	Arrowhead Landmark	San Bernardino	San Bernardino	40
18	Black Mountain	Riverside	San Bernardino	7,109

Total: 81,028

Botanical
Geological
Scenic
Scientific

The designation of Outstanding Natural Area is made by the Bureau of Land Management to ensure the preservation of scenic values and the features of an area still in their natural condition. However, recreational use of an Outstanding Natural Area to the maximum extent possible without damage to the natural values of the site, is a requirement of the Bureau. Consequently, Outstanding Natural Areas, which are selected through the planning system outlined in the Bureau of Land Management's Resource Management Plan, must be large enough to ensure that the area will not be adversely or permanently harmed by recreational activities. In addition, any such recreational use must be compatible with the original established purposes of the site. Despite the recreational requirement, the Bureau may sometimes designate an Outstanding Natural Area as a preserve for rare and endangered species.

Eighteen such Areas have been designated in the state with a total extent of 111,687 acres, and with resource values ranging from wildlife and vegetation to cultural and historic.

	Natural Area	County	Acres
1	Eureka Valley Dunes	Inyo	5,374
2	Darwin Falls	Inyo	6,000
3	Trona Tufa Pinnacles	San Bernardino	6,360
4	Clark Mountain	San Bernardino	4,480
5	Cinder Cones	San Bernardino	4,783
6	Rainbow Basin	San Bernardino	3,840
7	Mid Hills	San Bernardino	570
8	Caruthers Canyon	San Bernardino	4,200
9	Fort Piute	San Bernardino	410
10	Piute Range	San Bernardino	760
11	Bigelow Cholla	San Bernardino	80
12	Kelso Dunes	San Bernardino	11,590
13	Mopah Springs	San Bernardino	160
14	Chuckwalla Valley	Riverside	2,040
15	Jacumba Mountain	San Diego	32,000
16	San Sebastian Marsh	Imperial	6,400
17	Imperial Dunes	Imperial	22,000
18	Indian Springs	Imperial	640
		Total:	111,687

Note: Protected Natural Areas - National Park Service
Experimental Ecological Areas - National Park Service
Public Use Natural Areas - U.S. Fish and Wildlife Service

Three federally designated natural "Areas," designed for the preservation of natural diversity in the country, have no established sites in California. The National Park Service's Protected Natural Areas (1981) are defined as "sites given extra protection and surveillance because they possess rare, unique, or outstanding geologic, biologic, or other resources typically other than the park's foremost natural features." Protected Natural Areas are designated by the Park Superintendent to decrease the likelihood of inadvertent detrimental impact to unique sites within a National Park.

The National Park Service has also established the designation, Experimental Ecological Area (1981). These areas are specific tracts to be set aside and managed for approved manipulative research such as cutting and burning. Such research, either by park biologists or outside scientists, must be compatible with the Park's Resource Management Plan, and the areas are to be established on sites which have been disturbed.

U. S. Fish and Wildlife Service has established the designation of Public Use Natural Area (1980), a relatively undisturbed ecosystem or subsystem that is available for use by the public with certain restrictions for protecting the area. The designation is used only on the National Wildlife Refuge System, and is designed to: 1) Assure the preservation of a variety of significant natural areas for public use which, considered together, illustrate the diversity of the National Refuge System natural environments; 2) Preserve those environments that are essentially unmodified by human activity for future use. Nominations for selection are made by the refuge manager, and designation follows approval by the regional and Washington office. Where applicable, the guidelines for Public Natural Areas follow those of the Service's Research Natural Areas

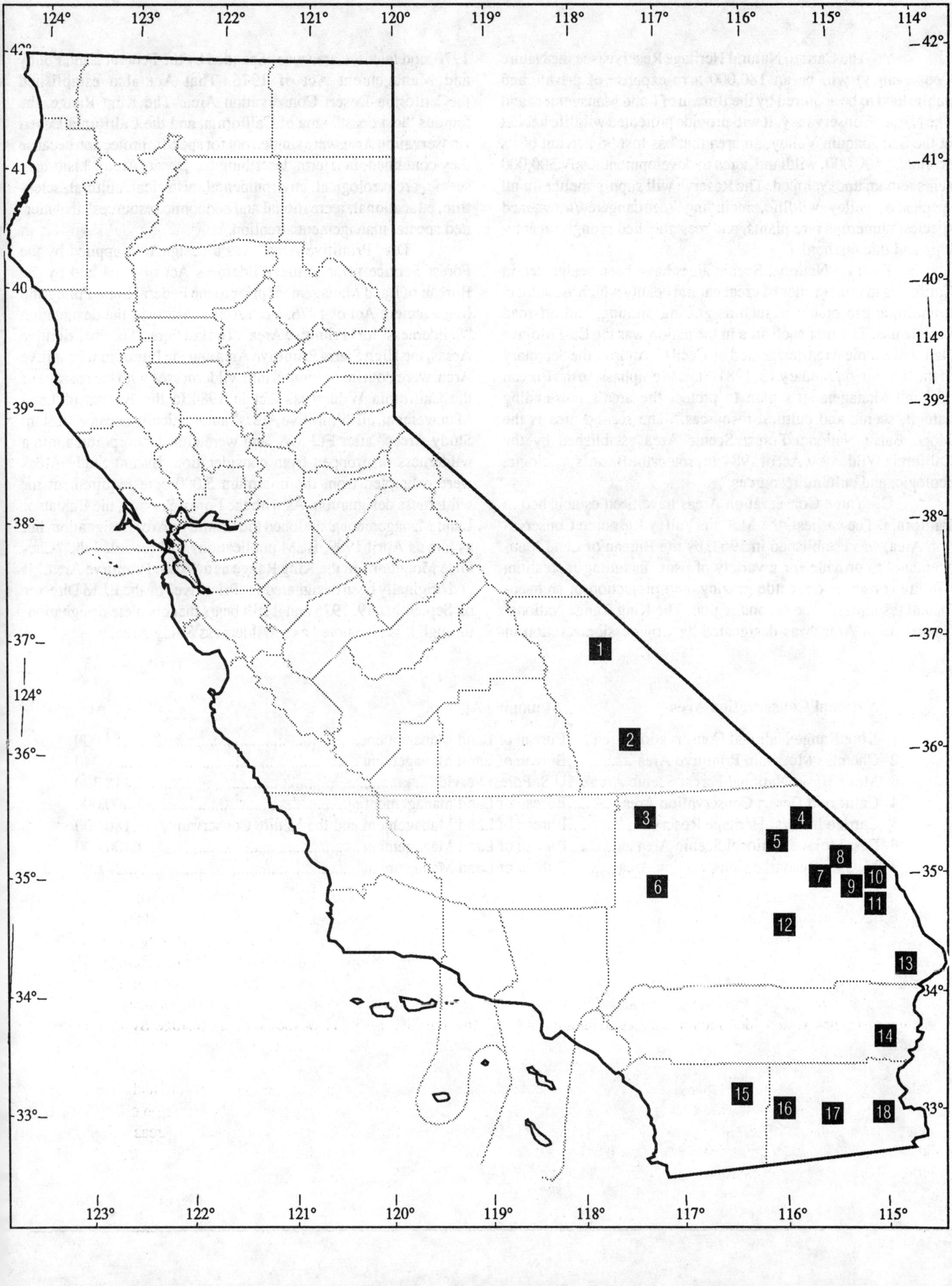

124° 123° 122° 121° 120° 119° 118° 117° 116° 115° 114°

42° 41° 40° 39° 38° 37° 36° 35° 34° 33°

1 2 3 4 5 6 7 8 9 10 11 12 13 14 15 16 17 18

123° 122° 121° 120° 119° 118° 117° 116° 115°

A - The Carrizo Natural Heritage Reserve (see the Nature Conservancy) will be an 180,000 acre expanse of private and public land to be acquired by the Bureau of Land Management and The Nature Conservancy. It will provide protected wildlife habitat in the San Joaquin Valley, an area that has lost 96 percent of its original 7,500,000, wildland acres to development. Only 300,000 acres remain undeveloped. The Reserve will supply shelter for all species of valley wildlife, including 7 endangered/threatened species, numerous rare plants, and reestablished pronghorn antelope and tule elk herds.

B - Two National Scenic areas have been designated in California to protect sites of great natural beauty which are subject to multiple use practices such as grazing, mining, and off-road vehicle use. The first such area in the nation was the East Mojave National Scenic Area designated by Cecil C. Andrus, the Secretary of the Interior, on January 13, 1981, "to add emphasis to the Bureau of Land Management's plan to protect the area's outstanding natural, scenic and cultural resources." The second area is the Mono Basin National Forest Scenic Area established by the California Wilderness Act of 1984, to protect the Basin's "geologic, ecologic, and cultural resources."

C - Three Conservation Areas have been established in California. The earliest, the McCain Valley Resource Conservation Area, was established in 1963, by the Bureau of Land Management "to provide for a variety of uses, including recreation, wildlife conservation, cattle grazing, and protection of archaeological resources." The second region, The King Range National Conservation Area, was designated by Congressional action in 1970, and later the area was expanded by the Federal Land Policy and Management Act of 1976. That Act also established theCalifornia Desert Conservation Area. The King Range, the famous "lost coast" area of California, and the California Desert Conservation Area were singled out for specific protection because they contained, in a term describing the Desert Area, "historical, scenic, archaeological, environmental, historical, cultural, scientific, educational, recreational and economic resources" that merited special management attention.

D - "Primitive Area" was a designation applied by the Forest Service prior to the Wilderness Act of 1964 and by the Bureau of Land Management prior to the Federal Land Policy and Management Act of 1976. Both Acts substituted the designation "Wilderness" for "Primitive Area". The last Forest Service Primitive Areas, the High Sierra Primitive Area and the Immigrant Primitive Area, were both incorporated into wildernesses with the passage of the California Wilderness Act in 1984.In the Bureau of Land Management, all "Primitive Areas" automatically became "Instant Study Areas" after FLPMA, and were either incorporated into a wilderness or dropped from consideration. Instant Study Areas were exempted from the minimum 5,000 acre requirement for wilderness designation. As with the Forest Service, the Bureau of Land Management abandoned the Primitive Area designation, but as late as April 1989, BLM publications still recorded the "Chemise Mountain" in the King Range as a "BLM Primitive Area." It had originally been designated as "Primitive" by the BLM Director on September 19, 1975 and it still bears that obsolete designation, through it is also noted as a Wilderness Study Area.

	National Conservation Area	Managing Agency	Acres
1	King Range National Conservation Area	Bureau of Land Management	54,000
2	Chemise Mountain Primitive Area	Bureau of Land Management	4,340
3	Mono Basin National Forest Scenic Area	U.S. Forest Service	118,300
4	California Desert Conservation Area	Bureau of Land management	2,500,000
5	Carrizo Natural Heritage Reserve	Bureau of Land Management and the Nature Conservancy	180,000
6	East Mojave National Scenic Area	Bureau of Land Management	1,400,000
7	McCain Resource Conservation Area	Bureau of Land Management	38,692

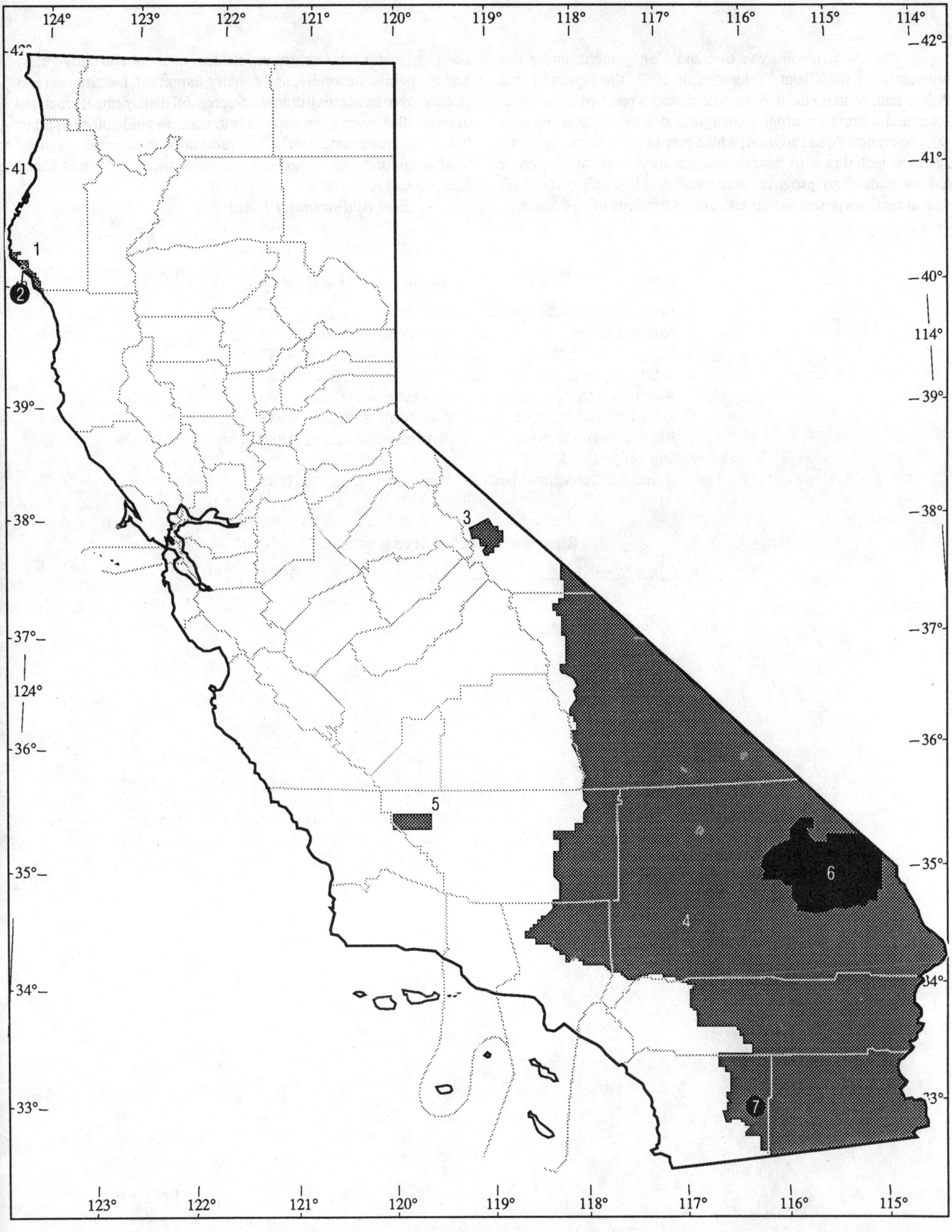
124°
123°
122°
121°
120°
119°
118°
117°
116°
115°
114°
42°
41°
40°
39°
38°
37°
36°
35°
34°
33°
1
2
3
4
5
6
7

In 1986, the Bureau of Land Management, under the provisions of the Clean Water Act of 1972, the Federal Land Policy and Management Act of 1976, and a patchwork of other laws and executive orders, inaugurated the Riparian Habitat Demonstration Area Program, which may involve intensive management techniques to restore and maintain important riparian habitat sites. The program was established, chiefly, for dual educational purposes: for on-site demonstrations of riparian values and uses for school children, academic groups and individuals, and the public. Secondly, and equally important, the sites were to be employed as learning models. Successful management practices at these sites were to be worked into plans to guide other riparian habitat improvement efforts. The program complements ongoing restoration projects throughout the state, such as those at Clark Canyon and Ash Creek.

See: Bibliography 114, 198.

	Area	County	Established
1	Fitzhugh and Cedar Creeks	Modoc	1987
2	Nooning Creek	Humboldt	1987
3	Long Valley Creek	Tulare	1987
4	Sand Canyon	Kern	1987
5	Afton Canyon	San Bernardino	1988
6	Piute Creek	San Bernardino	1988
7	Big Morongo Canyon	Riverside	1988
8	San Felipe Creek and San Sebastian Marsh	Imperial	1988

Riparian Showcase Area

	Area	County	Established
A	Cedar Creek	Lassen	1987

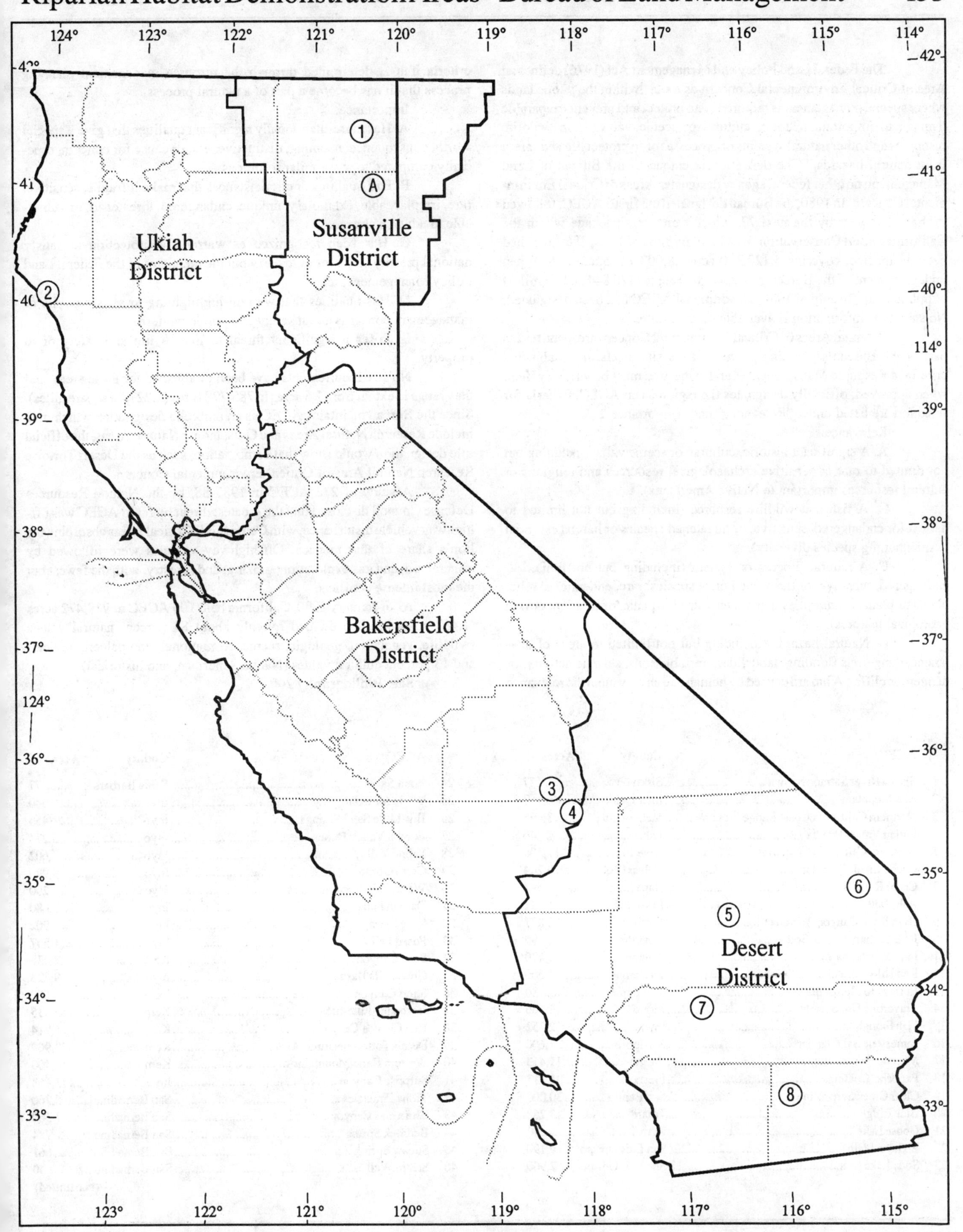
124°
123°
122°
121°
120°
119°
118°
117°
116°
115°
114°
42°
41°
40°
39°
38°
37°
36°
35°
34°
33°
Ukiah
District
Susanville
District
Bakersfield
District
Desert
District
1
A
2
3
4
5
6
7
8

The Federal Land Policy and Management Act (1976) defines an Area of Critical Environmental Concern as a site "within the public lands where special management is required . . .to protect and prevent irreparable damage to important historic, cultural or scenic values, fish, wildlife resources and other natural systems or processes or to protect life and safety from natural hazards." The definition is unique to the Bureau of Land Management; no other federal agency designates Areas of Critical Environmental Concern. In 1980, the Bureau designated the first 3 ACEC, followed in that same year by the next 77, which came as a package within the California Desert Conservation Area Plan. By June of 1988, 286 Areas had been designated, covering 5,122,380 acres (8,003.72 square miles), just under 2 percent of the Bureau's 269,340,030 acres (420,844 square miles) of public land. (In July of 1988, 28 additional ACEC had been designated. No statistical information is available on those sites.)

Potential Areas of Critical Environment Concern are identified as part of the Bureau's planning process. If the site meets the established criteria, a Resource Management Plan for the area must be written which, when approved, officially designates the region as an ACEC. Criteria for selection are listed under "Relevance" and "Importance":

Relevance:

A. A significant historic, cultural, or scenic value (including but not limited to rare or sensitive archaeological resources and religious or cultural resources important to Native Americans).

B. A fish and wildlife resource (including but not limited to habitat for endangered, sensitive, or threatened species or habitat essential to maintaining species diversity).

C. A natural process or system (including but not limited to endangered, sensitive, or threatened plant species; rare, endemic, or relict plants or plant communities that are terrestrial, aquatic, or riparian; or rare geological features).

D. Natural hazards (including but not limited to areas of avalanche, dangerous flooding, landslides, unstable soils, seismic activity, or dangerous cliffs). A hazard caused by human action may meet the relevance criteria if it is determined through the resource management planning process that it has become a part of a natural process.

Importance:

A. Has more-than-locally significant qualities that give it special worth, consequence, meaning, distinctiveness, or cause for concern, especially compared to any similar resources.

B. Has qualities or circumstances that make it fragile, sensitive, rare, irreplaceable, exemplary, unique, endangered, threatened, or vulnerable to adverse change.

C. Has been recognized as warranting protection to satisfy national priority concerns or to carry out the mandates of the Federal Land Policy Management Act.

D. Has qualities that warrant highlighting to satisfy public or management concerns about safety and public welfare.

E. Poses a significant threat to human life and safety or to property.

No size requirements have been established for these areas, and they range in extent from 2.5 acres to 785,778 acres (1,227.78 square miles). Since the Bureau maintains ACEC as an umbrella designation which may include Research Natural Areas and Outstanding Natural Areas, the official title designation would show that combination, such as the Desert Tortoise Research Natural Area of Critical Environmental Concern.

A study of 273 ACEC in 1987-88, by the Natural Resources Defense Council disclosed that the greatest threat by far to ACEC was off-highway vehicle disturbance, with California's Critical Areas absorbing the lion's share of that menace. Off-highway vehicles were followed by mineral, oil, and gas exploration; grazing; and forestry, with the fewest but the most intense conflicts.

As of January 1990, California held 105 ACEC of 919,472 acres (1,437 square miles) divided roughly about 87 percent natural values (wildlife, vegetation, geologic, scenic, recreational, and paleontological) and 13 percent cultural values (Native American, and historical).

See: Bibliography 200.

	ACEC	County	Acres
1	Shasta River Salmon Spawn	Siskiyou	127
2	Red Mountain	Mendocino	6,957
3	Northern California Coast Range Preserve	Mendocino	3,695
4	Indian Valley Brodia	Lake	40
5	Northern California Chaparral	Lake	11,206
6	Cache Creek Corridor	Colusa/Yolo	8,204
7	Cedar Roughs	Napa	5,597
8	Ash Valley	Lassen	1,121
9	High Rock Canyon, Nevada*	Washoe,	12,877
10	Ione Tertiary Oxisol Soil	Amador	90
11	Ione Manzanita	Amador	120
12	Red Hills	Tuolomne	4,500
13	El Dorado Manzanita	Tuolomne	80
14	Travertine Hot Springs	Mono	160
15	Fish Slough	Inyo	32,352
16	Limestone Salamander	Mariposa	1,600
17	Moreno Paleontological	Fresno	11,413
18	Panoche-Coalinga	Fresno	31,717
19	Clear Creek Serpentine	San Benito	30,000
20	Blue Ridge	Tulare	3,268
21	Goose Lake	Kern	40
22	Elkhorn Plain	San Luis Obispo	9,190
23	Soda Lake	San Luis Obispo	2,960
24	Point Sal	Santa Barbara	77
25	White Mountain City	Inyo	640
26	Big-Little Sand Springs	Inyo	450
27	Eureka Valley Dunes	Inyo	5,274
28	Saline Valley	Inyo	7,602
29	Cerro Gordo	Inyo	9,990
30	Warm Sulfur Springs	Inyo	290
31	Darwin Falls	Inyo	2,680
32	Rose Spring	Inyo	902
33	Fossil Falls	Inyo	1,547
34	Sand Canyon	Kern	2,338
35	Great Falls Basin	Inyo	9,723
36	Short Canyon	Kern	600
37	Jawbone-Butterbredt	Kern	155,435
38	Last Chance Canyon	Kern	5,914
39	Desert Tortoise Natural Area	Kern	23,909
40	Western Rand Mountains	Kern	16,400
41	Surprise Canyon	Inyo	13,168
42	Trona Pinnacles	San Bernadino	6,360
43	Christmas Canyon	San Bernadino	7,560
44	Bedrock Spring	San Bernadino	784
45	Squaw Spring	San Bernadino	661
46	Steam Well	San Bernadino	40

(continued)

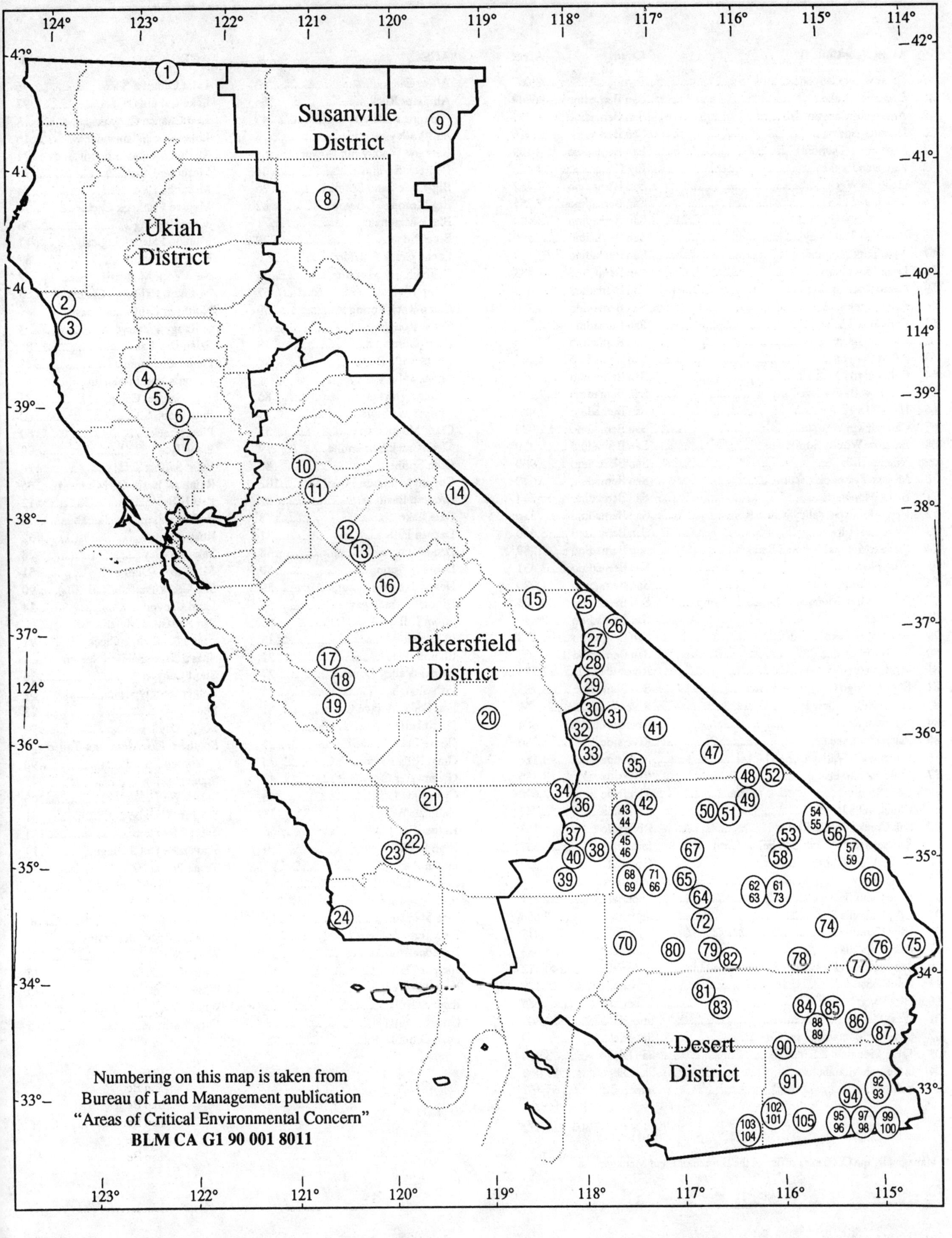

Susanville District

Ukiah District

Bakersfield District

Desert District

Numbering on this map is taken from
Bureau of Land Management publication
"Areas of Critical Environmental Concern"
BLM CA G1 90 001 8011

	ACEC (continued)	County	Acres
47	Greenwater Canyon	Inyo	3,067
48	Grimshaw Lake	San Bernadino	96049
49	Amargosa Canyon	San Bernadino	9,299
50	Denning Spring	San Bernadino	416
51	Salt Creek (Dumont)	San Bernadino	2,109
52	Kingston Range	Inyo	14,452
53	Halloran Wash	San Bernadino	1,862
54	Mesquite Lake	San Bernadino	7,251
55	Clark Mountain	San Bernadino	23,400
56	Dinosaur Trackway	San Bernadino	590
57	New York Mountains	San Bernadino	62,720
58	Dead Mountains	San Bernadino	21,853
59	Camp Rock Spring	San Bernadino	663
60	Piute Creek	San Bernadino	4,320
61	Fort Soda-Mojave Chub	San Bernadino	6,770
62	Cronese Basin	San Bernadino	7,760
63	Afton Canyon	San Bernadino	4,904
64	Calico Early Man Site	San Bernadino	930
65	Rainbow Basin/Owl Canyon	San Bernadino	2,158
66	Harper Dry Lake	San Bernadino	480
67	Camp Irwin	San Bernadino	2,020
68	Barstow Woolly Sunflower	San Bernadino	320
69	Kramer Hills	San Bernadino	960
70	Mojave Fishhook Cactus	San Bernadino	640
71	Black Mountain	San Bernadino	5,304
72	Upper Johnson Valley Yucca Rings	San Bernadino	310
73	Mesquite Hills	San Bernadino	5,640
74	Marble Mountains Fossil Bed	San Bernadino	289
75	Whipple Mountains	San Bernadino	3,431
76	Mopah Spring	San Bernadino	1,320
77	Patton's Iron Mountain Divisional Camp	San Bernadino	3,606
78	Dale Lake	San Bernadino	2,380
79	Soggy Dry Lake	San Bernadino	278
80	Juniper Flats	San Bernadino	3,107
81	Whitewater Canyon	Riverside	12,785
82	Big Morongo Canyon Riverside	San Bernadino	3,705
83	Edom Hill-Willow Hole	Riverside	1,760
84	Alligator Rock	Riverside	7,684
85	Palen Dry Lake	Riverside	3,386
86	Chuckwalla Valley Dune Thicket	Riverside	3,126
87	Mule Mountains	Riverside	3,886
88	Corn Spring	Riverside	2,690
89	Chuckwalla Bench	Riverside	52,749
90	Salt Creek Pupfish-Rail	Riverside	4,253
91	San Sebastian Marsh-San Felipe Creek	Imperial	6,337
92	Singer Geoglyphs	Imperial	1,253
93	Indian Pass	Imperial	1,920
94	Lake Cahuilla 2	Imperial	1,214
95	Lake Cahuilla 3	Imperial	2,554
96	Lake Cahuilla 5	Imperial	5,412
97	Lake Cahuilla 6	Imperial	4,483
98	Southern East Mesa Flat-Tailed Horned Lizard	Imperial	40,712
99	Plank Road	Imperial	129
100	Pilot Knob	Imperial	685
101	West Mesa	Imperial	17,400
102	Coyote Mountains Fossil Site	Imperial	640
103	Table Mountain	San Diego	3,960
104	In-ko-pah Mountains	San Diego	17,060
105	Yuha Basin	Imperial	64,462

Total: 919,472

* Managed by the California office of the Bureau of Land Management

ACEC	#
Afton Canyon	63
Alligator Rock	84
Amargosa Canyon	49
Ash Valley	8
Barstow Woolly Sunflower	68
Bedrock Springs	44
Big-Little Sand Springs	26
Big Morongo Canyon	82
Black Mountain	71
Blue Ridge	20
Cache Creek Corridor	6
Calico Early Man Site	64
Camp Irwin	67
Camp Rock Spring	59
Cedar Roughs	7
Cerro Gordo	29
Chuckwalla Bench	89
Chuckwalla Valley Dune Thicket	86
Christmas Canyon	43
Clark Mountain	55
Clear Creek Serpentine	19
Corn Spring	88
Coyote Mountain Fossil Site	102
Cronese Basin	62
Dale Lake	78
Darwin Falls	31
Dead Mountain	58
Denning Spring	50
Desert Tortoise Natural Area	39
Dinosaur Trackway	56
Edom Hill-Willow Hole	83
El Dorado Manzanita	13
Elkhorn Plain	22
Eureka Valley Dunes	27
Fish Slough	15
Fort Soda-Mojave Chub	61
Fossil Falls	33
Goose Lake	21
Great Falls Basin	35
Greenwater Canyon	47
Grimshaw Lake	48
Halloran Wash	53
Harper Dry Lake	66
High Rock Canyon	9
Indian Pass	93
Indian Valley Brodia	4
In-ko-pah Mountains	104
Ione Manzanita	11
Ione Tertiary Oxisol Soil	10
Jawbone-Butterbredt	37
Juniper Flats	80
Kingston Range	52
Kramer Hills	69
Lake Cahuilla 2	94
Lake Cahuilla 3	95

ACEC	#
Lake Cahuilla 5	96
Lake Cahuilla 6	97
Last Chance Canyon	8
Limestone Salamander	16
Marble Mountain Fossil Bed	74
Mesquite Hills	73
Mesquite Lake	54
Mojave Fishhook Cactus	70
Mopah Spring	76
Moreno Paleontological	17
Mule Mountain	87
New York Mountains	57
Northern California Chaparral	5
Northern California Coast Range Preserve	3
Palen Dry Lake	85
Panoche-Coalinga	18
Patton's Iron Mountain Divisional Camp	77
Pilot Knob	100
Piute Creek	60
Plank Road	99
Point Sal	24
Rainbow Basin-Owl Canyon	65
Red Hills	12
Red Mountain	2
Rose Spring	32
Saline Valley	28
Salt Creek (Dumont)	51
Salt Creek Pupfish/Rail	90
Land Canyon	34
San Sebastian Marsh and San Felipe Creek	91
Shasta River Salmon Spawn	1
Short Canyon	36
Singer Geoglyphs	92
Soda Lake	23
Soggy Dry Lake	79
Southern East Mesa Flat-Tailed Horned Lizard	98
Squaw Spring	45
Steam Well	46
Surprise Canyon	41
Table Mountain	103
Travertine Hot Springs	14
Trona Pinnacles	42
Upper Johnson Valley Yucca Rings	72
Warm Sulfur Springs	30
Western Rand Mountains	40
West Mesa	101
Whipple Mountains	75
White Mountain City	25
Whitewater Canyon	81
Yuha Basin	105

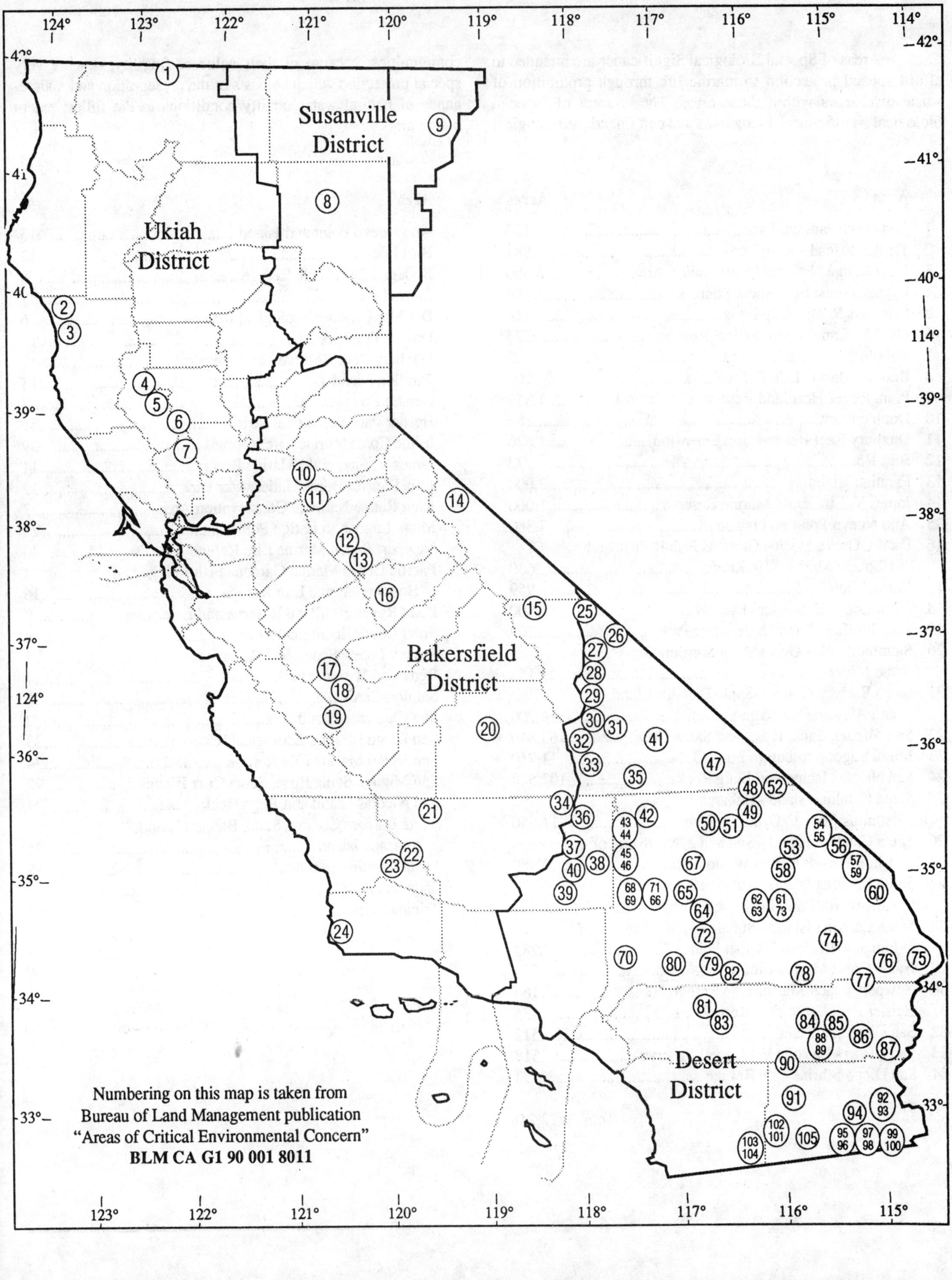

Susanville District

Ukiah District

Bakersfield District

Desert District

Numbering on this map is taken from
Bureau of Land Management publication
"Areas of Critical Environmental Concern"
BLM CA G1 90 001 8011

Areas of Special Biological Significance are intended to afford special protection to marine life through prohibition of waste discharges within those areas. The concept of "special biological significance" recognizes that certain unique biological communities, because of their value or fragility, deserve very special protection which consists of the preservation and maintenance of natural water quality conditions to the fullest extent practical.

	Area	Acres
1	Redwood National Park	4,160
2	Trinidad Head - Kelp Beds	1,581
3	King Range National Conservation Area	3,680
4	Pygmy Forest Ecological Staircase	259
5	Saunders Reef - Kelp Beds	618
6	Del Mar Landing Ecological Reserve	77
7	Gerstle Cove	2
8	Bodega Marine Life Refuge	200
9	Point Reyes Headland Reserve and Extension	1,359
10	Double Point	86
11	Duxbury Reef Reserve and Extension	1,626
12	Bird Rock	72
13	Farallon Island	2,000
14	James V. Fitzgerald Marine Reserve	1,006
15	Año Nuevo Point and Island	7,360
16	Pacific Grove Marine Gardens Fish Refuge and Hopkins Marine Life Refuge	680
17	Carmel Bay	959
18	Point Lobos Ecological Reserve	800
19	Julia Pfeiffer Burns Underwater Park	709
20	Salmon Creek - Ocean Area Surrounding the Mouth	1,152
21	Santa Barbara County, Santa Barbara Island, and Anacapa Island	14,000
22	San Miguel, Santa Rosa, and Santa Cruz Islands	163,840
23	Mugu Lagoon to Latigo Point	11,710
24	San Nicolas Island and Begg Rock	102,528
25	Santa Catalina Island - Subarea 1, Isthmus Cove to Catalina Head	11,650
26	Santa Catalina Island - Subarea 2, North end of Little Harbor to Ben Weston Point	2,989
27	Santa Catalina Island - Subarea 3, Farnsworth Bank Ecological Reserve	14
28	Santa Catalina Island - Subarea 4, Pinnacle Rock to Jewfish Point	3,283
29	Irvine Coast Marine Life Refuge	1,024
30	Newport Beach Marine Life Refuge	166
31	Heisler Park Ecological Reserve	1,536
32	San Clemente Island	80,512
33	San Diego - La Jolla Ecological Reserve	518
34	San Diego Marine Life Refuge	92
		Total: 422,248

Area	#
Año Nuevo Point and Island	15
Bird Rock	12
Bodega Marine Life Refuge	8
Carmel Bay	7
Del Mar Landing Ecological Reserve	6
Double Point	10
Duxbury Reef Reserve and Extension	11
Farallon Island	13
Gerstle Cove	7
Heisler Park Ecological Reserve	31
Irvine Coast Marine Life Refuge	29
James V. Fitzgerald Marine Reserve	14
Julia Pfeiffer Burns Underwater Park	19
King Range National Conservation Area	3
Mugu Lagoon to Latigo Point	23
Newport Beach Marine Life Refuge	30
Pacific Grove Marine Gardens Fish Refuge and Hopkins Marine Life Refuge	16
Point Reyes Headland Reserve and Extension	9
Point Lobos Ecological Reserve	18
Pygmy Forest Ecological Reserve	4
Redwood National Park	1
Salmon Creek	20
San Clemente Island	32
San Diego La Jolla Ecological Reserve	33
San Diego Marine Life Refuge	34
San Miguel, Santa Rosa, Santa Cruz Islands	22
San Nicolas Island and Begg Rock	24
Santa Barbara County, Santa Barbara Island, Anacapa Island	21
Santa Catalina Island	25, 26, 27, 28
Saunders Reef	5
Trinidad Head	2

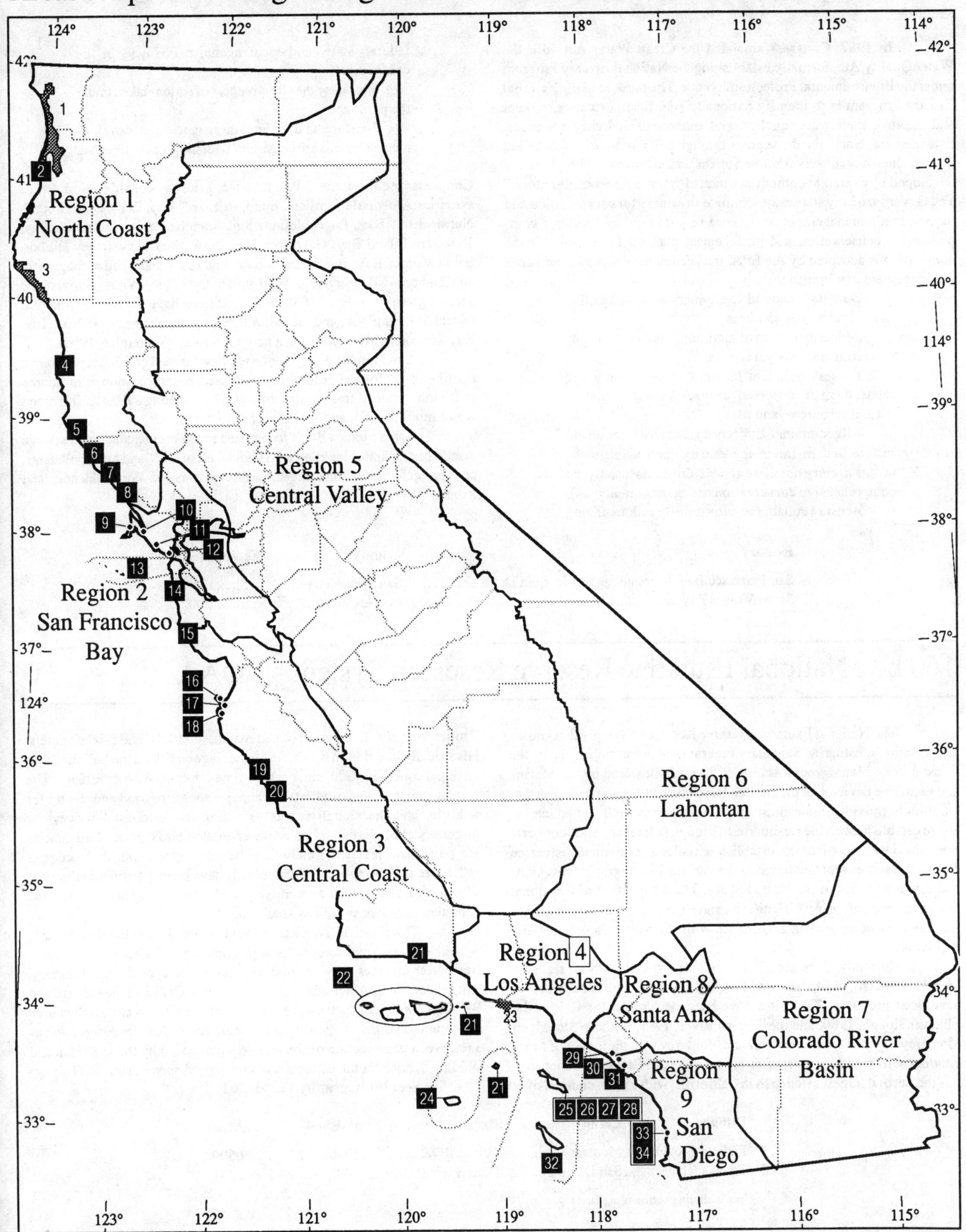
124°
123°
122°
121°
120°
119°
118°
117°
116°
115°
114°
42°
41°
40°
39°
38°
37°
36°
35°
34°
33°
1
2
3
4
5
6
7
8
9
10
11
12
13
14
15
16
17
18
19
20
21
22
23
24
25
26
27
28
29
30
31
32
33
34
Region 1
North Coast
Region 2
San Francisco
Bay
Region 3
Central Coast
Region 4
Los Angeles
Region 5
Central Valley
Region 6
Lahontan
Region 7
Colorado River
Basin
Region 8
Santa Ana
Region
9
San
Diego

160 a National Estuary Program - Environmental Protection Agency

In 1987, Congress amended the Clean Water Act with the Water Quality Act, formally establishing the National Estuary Program under the Environmental Protection Agency. The purpose of the National Estuary Program is to identify nationally significant estuaries, protect and improve their water quality, and enhance their living resources. Estuaries are to achieve these goals through collaborative efforts called "comprehensive conservation and management plans." The plans are developed by oversight committees termed "management conferences." The Governor of any state may nominate an estuary for consideration and request that a management conference be convened to develop a comprehensive conservation and management plan for the estuary. If the nomination is accepted by the EPA, the agency administrator convenes a management conference to:

1-assess trends in the estuary's water quality, natural resources and uses;

2-identify causes of environmental problems by collecting and analyzing data;

3-assess pollutant loadings in the estuary and relate them to observed changes in water quality, natural resources, and uses;

4-Recommend and schedule priority actions to restore and maintain the estuary, and identify the means to carry out these actions(this is essentially the comprehensive conservation and management plan).

5-ensure collaboration on priority actions among federal, state, and local agencies involved in the conference;

6-monitor the effectiveness of actions taken under the plan; and

7-ensure that federal assistance and development programs are consistent with the goals of the plan.

Congress required the EPA to give priority consideration to 12 estuaries: Albemarle-Pimlico Sound, Buzzards Bay, Long Island Sound, Narragansett Bay, Puget Sound, San Francisco Bay, Delaware Bay, Delaware Inland Bay, Galveston Bay, New York-New Jersey Harbor, Santa Monica Bay, and Sarasota Bay. The first 6 bays had management conferences called early in 1986 under the Clean Water Act and the second group in July of 1988. The conferees have 5 years in which to complete their plans , and the restoration projects for both San Francisco Bay and Santa Monica Bay are now underway (September 1990).

San Francisco Bay conferees have agreed on 5 basic problems: decline of biological resources, increased point and nonpoint source pollution, reduced fresh water inflow and resulting salinity, increased water modification, and intensified land use.

Santa Monica Bay's recognized problems include: major sewage discharges, sediment building with high levels of DDT and other pollutants, sewage spills in Ballona Creek, severe pathenogen contamination, and habitats

	Estuary	County	Established
1	San Francisco Bay, Sacramento/San Joaquin Delta	San Francisco+	1986
2	Santa Monica Bay	Los Angeles	1988

160 b National Estuarine Reserve Research System - NOAA

The National Estuarine Reserve Research System (also known as the National Estuarine Sanctuary Program), which is authorized by the Coastal Zone Management Act of 1972, is administered by the Marine and Estuarine Division of the National Oceanic and Atmospheric Administration to provide opportunities for long-term research and education and to establish estuarine sanctuaries. Through federal and state cooperation, the Division plans to establish a national system of estuarine research reserves representative of the various biogeographical regions and estuarine types in the United States. The designation of a national estuarine research reserve signifies a state's commitment to long-term management of an area in accordance with the System's policies and regulations.

Currently there are 17 research reserves (2 of those reserves hold 4 sites each), in 14 states and Puerto Rico, and 3 additional reserves have been proposed. Two sites have been designated for California: Elkhorn Slough, 1980, and the Tijuana River, 1981. Elkhorn Slough on Monterey Bay between Santa Cruz and Monterey with the harbor of Moss Landing at its mouth, is most often referred to as an Estuarine Sanctuary. It is the second largest salt marsh in California, with a main channel over 7 miles long and 2,500 acres of salt marsh, mudflats, and tidal channels. Historically, the slough was a freshwater lagoon before the Salinas River changed course in 1908, since which time it has become an estuary. The waters, marshes, and mudflats are a major source of food and shelter for wildlife, and the slough serves as a link in the coastal flyway for migratory shore birds and other water-related birds. Many bird species are permanent residents, including the endangered California clapper rail. More than 90 species of water birds have been observed in the area. The slough also serves as a nursery and a feeding area for sport and commercial fishes as well as shellfish.

The Tijuana River, associated with the Border Field State Park, is a major waterfowl refuge. Its approximately 2,531 acres of salt and freshwater marshes provide food and resting sites for the black-necked stilt, avocet, teal, American widgeon, pelican, California least tern, and Beldings Savannah sparrow. The waters of the estuary are a rich source for commercial fishes, as well as for shellfish. Before its designation as a reserve, a large section of the area was purchased by the U.S. Fish and Wildlife Service for the protection of endangered species and their habitat.

See: Bibliography 12, 93, 202

	Estuary	County	Stream	Established	Acres
A	Elkhorn Slough	Monterey	Salinas River	1980	2,500*
B	Tijuana River	San Diego	Tijuana River	1981	2,531

*not including surrounding lands

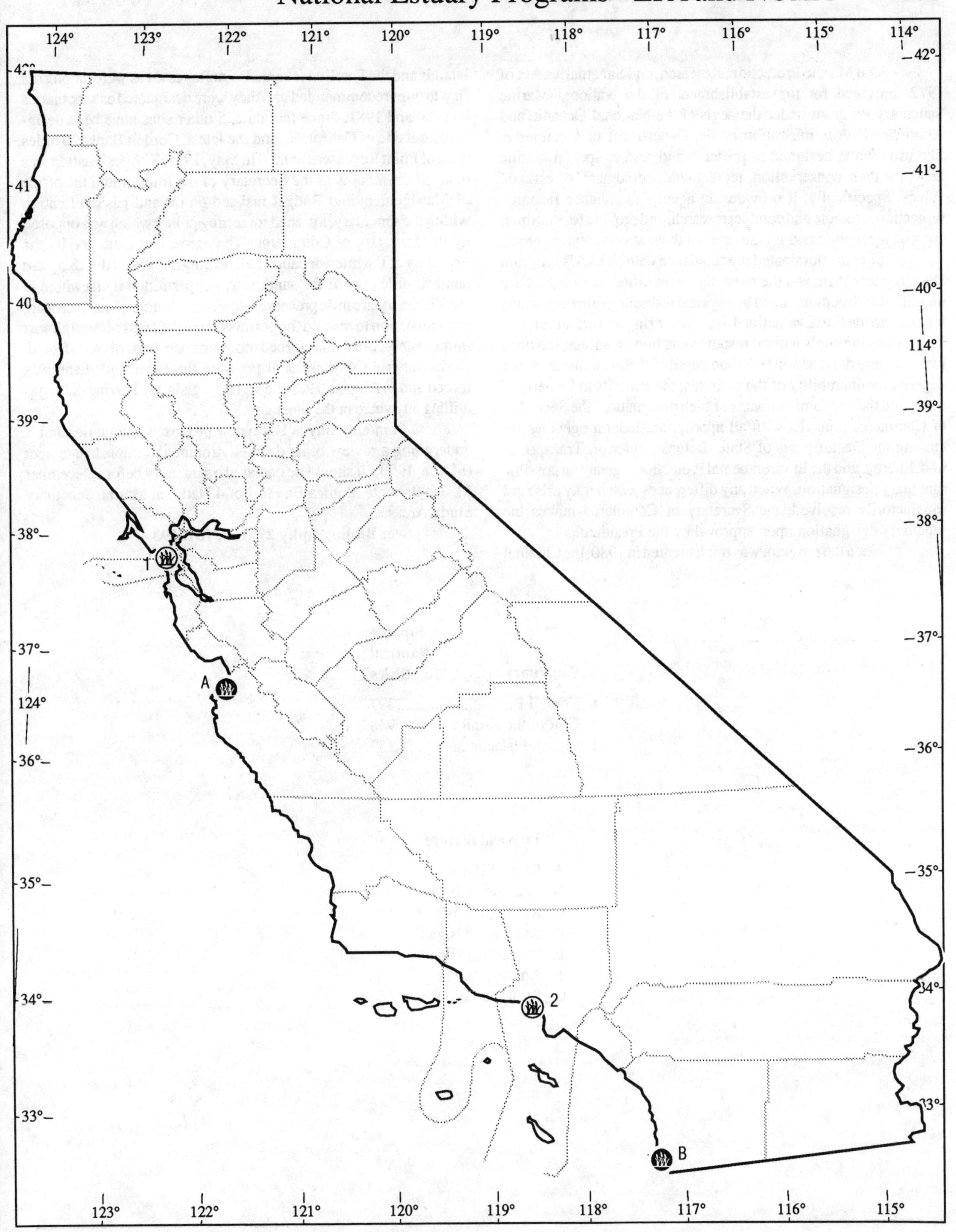
124°
123°
122°
121°
120°
119°
118°
117°
116°
115°
114°
42°
41°
40°
39°
38°
37°
36°
35°
34°
33°
1
2
A
B

The Marine Protection, Research, and Sanctuaries Act of 1972, provided for the establishment of the National Marine Sanctuary Program under the aegis of the National Oceanic and Atmospheric Administration of the Department of Commerce. The program is designed to preserve and restore special marine areas for their conservation, recreational, ecological, or esthetic values. Specifically, it provides an agency to enhance resource protection, promote and conduct research, and provide for multiple, compatible, public and private uses of these special marine areas.

Sites are nominated in accordance with NOAA's Program Development Plan, and the most highly qualified marine sites are identified and recommended by regional resource evaluation teams for entry to the Site Evaluation List. After stringent further review, which rates the site's natural resource and human values, the need for sanctuary designation based on possible threats to the area, the degree of vulnerability of the resource, the benefits to be derived from, and the economic impacts of such designation, the Secretary of Commerce consults with all appropriate federal agencies including the Departments of State, Defense, Interior, Transportation, Energy, and the Environmental Protection Agency on possible sanctuary designation. When any differences which may arise are satisfactorily resolved, the Secretary of Commerce makes the sanctuary designation upon approval by the President.

When the program was implemented in 1980, the Channel Islands and the Farallon Islands/Point Reyes areas were among the first groups recommended and they were designated as sanctuaries in 1980 and 1981. Since that date, 5 other sites have been designated outside of California, and the latest, Cordell Bank, 20 miles west of Point Reyes was named in May, 1989. The designation was delayed for months as the Secretary of the Interior and the office of Management and Budget insisted on oil and gas exploration within the sanctuary, a precedent setting policy which was opposed by the Secretary of Commerce. The agreement extracted by the Secretary of the Interior banned oil drilling only from the 18 square nautical mile core of the sanctuary, but permitted it elsewhere in the 397.05 square mile preserve. However, Congress was incensed, and legislation to rescind the permission to drill was offered almost immediately after the forced compromise had been effected. Under intense Congressional pressure the President's hand was forced and in August, 1989, he signed legislation barring oil or gas drilling anywhere in the sanctuary.

Monterey Bay is 1 of 6 new proposed sanctuaries and is undergoing a review of its draft environmental impact statement (March, 1990). It should be declared a sanctuary before December 31, 1990. Santa Monica Bay is 1 of 4 National Marine Sanctuary Study Areas.

See: Bibliography 27, 174, 181, 193

	Sanctuary	**Square Nautical Miles**
1	Cordell Bank	397
2	Gulf of the Farallones	948
3	Channel Islands	1,252

Physical feature

A Cordell Bank
B Farallon Islands
C San Miguel Island
D Santa Rosa Island
E Santa Cruz Island
F Anacapa Island
G Santa Barbara Island

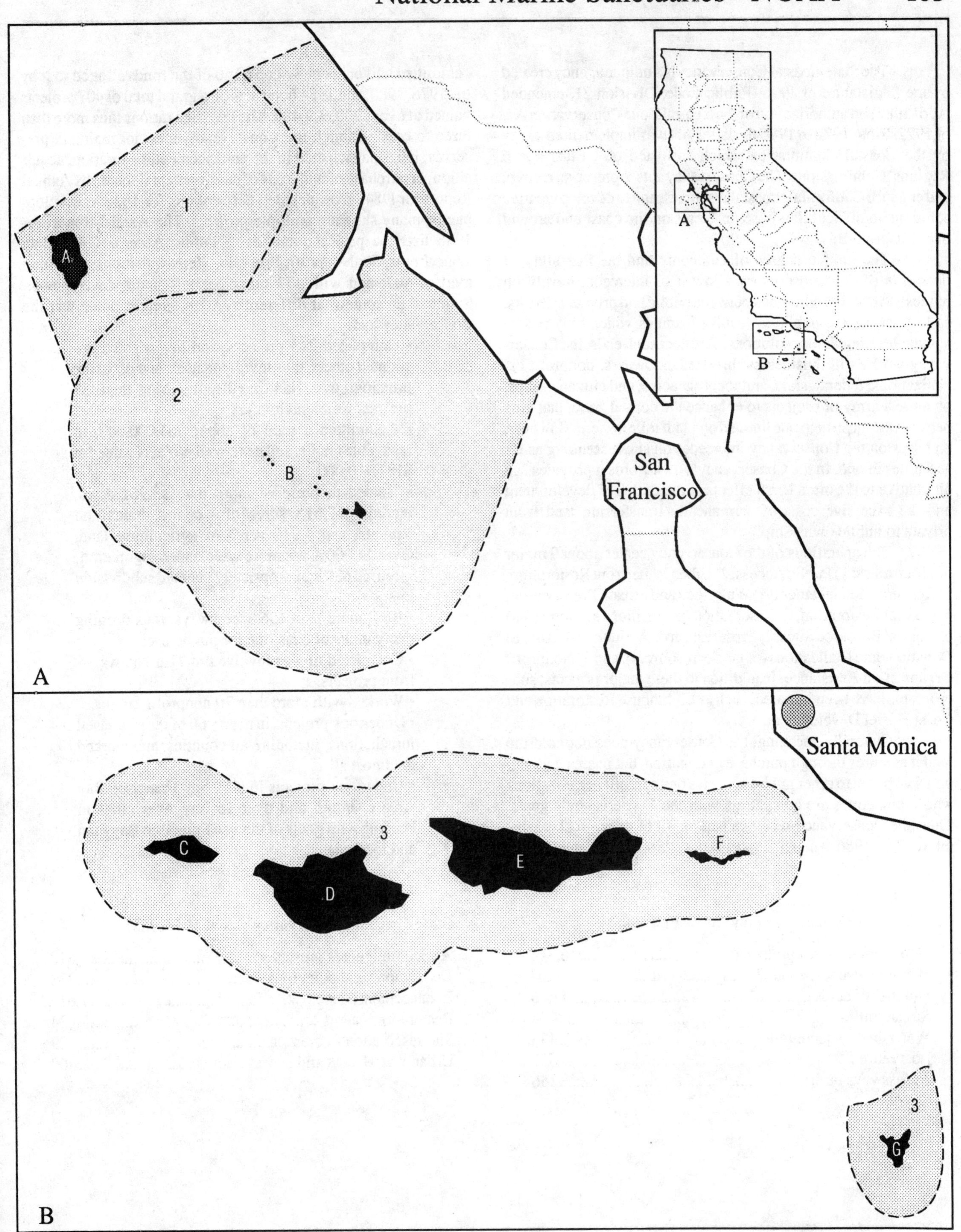
1
A
2
B
San
Francisco
A
B
A
Santa Monica
3
C
D
E
F
3
G
B

The State Coastal Conservancy is a unique agency created by the Legislature in 1976 (Public Code, Division 21, amended April 1982) on authorization of the Coastal Zone Conservation Act of 1972. From 1972 to 1976 the latter Act was implemented solely by the Coastal Commission which consisted of 1 State and 6 Regional Commissions. The Conservancy acts to preserve, restore, and enhance California's coastal resources, and to develop creative solutions to difficult land use problems on the coast and around San Francisco Bay.

The unique degree of autonomy and the flexibility of operation of the Conservancy allow it to intercede, mostly on request, in conflict situations between public and private interests, and to act for the public to resolve disputes which may not be amenable to regulatory solutions. To effect such ends, the Conservancy works with businesses; private landowners; nonprofit organizations; federal, state, and local agencies; and citizen groups, on a great array of projects to enhance the coastal zone, that area between the mean high tide line and one-half mile inshore. However, on occasion the Conservancy intercedes on sites extending as far as 5 miles inland. In the Conservancy's own words it provides "an alternative to the often harsh effects of regulation of development and the excessive costs of permanently transferring land from private to public ownership."

The operations of the Conservancy center about 9 major program areas: 1) Public Access, 2) Urban Waterfront Restoration, 3) Resource Enhancement, 4) Watershed and Stream Preservation, 5) Coastal Restoration, 6) Reservation (acquisition for future sale) of Coastal Resource Sites, 7) Preservation of Agricultural Land, 8) Donations and Dedications (of land or resources), and 9) Nonprofit Organization Assistance. In addition to these major projects, subprograms have been initiated, such as Lighthouse Restoration and Coast Hostel Development

Generally speaking, the Conservancy does not hold the land it acquires through purchase or donation, but passes it along, on a sale basis, to other public agencies or nonprofit organizations whose objectives are compatible with the Conservancy's goals. Thus, the Conservancy recovers part or all of its original expenditure. The 1986 Annual Report states, "So far, the agency has committed and or spent $74,667,150 of the funds allotted to it by the 1976, 1980, and 1984 bond acts on a grand total of 407 projects valued at over $200,609,480. The leverage ratio is thus more than three to one." Though the Conservancy does not maintain preserves, it has participated in the enhancement, restoration, acquisition, or purchase of hundreds of environmental tracts. Its Annual Report for 1984 gives detailed descriptions for 162 site developments, many showing multiple projects. The Annual Report for 1986 lists the project totals as: "Public Access, 114; Urban Waterfronts, 50; Nonprofit Program, 76; Acquisitions, 82;" for a total of 407; and with 173 other projects under development. Summed up somewhat differently, a 1986 report noted that the Conservancy had:

- Completed 243 projects, and investigated uncounted others, thus involving itself in most of the principal coastal land use disputes. More than 337 projects were underway.
- Committed a total of about $64,000,000, in repayable grants, and generated projects valued at $181,710,000
- Helped to preserve more than 13,961 acres, including 7,615 acres of wetland dunes and habitats, and 1,810 acres of agricultural land. Over 14,000 acres more were under negotiation.
- Retired 639 inappropriately planned subdivision lots.
- Built more than 156 accessways, thus opening many miles of coastline for public use.
- Completed or was involved in 71 urban waterfront projects.
- Worked with more than 70 nonprofit groups.
- Undertook projects in nearly 60 of the 67 coastal jurisdictions, including all counties, and worked in almost all.
- Undertook projects in all 9 San Francisco Bay Area counties, and over 15 Bay Area cities in keeping with goals of the San Francisco Bay Plan and local plans.

State Coastal Conservancy (As of Fall 1988)

Non-wetland acres protected	6,346
Wetlands-Dunes, and habitats preserved	64
Number of acres	7,615
Stream miles	208
Watershed - square miles	433
Lots retired	639
Accessways built	156

State Coastal Conservancy (As of Fall 1988)

Agricultural acres preserved	1,810
Dedications and donations - acres	333
Enhancement - acres	7,615
Restoration - acres	3,634
Site reservation - acres	109
Urban waterfronts and access - acres	460

Four conservationists met in 1973 to map a strategy to prevent the completion of a 625 foot-high dam, the New Melones Dam, in the canyon of the Stanislaus River, one of the most beautiful and popular waterways of the west. In the 7 year struggle which ensued, the group drew together a network of environmental, legal, and political experts to fight for the river. That battle was lost.The New Melones Dam was capped in 1979. Nonetheless, those efforts helped to change much of the public attitude from a passive acceptance to a firm resistance to dams, based on the knowledge of the irreparable damage dams cause in living river systems.

Today, December 1990, membership in the Friends Of The River has grown to 10,000 with 3 regional chapters. It is the largest grassroots river conservation organization in the nation, "dedicated to the preservation of our rivers and streams, and to the conservation of our water and energy supplies." In more recent years, the organization's success may be measured in the protections added for California rivers since 1980: National Wild and Scenic status for the Tuolomne River in 1984, and the same designation for the Kern, the Kings, and the Merced Rivers in 1986. In 1984, the Friends led the successful drive to stop the environmentally destructive Through Delta Canal, and in1987, the association helped to gain passage of hydroelectric reform legislation in Congress. The Friends Of The River are currently engaged in conservation efforts on the Clavey, the Salt Caves project on the Upper Klamath, the Hobo project on the Lower Kern, the Devils Nose project on the North Fork of the Mokelumne, and the Auburn Dam on the Upper American Rivers.

In 1987, the organization recognized the need to develop a broader role for itself, a need to directly address underlying water policies and management practices responsible for river destruction. Consequently, 2 ambitious goals emerged: the formation of a new water agenda to research, articulate, and promote water policies that would eliminate new dams from the public agenda, and secondly, the 100 rivers campaign. That campaign is intended to set the groundwork for the eventual designation of more than 100 additions to the National Wild and Scenic Rivers System.

Friends Of The River
909 12th Street #207
Sacramento, CA, 95814
916-442-3155

Friends Of The River
Fort Mason Center, Building C
San Francisco, CA, 94123
415-771-0400

Natural Reserve System - University of California

The University of California's Natural Reserve System (formerly the Natural Land and Water Reserve System) was established in 1965 in response to the continuing loss and disruption of field teaching sites due to vandalism, theft, urbanization, and other developments. Since the university believes that only through field study can it produce competent environmental planners and citizens imbued with a land ethic, the Regents designated 7 university sites as the first NRS reserves. Since that date the system has grown to 30 sites including 3 candidate reserves (98,800 acres, 154 sqare miles).

The primary objective of the system is the preservation and management of the state's natural diversity to meet the university's teaching and research needs in disciplines that require field work. Each reserve functions as an outdoor classroom or laboratory, in the knowledge that an outdoor laboratory, once lost, can never be reconstructed. The habitats and biota it supported are permanently lost and its particular gene pool cannot be replaced. In this context, the NRS functions as an ecosystem library, an irreplacable storehouse of the state's biota and natural diversity.

To meet these objectives, three classes of reserves are envisioned for the future: 1) Campus teaching reserves, natural areas near campuses that meet the needs of regularly scheduled classes and student research with a minimum of travel; 2) Special habitat reserves for the study of representative samples of both widespread habitat types and distinctive ecosystems and features, such as pygmy forests, vernal pools, archaeological sites, and geological formations of special value for teaching and research; 3) Multipurpose ecological reserves, large ecologically diverse areas capable of supporting a wide variety of academic research and teaching at a single location.

Of the 178 major habitat types that have been identified in a statewide inventory by university faculty and NRS staff, the 30 existing reserves hold about 124 of those types.

See: Bibliography 140.

	Reserve	County	Associated Campus	Acres
1	Eagle Lake	Lassen	Candidate	26,985
2	Northern California Coast Range	Mendocino	Candidate	3,952
3	Pygmy Forest	Mendocino	Berkeley	70
4	Chickering American River	Placer	Berkeley	1,786
5	Bodega Marine	Sonoma	Davis	326
6	Stebbins Cold Canyon	Solano	Davis	277
7	Jepson Prairie	Solano	Davis	277
8	Año Nuevo Island	San Mateo	Santa Cruz	16
9	Younger Lagoon	Santa Cruz	Santa Cruz	49
10	Valentine Eastern Sierra	Mono	Santa Barbara	136
11	Fish Slough	Inyo-Mono	Candidate	1,606
12	Hastings Natural History Reservation	Monterey	Berkeley	1,930
13	Landels-Hill Big Creek	Monterey	Santa Cruz	3,848
14	Granite Mountains	San Bernardino	Riverside	1,850
15	Sacramento Mountains	San Bernardino	Riverside	591
16	Coal Oil Point	Santa Barbara	Santa Barbara	117
17	Carpenteria Salt Marsh	Santa Barbara	Santa Barbara	120
18	Santa Cruz Island	Santa Barbara	Santa Barbara	54,488
19	San Joaquin Freshwater Marsh	Orange	Irvine	202
20	Etiwanda Wash	San Bernardino	Riverside	157
21	Box Springs	Riverside	Riverside	160
22	Burns Piñon Ridge	San Bernardino	Irvine	265
23	Motte Rimrock	Riverside	Riverside	237
24	James San Jacinto Mountains	Riverside	Riverside	29
25	Philip. L. Boyd Deep Canyon Research Center	Riverside	Riverside	5,736
26	Dawson Los Monos Canyon	San Diego	San Diego	133
27	Ryan Oak Glen	San Diego	San Diego	15
28	Scripps Shoreline-Underwater	San Diego	San Diego	80
29	Elliot Chaparral	San Diego	San Diego	183
30	Kendall-Frost Mission Bay Marsh	San Diego	San Diego	16

Reserve	#
Año Nuevo Island	8
Bodega Marine	5
Box Springs	21
Burns Piñon Ridge	22
Carpenteria Salt Marsh	17
Chickering American River	4
Coal Oil Point	16
Dawson Los Monos Canyon	26
Eagle Lake	1
Elliot Chaparral	29
Etiwanda Wash	20
Fish Slough	11
Granite Mountains	14
Hastings Natural History Reservation	12
James San Jacinto Mountains	24
Jepson Prairie	7
Kendall Frost Mission Bay Marsh	30
Landels-Hill Big Creek	13
Motte Rimrock	23
Northern California Coast Range	2
Philip. L. Boyd Deep Canyon Research Center.	25
Pygmy Forest	3
Ryan Oak Glen	27
Sacramento Mountains	15
San Joaquin Freshwater Marsh	19
Santa Cruz Island	18
Scripps Shoreline Underwater	28
Stebbins Cold Canyon	6
Valentine Eastern Sierra	10
Younger Lagoon	9

(continued)

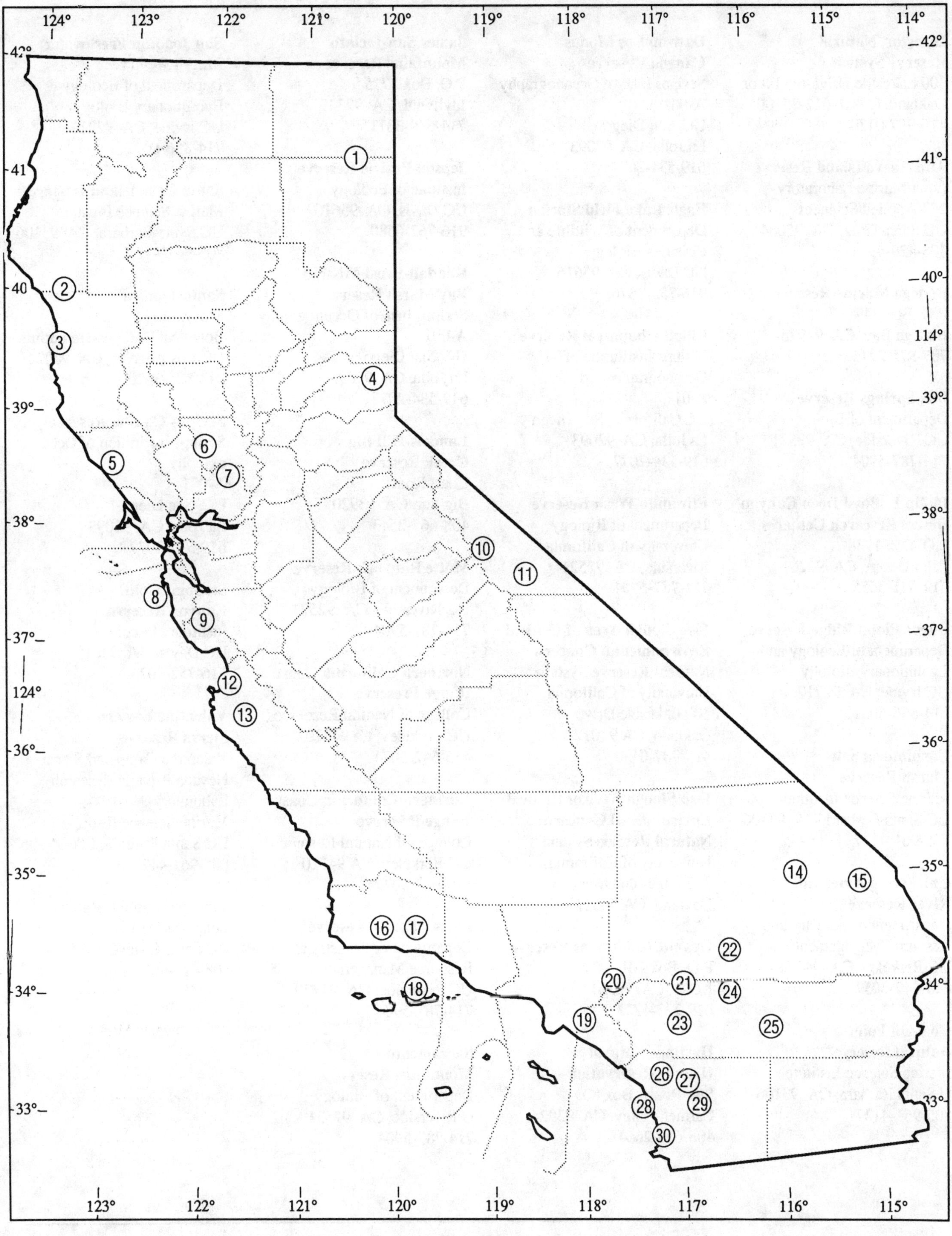

124° 123° 122° 121° 120° 119° 118° 117° 116° 115° 114°

42° 41° 40° 39° 38° 37° 36° 35° 34° 33°

1 2 3 4 5 6 7 8 9 10 11 12 13 14 15 16 17 18 19 20 21 22 23 24 25 26 27 28 29 30

114° 124°

123° 122° 121° 120° 119° 118° 117° 116° 115°

Director, Natural Reserve System
300 Lakeside Drive-6th Floor
Oakland, CA 94612-35500
415-987-0150

Año Nuevo Island Reserve
Long Marine Laboratory
273 Applied Sciences
UC Santa Cruz, CA 95064
408-429-4971

Bodega Marine Reserve
P.O. Box 247
Bodega Bay, CA 94923
707-875-2211

Box Springs Reserve
Department of Biology
UC Riverside, CA 92521
714-787-5904

Philip L. Boyd Deep Canyon Desert Research Center
P.O. Box 1738
Palm Desert, CA 92261
619-341-3655

Burns Pinon Ridge Reserve
Department of Ecology and Evolutionary Biology
UC Irvine, CA 92717
714-856-6031

Carpinteria Salt Marsh Reserve
Marine Science Institute
UC Santa Barbara, CA 93106
805-961-4127

Chickering American River Reserve
Department of Forestry and Resource Management
UC Berkeley, CA 94720
415-642-5057

Coal Oil Point Natural Reserve
Marine Science Institute
U Santa Barbara, CA 93106
805-961-4127

Dawson Los Monos Canyon Reserve
Scripps Inst. of Oceanography
A-001
UC, San Diego
La Jolla, CA 92093
619-534-2077

Eagle Lake Field Station
Department of Wildlife and Fisheries Biology
UC Davis, CA 95616
916-752-3576

Elliott Chaparral Reserve
Scripps Institution of Oceanography
A-011
UC California, San Diego
La Jolla, CA 92093
619-534-2077

Etiwanda Wash Reserve
Department of Biology
University of California
Riverside, CA 92521
714-787-5904

Fish Slough Area of Critical Environmental Concern Natural Reserve System
University of California
300 Lakeside Drive
Oakland, CA 94012
415-987-0150

Fish Slough Area of Critical Environmentl Concern Natural ReserveSystem
University of California
300 Lakeside Drive
Oakland, CA 94612

Granite Mountains Reserve
P.O. Box 101
Kelso, CA 92351
619-733-4222

Hastings Natural History Reservation
Star Route, Box 80
Carmel Valley, CA 93924
408-659-2664

James San Jacinto Mountains Reserve
P.O. Box 1775
Idyllwild, CA 92349
714-659-3811

Jepson Prairie Reserve
Institute of Ecology
UC Davis, CA 95616
916-752-6580

Kendall-Frost Mission Bay Marsh Reserve
Scripps Inst. of Oceanography
A-001
UC, San Diego
La Jolla, CA 92093
619-534-2077

Landels-Hill Big Creek Reserve
Coast Route
Big Sur, CA 93920
408-667-2543

Motte Rimrock Reserve
Department of Biology
UC Riverside, CA 92521
714-787-5904

Northern California Coast Range Preserve
College of Natural Resources
UC Berkeley, Ca 94720
415-642-5074

Northern California Coast Range Preserve
College Of Natural Resources
UC Berkeley, CA 94720
415-642-5079

Pygmy Forest Reserve
Department of Forestry and Resource Management
UC Riverside, CA 94720
714-787-5904

Sacramento Mountains Reserve
Department of Biology
U Riverside, CA 92521
714-787-5904

San Joaquin Freshwater Marsh Reserve
Department of Ecology & Evolutionary Biology
UC Irvine, CA 92717
714-856-6031

Santa Cruz Island Reserve
Marine Science Institute
UC Santa Barbara, CA 93106
805-961-4127

Santa Monica Mountains Reserve
Botanical Gardens-Herbarium
UC Los Angeles, CA 90024
213-825-3620

Scripps Coastal Reserve
Scripps Institution of Oceanography
A-001
UC, San Diego
La Jolla, CA 92093
619-534-2077

Stebbins Cold Canyon Reserve
Institute of Ecology
UC Davis, CA 95616
916-752-3026

Valentine Eastern Sierra Reserve
Valentine Camp and Sierra Nevada Aquatic Research Laboratory
Marine Science Institute
UC Santa Barbara, CA 93106
805-961-4127

Younger Lagoon Reserve
Long Marine Laboratory
UC Santa Cruz, CA 95064
408-429-4971

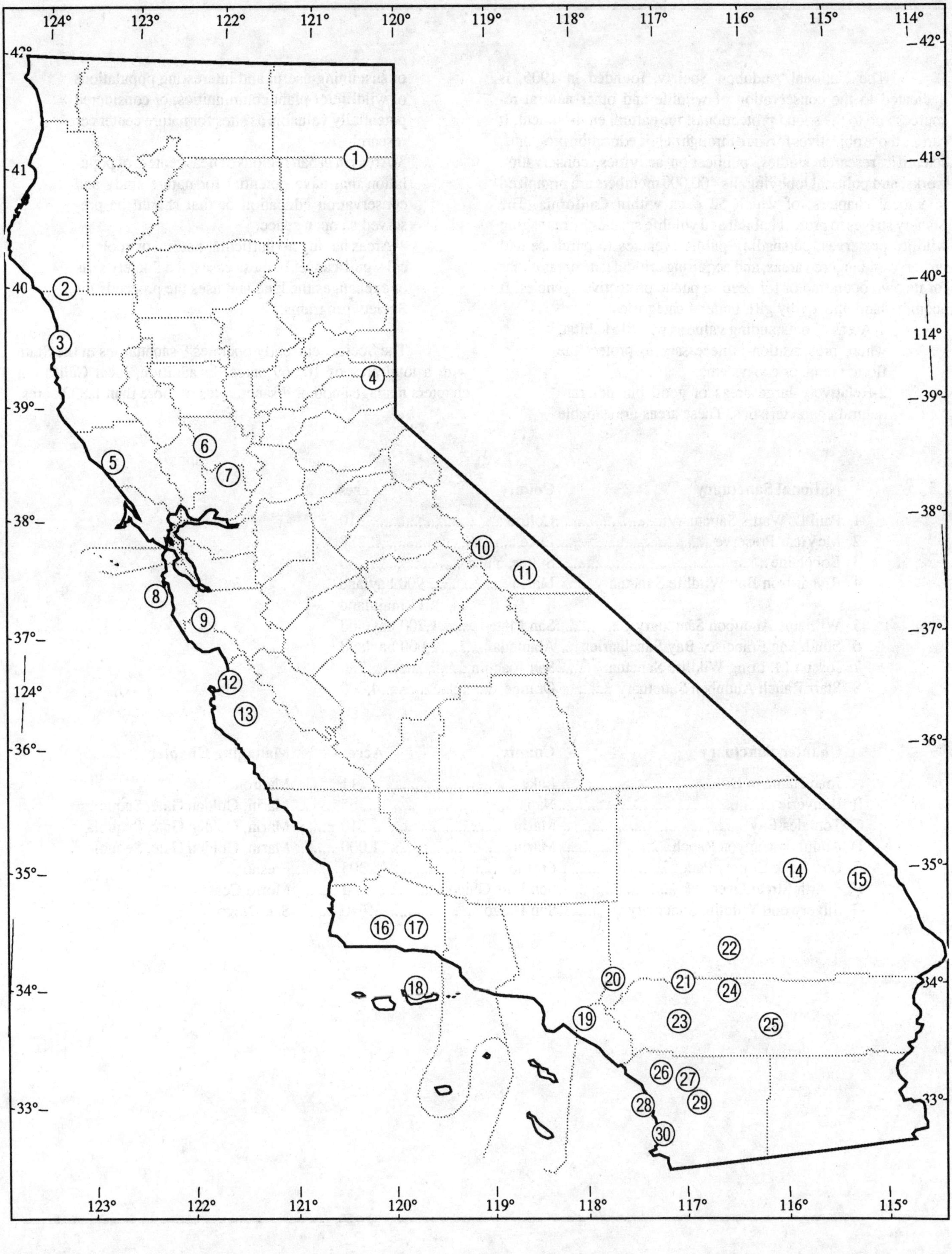
124°
123°
122°
121°
120°
119°
118°
117°
116°
115°
114°
42°
41°
40°
39°
38°
37°
36°
35°
34°
33°
1
2
3
4
5
6
7
8
9
10
11
12
13
14
15
16
17
18
19
20
21
22
23
24
25
26
27
28
29
30

The National Audubon Society, founded in 1905, is dedicated to the conservation of wildlife and other natural resources and to the sound protection of the natural environment. It carries those objectives forward through public education programs, scientific research studies, publication activities, conservation works, and political lobbying. Its 400,000 members are organized into local chapters, of which 52 exist within California. The Society strives to protect habitat and wildlife species by managing wildlife preserves, persuading public agencies to purchase and preserve endangered areas, and acquiring critical land areas either for its own operation or for deed to public protective agencies. It acquires land, mostly by gift, under 4 categories:

1-Areas of outstanding value as wildlife habitat, where preservation is necessary to protect the flora, fauna, or ecosystems.

2-Relatively large areas of good but not rare natural characteristics. These areas are capable of sustaining useful and interesting populations of wildlife or plant communities, or considered potentially valuable as sites for nature centers or research.

3-Areas in urban regions or near centers of population that have potential for nature study and conservation education or that should be preserved as open space.

4-Areas having no particular natural or ecological significance. In these cases, the Society sells or exchanges the land and uses the proceeds for Society programs.

The Society currently operates 8 sanctuaries in the State with a total area of 10,069 acres. In addition, local California chapters manage another 7 sanctuaries of more than 2,832 acres.

	National Sanctuary	County	Acres
1	Paul L. Wattis Sanctuary	Colusa	510
2	McVicar Preserve	Lake	238
3	Bobelaine	Sutter, Yuba	430
4	Richardson Bay Wildlife Sanctuary	Marin	900 bayland 11 mainland
5	Williams Audubon Sanctuary	San Mateo	1,200 bayland
6	South San Francisco Bay Sanctuaries	Alameda	2,000 bayland
7	Joseph M. Long Wildlife Sanctuary	San Joaquin	780
8	Starr Ranch Audubon Sanctuary	Orange	4,000

	Chapter Sanctuary	County	Acres	Managing Chapter
A	Joan Hamann Dole	Lake	11	Madrone
B	Bouverie	Napa	365	Marin, Golden Gate, Sequoia
C	Tomales Bay	Marin	510	Marin, Golden Gate, Sequoia
D	Audubon Canyon Ranch	Marin	1,000	Marin, Golden Gate, Sequoia
E	Lost Lake County Park	Fresno	305	Fresno
F	Fourth Street Overlook	San Luis Obispo	1	Morro Coast
G	Silverwood Wildlife Sanctuary	San Diego	640	San Diego

124° 123° 122° 121° 120° 119° 118° 117° 116° 115° 114°

42° 41° 40° 39° 38° 37° 36° 35° 34° 33°

1 2 3 4 5 6 7 8

A B C D E F G

National Sanctuaries

Chapter Sanctuaries

In 1917, 2 committees of the Ecological Society of America merged to form the Nature Conservancy. Since that time the Conservancy has acquired 8,690 parcels of ecologically important and sensitive land, 5,121,522 acres, an area slightly smaller than the state of Massachusetts. These sites are scattered throughout the 50 states, with others in Canada, Latin America, and the Caribbean. The Conservancy manages more than 900 preserves, the largest privately owned nature preserve system in the world.

The Conservancy made its first California acquisition in 1958, when a few tenacious volunteers, after protracted negotiations, purchased 4,000 acres of old growth Douglas fir forest and an undisturbed tributary of the Eel River, now known as the Northern California Coast Range Preserve. By 1977, the escalating number and size of the California projects prompted the initiation of the California Field Office, whose first project was the extraordinarily ambitious acquisition of 90 percent of Santa Cruz Island (54,488 acres). Since 1958, the Conservancy has been involved in 209 California land acquisition projects, protecting 321,786 acres.

The exacting standards of the Conservancy require that the land for acquisition be of the highest biological significance. Accordingly, the Conservancy launched the California Critical Areas Program, a plan to identify and safeguard the State's most endangered biological species, and has attempted to acquire lands holding such specimens. Since the Conservancy makes frequent transfers of its properties to other environmental protection agencies, such as the University of California and county park systems (for instance: Bumpy Camp Nature Preserve, 160 acres, to Napa College; Jacks Peak Regional Park, 525 acres, to Monterey County Parks and Recreation Department; and Robert Lee Sims Preserve, 323 acres, to U.S. Fish and Wildlife Service) the number of preserves held varies annually. Presently, however, the Conservancy holds, manages, or has placed easements on 52 California areas ranging in size from the tiny Bunnell Vernal Pools, 1 acre, and Spindrift Point, 3.47 acres, to the immense 180,000 acre Carizzo Plain (90,361 acres now held cooperatively with other agencies, 180,000 acres by 1997).

In recent months, the California Conservancy has collaborated with other public agencies to protect 7 threatened areas: Ten-Mile Dunes in Mendocino County, Manzanita Hill in Amador County, Grimshaw Lake in Inyo County, Mission Creek in Riverside County, Coal Canyon in Los Angeles County, and Rice Canyon and Spooner's Mesa in San Diego County. It is also in the process of acquiring 6 sites as preserves: Fall River in Shasta County, Sacramento River areas in Yuba County, Haystack Mountain in Merced County, Cambria Pines in San Luis Obispo County, Guadalupe Dunes in Santa Barbara County, and San Diego Vernal Pools in San Diego County.

While none of the easement areas allow for visitation, practically all of the preserves do permit on-site field research, study, and visit by prior arrangement. Ten preserves may be visited on a "walk-on" basis: McCloud River, Northern California Coast Range, Fairfield Osborn, Ring Mountain, Kaweah Oaks, Creighton Ranch, Kern River, Big Morongo Canyon, Santa Rosa Plateau, and Coachella Valley.

See: Bibliography 67.

	Preserve	County	Acres
1	Lanphere-Christensen Dunes	Humboldt	456
2	McCloud River	Shasta	2,330
3	Inks Creek Ranch	Tehama	40
4	Gray Davis-Dye Creek	Tehama	37,540
5	Kopta	Tehama	706
6	Vina Plains	Tehama	1,950
7	North Coast Range	Mendocino	7,320
8	Pygmy Forest	Mendocino	632
9	Sacramento River Oxbow	Glenn	94
10	Boggs Lake	Lake	182
11	Russian River	Sonoma	168
12	Fairfield Osborn	Sonoma	210
13	Bunnell Vernal Pools	Sonoma	1
14	Jepson Prairie	Solano	1,566
15	Cosumnes River	Sacramento	1,454
16	Bishop Pine	Marin	400
17	Spindrift Point	Marin	347
18	Ring Mountain	Marin	377
19	Bennett Juniper	Tuolumne	372
20	Bonny Doon	Santa Cruz	526
21	Struve Pond	Santa Cruz	20
22	Flying M Ranch	Merced	2,400
23	Elkhorn Slough	Monterey	432
24	Landels-Hill Big Creek	Monterey	3,848
25	Kaweah Oaks	Tulare	311
26	Creighton Ranch	Tulare	3,280
27	Pixley Vernal Pools	Tulare	40
28	Carrizo Plain	San Luis Obispo	90,361*
29	Paine Wildflower-Semitropic Ridge	Kern	1,133
30	Williams Wildlife	Kern	40
31	Kern River	Kern	1,194
32	Amargosa River	Inyo	1,320
33	Lokern	Kern	1,994
34	Hibberd	San Luis Obispo	1,470
35	Kern Lake	Kern	83
36	Sand Ridge	Kern	117
37	Desert Tortoise	Kern	24,577
38	Nipomo Dunes	San Luis Obispo	2,550
39	Santa Cruz Island	Santa Barbara	54,488

(continued)

* 180,000 planned

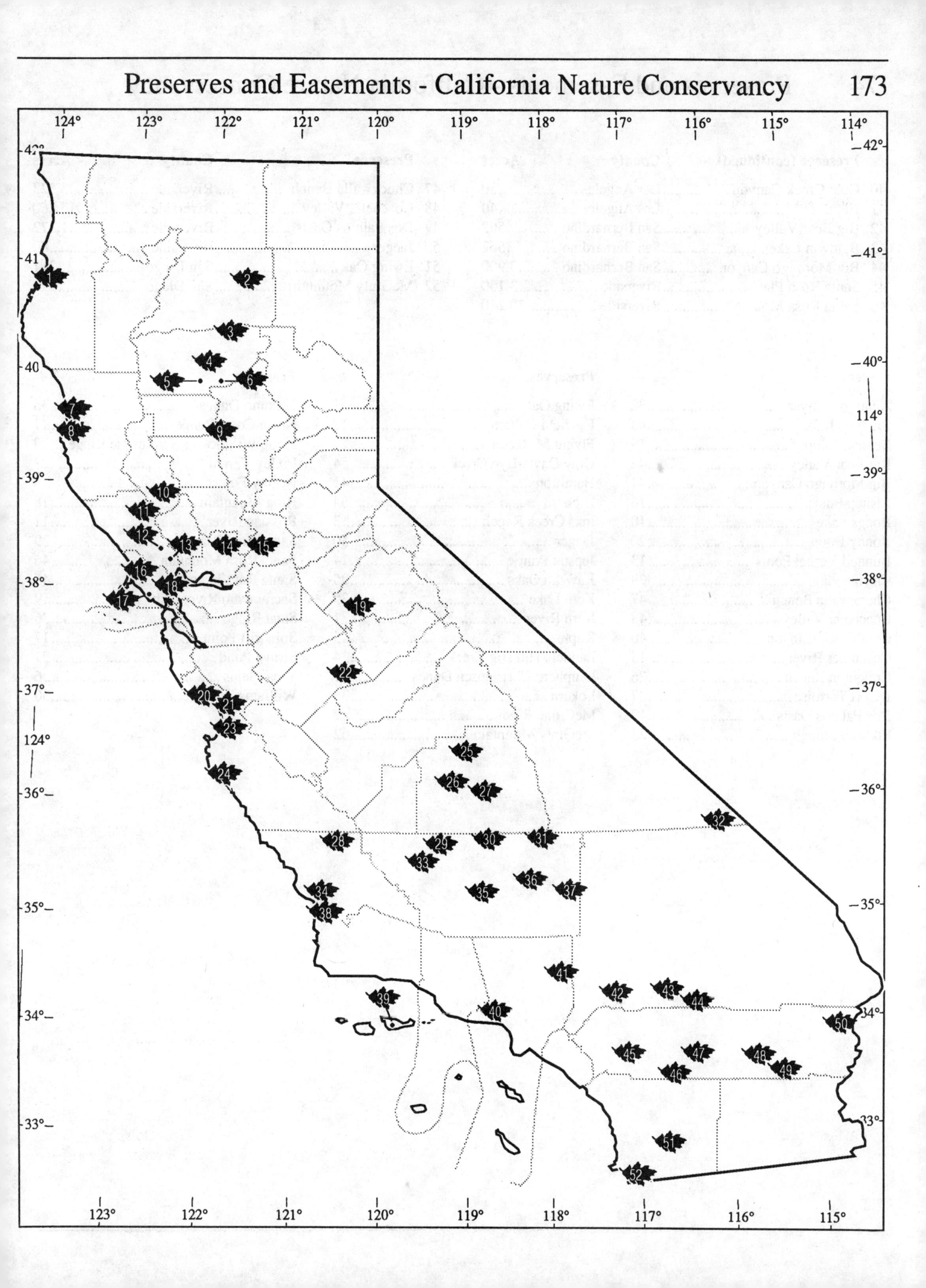
124° 123° 122° 121° 120° 119° 118° 117° 116° 115° 114°
42° 41° 40° 39° 38° 37° 36° 35° 34° 33°
1 2 3 4 5 6 7 8 9 10 11 12 13 14 15 16 17 18 19 20 21 22 23 24 25 26 27 28 29 30 31 32 33 34 35 36 37 38 39 40 41 42 43 44 45 46 47 48 49 50 51 52
123° 122° 121° 120° 119° 118° 117° 116° 115°

	Preserve (continued)	County	Acres
40	Cold Creek Canyon	Los Angeles	30
41	Hamilton	Los Angeles	40
42	Big Bear Valley	San Bernardino	562
43	Baldwin Lake	San Bernardino	562
44	Big Morongo Canyon	San Bernardino	3,900
45	Santa Rosa Plateau	Riverside	3,100
46	Santa Rosa Mountain	Riverside	10,240
47	Chuckwalla Bench	Riverside	3,523
48	Coachella Valley	Riverside	13,260
49	Dos Palmos Oasis	Riverside	1,372
50	Jaeger	Riverside	80
51	Ewing Oak	San Diego	211
52	McGinty Mountain	San Diego	914

Preserve	#
Amargosa River	32
Baldwin Lake	43
Bennett Juniper	19
Big Bear Valley	42
Big Morongo Canyon	44
Bishop Pine	16
Boggs Lake	10
Bonny Doon	20
Bunnell Vernal Pools	13
Carrizo Plain	28
Chuckwalla Bench	47
Coachella Valley	48
Cold Creek Canyon	40
Cosumnes River	15
Creighton Ranch	26
Desert Tortoise	37
Dos Palmos Oasis	49
Elkhorn Slough	23
Ewing Oak	51
Fairfield Osborn	12
Flying M Ranch	22
Gray Davis-Dye Creek	4
Hamilton	41
Hibberd	34
Inks Creek Ranch	3
Jaeger	50
Jepson Prairie	14
Kaweah Oaks	25
Kern Lake	35
Kern River	31
Kopta	5
Landels-Hill Big Creek	24
Lanphere-Christensen Dunes	1
Lokern	33
McCloud River	2
McGinty Mountain	52
Nipomo Dunes	38
North Coast Range	7
Paine Wildflower-Semitropic Ridge	29
Pixley Vernal Pools	27
Pygmy Forest	8
Ring Mountain	18
Russian River	11
Santa Cruz Island	39
Santa Rosa Mountain	46
Santa Rosa Plateau	45
Sacramento River Oxbow	9
Sand Ridge	36
Spindrift Point	17
Struve Pond	21
Vina Plains	6
Williams Wildlife	30

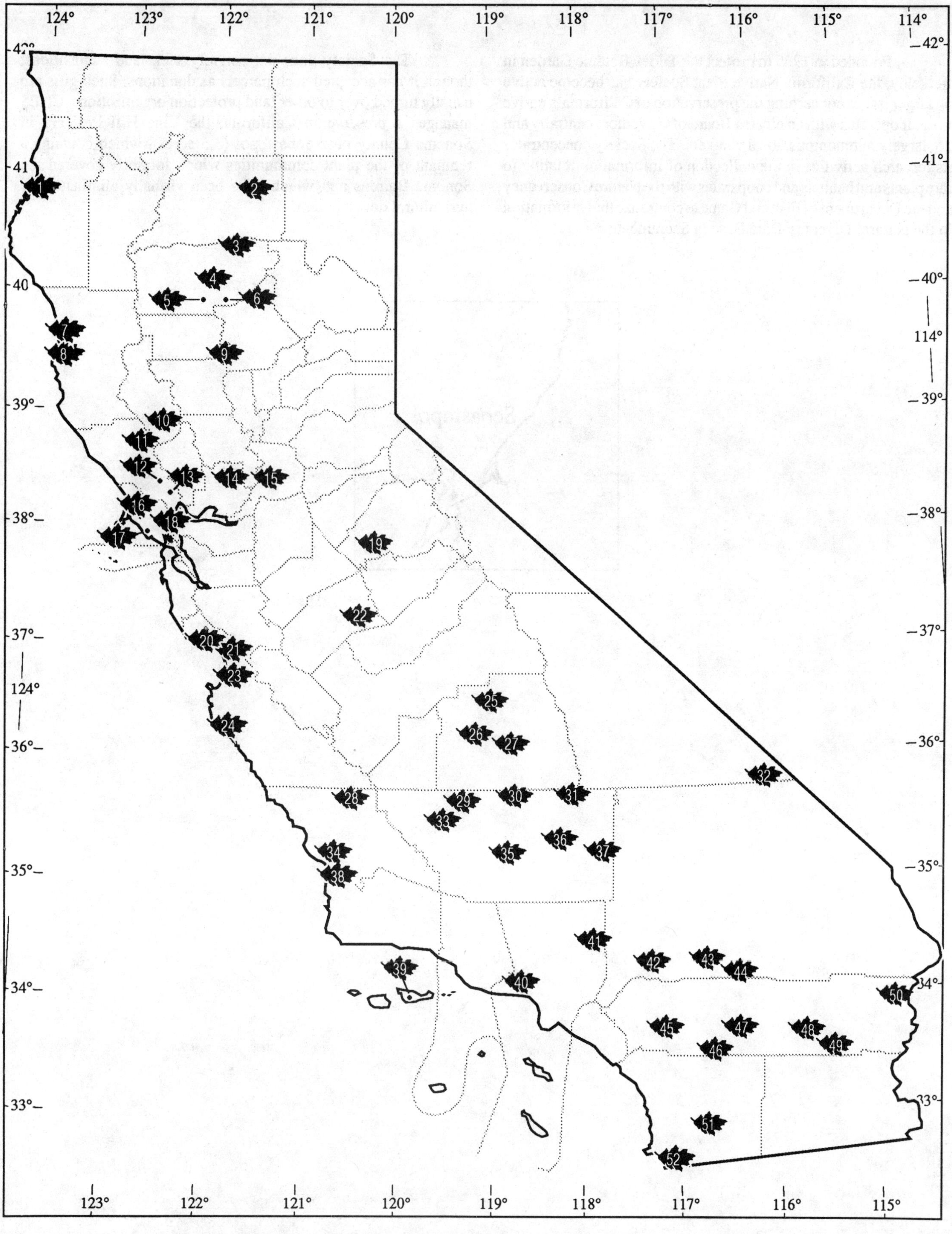
124°
123°
122°
121°
120°
119°
118°
117°
116°
115°
114°
42°
41°
40°
39°
38°
37°
36°
35°
34°
33°
1
2
3
4
5
6
7
8
9
10
11
12
13
14
15
16
17
18
19
20
21
22
23
24
25
26
27
28
29
30
31
32
33
34
35
36
37
38
39
40
41
42
43
44
45
46
47
48
49
50
51
52

Founded in 1965 to protect the Tilden Botanic Garden in Berkeley, the California Native Plant Society has become active on larger issues concerning the preservation of California's native flora. It operates with an elected Board of Governors centrally and 26, largely autonomous, local chapters. The Society concentrates its research activities in the collection of information relating to rare plants and habitats and cooperates with the Nature Conservancy and the Department of Fish and Game to centralize that information in the Natural Diversity Data Base in Sacramento.

The Society does not actively seek land acquisitions, though it has accepted such parcels as donations. Such gifts are usually turned over to other land protection organizations. CNPS manages a preserve in California, the Vine Hill Preserve in Sonoma County near Sebastopol (3.5 acres), which contains a remnant of the plant communities which formerly covered the Sonoma Barrens and which have been virtually eliminated by agricultural development.

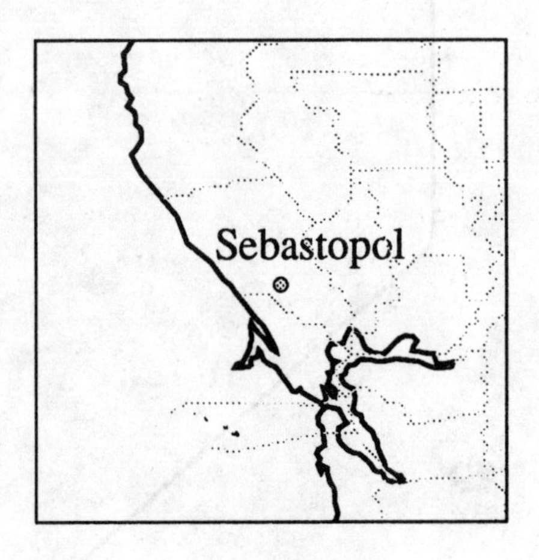

The objectives of the National Wildlife Federation, founded in 1936, are "to encourage the intelligent management of the life sustaining resources of the earth — its productive soil — and its dependent wildlife — and to promote and encourage the knowledge and appreciation of these resources; their interrelationship and wise use, without which there can be little hope for a continuing abundant life." The Federation claims a membership in excess of 4.5 million, with headquarters in Washington, D.C.; 13 regional offices (the Western Regional Office, Gardnerville, Nevada, directs operations for Arizona, California, Guam, Hawaii, and Nevada); and affiliate organizations in each state. Federation activities include 1) conservation education through publication (Journals: *National Wildlife, International Wildlife, Ranger Rick)*, school workshops, and nature centers; 2) legal consultation, including lobbying and legislative activity; 3) research, including direct research programs and graduate scholarships; and 4) wildlife habitat protection, including land acquisitions and backyard designations as wildlife preserves.

Land for acquisition is based upon its value as wildlife habitat according to a tripartite category: 1) property of high wildlife habitat or natural resource value for conservation areas, 2) property of limited wildlife habitat or natural resource value for lease or concession agreement for revenue, and 3) property of limited or no wildlife habitat or natural resource value but with some economic value for resale. The Federation manages 16 properties in the United States with 1 in California, the Lava Lakes Wildlife Area and Nature Center in Siskiyou County.

With 140,000 miles of rail line already out of service, and the abandonment of 3,000 to 4,000 miles of track each year, the Rails to Trails Conservancy was organized in 1985 in an effort to convert some of these right-of-way corridors to recreational trails. The Conservancy is aided in these efforts by a section of the National Trails System Act of 1968, which provides for the "banking" of abandoned lines for possible future restoration of rail service. It notes that the Secretary of Transportation, the Chairman of the Interstate Commerce Commission, and the Secretary of the Interior "shall encourage state and local agencies and private interests to establish appropriate trails...to preserve established railroad right-of-way for [possible] future reactivation of rail service." In addition, a California bill is before the legislature to inventory rail lines so that routes may be preserved in the event of abandonment. The legislation hopes to avoid a repetition of the insanity which destroyed the old Pacific Electric "Red Car" lines, dooming the Los Angeles region to its current car induced poisonous air situation.

One hundred and sixty rail corridors have already been converted by the RTC and local agencies into parks, bikeways, and foot trails in 31 states, and more than 150 other conversions are being developed. California holds 22 such trails, the majority of which are short stretches under 5 miles, but with 2 over 20 miles and the remainder between 5 and 10 miles.

	Trail	**Town or City**	**Length (miles)**
1	Hammond Trail	McKinleyville	3.0
2	Bizz Johnson Trail	Susanville to Westwood	24.5
3	Paradise Memorial Trailway	Paradise	5.0
4	Truckee River Bike Trail	Tahoe City	3.5
5	Sacramento-RioLinda Bikeway	Sacramento to Rio Linda	8.0
6	Larkspur to Corte Madera Bike Path	Larkspur to Corte Madera	5.0
7	Tiburon Lineal Park	Tiburon	3.2
8	Mill Valley Bike Trail	Corte Madera tunnel to Coyote Creek	2.5
9	Old Railroad Grade	Mill Valley to Mt. Tamalpais	9.0
10	Lafayette/Moraga Trail	Lafayette to Moraga	7.6
11	San Ramon Valley Iron Horse Trail	Alamo to San Ramon	21.0
12	Shepard Canyon Trail	Oakland	3.0
13	Monterey Peninsula Rec Trail	Monterey to Pacific Grove	2.3
14	Mt. Lowe Railroad Trail	Rubio Canyon to Mt. Lowe	8.0
15	Ojai Valley Trail	Ventura to Ojai	9.5
16	Duarte Bike Trail	Duarte	1.5
17	Hermosa Beach Trail	Hermosa Beach	2.0
18	Juanita Cook Trail	Fullerton	2.0
19	Sally Pekarek Trail	Fullerton	4.0*
20	Bud Turner Trail	Fullerton	3.8
21	Alton Avenue Bike Trail	Santa Ana	0.4
22	Silver Strand Bikeway	Coronado to Imperial Beach	7.0

*1 mile on former railroad.

124° 123° 122° 121° 120° 119° 118° 117° 116° 115° 114°

42° 41° 40° 39° 38° 37° 36° 35° 34° 33°

1
2
3
4
5
6, 7, 8, 9
10,11
12
13
14
15
16
17
18, 19, 20
21
22

123° 122° 121° 120° 119° 118° 117° 116° 115°

Founded in 1892, with John Muir, a preeminent naturalist and writer, as first president, the Sierra Club began its career with a small band of conservation enthusiasts determined to protect the magnificent wilderness areas of the Sierra Nevada, particularly Yosemite and its environs. Today, about half a million strong, the club's volunteers maintain one of the strongest, most effective conservation programs in the world. The bulk of the club's programs are carried forward by 58 chapters across the United States and Canada (13 in California).The Club is involved with the protection of the earth throughout the planet, with direct action only on the North American continent. As a grass roots organization, the members determine policies and activities, and are assisted in these endeavors by 14 professionally staffed field offices (2 in California) and a major political watch-dog office in Washington, D.C. Headquarters is located in San Francisco, California, and staff operations are divided into departments: Conservation, Outings, Membership, Development, Public Affairs, and Volunteer Development. Four other organizations are closely associated with the club, but with separate legal entities: 1) Sierra Club Legal Defense Fund which manages environmental litigation on behalf of the club and other conservation organizations, 2) Sierra Club Political Committee which directs the club's electoral activities, 3) The Sierra Club Foundation which provides funding for programs other than political lobbying, such as expenses related to testimony, research, education, and publication, and 4) Sierra Club Books, a publishing house specializing in the printing of environmental and nature studies. The Club does not hold or manage preserves, but it does own several small parcels set aside primarily for recreational purposes.

The purposes of the Sierra Club have been formally enunciated: to explore, enjoy and protect the wild places of the earth; to practice and promote the responsible use of the earth's ecosystems and resources; to educate and enlist humanity to protect and restore the quality of the natural and human environment; and to use all lawful means to carry out these objectives.

See: Bibliography 26, 33.

	Chapter	**Founded**
1	Redwood Chapter	1958
2	Mother Lode Chapter	1939
3	San Francisco Bay Chapter	1924
4	Loma Prieta Chapter	1933
5	Tehipite Chapter	1953
6	Ventana Chapter	1963
7	Santa Lucia Chapter	1968
8	Kern Kaweah Chapter	1952
9	Toiyabe Chapter	1957
10	Los Padres Chapter	1952
11	Angeles Chapter	1911
12	San Gorgonio Chapter	1932
13	San Diego Chapter	1948

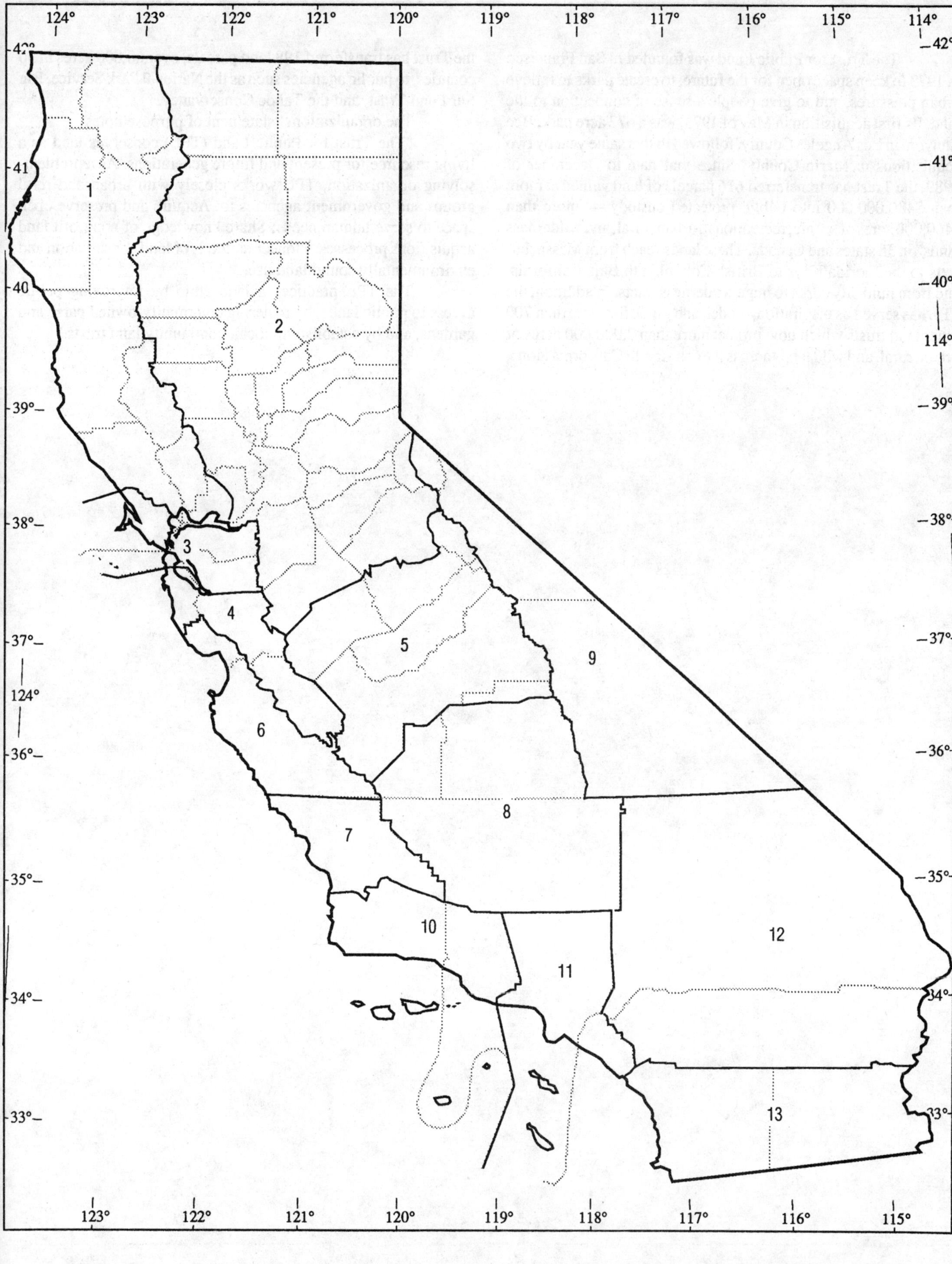

124° 123° 122° 121° 120° 119° 118° 117° 116° 115° 114°

42° 41° 40° 39° 38° 37° 36° 35° 34° 33°

1 2 3 4 5 6 7 8 9 10 11 12 13

123° 122° 121° 120° 119° 118° 117° 116° 115°

The Trust for Public Land was founded in San Francisco in 1973 to keep space open for the future, to create parks to relieve urban pressures, and to give people a sense of connection to the land. Its first acquisition in May of 1973, was a 672 acre park, Bee Canyon, in Los Angeles County, followed in that same year by two acquisitions in Marin County. Since that date to December of 1989, the Trust has transferred 616 parcels of land valued at more than $480,000,000 into public protected custody — more than "460,000 acres of scenic, recreational, urban, rural, and wilderness lands" in 35 states and Canada. These lands reach from Massachusetts to the Florida Keys to British Columbia to Baja California, and from mini city parks to huge wilderness tracts. In addition, the TPL has served as inspiration, model, and guide for more than 700 other land trusts which now protect more than 2,000,000 acres of urban, rural, and wild lands across the country. In California alone, the Trust has transferred 198 land parcels, over 60,000 acres in 30 counties, to public agencies such as the National Park Service, Big Sur Land Trust, and the Tahoe Conservancy.

The organizations' statement of purpose notes:

"The Trust for Public Land (TPL) conserves land as a living resource for present and future generations. As a problem-solving organization, TPL works closely with urban and rural groups and government agencies to: Acquire and preserve open space to serve human needs; Share knowledge of nonprofit land acquisitions processes; Pioneer methods of land conservation and environmentally sound land use.

The TPL practices a land ethic by improving public access to public lands, by fostering community-owned parks and gardens, and by establishing local, community land trusts."

The Planning & Conservation League, a non-profit statewide alliance of citizens and more than 100 conservation organizations united to protect and restore the quality of California's environment through legislation and administrative action, celebrated its 25th anniversary in 1990. In 1965, a small band of Bay Area conservationists recognized the need for an office to advocate protection of the California environment before the legislature. The legislature at that period was almost totally oblivious of the pollution, smog, coastal destruction, wetland encroachment, uncontrolled growth and the myriad other environmental problems which beset the state. Consequently, members of the California Roadside Council, an organization devoted to the elimination of highway billboards, and the beautification of the state's roadways and cities, along with the Sierra Club and several other conservation groups, formed the Planning & Conservation League.

In the past 25 years the PCL has been a major participant in the enactment of the California Environmental Quality Act, the state clean water and air acts, recycling legislation, the oil spill prevention bill, rollcall votes in legislative committees, and the defeat of the Through Delta Canal. In addition, it has been a sponsor of the June 1988, Proposition 70, The California Wildlife, Coastal, and Parklands Bond Act; the November 1988, Proposition 99, Tobacco Tax Initiative; and the June 1990, Propositions 116 and 117, for public rail transportation and the elimination of trophy hunting of the California Mountain Lion.

Future programs planned by the PCL include wildlife protection and park acquisition, maintenance of agricultural land and open space, development of public transport, growth planning, air quality enhancement, and water conservation.

Planning & Conservation League
909 12th Street #203
Sacramento, CA, 95814
916-444-8726

Injured and orphaned wildlife require very special care, handling, and diet, which are not available in the average household. It is a mistake for an untrained person to attempt to care for any wild creature, particularly an injured one. A call to a licensed rehabilitation center will provide information and the best hope for the full recovery and eventual release of the creature to the wild.

REGION 1
DF&G
601 Locust Avenue
Redding, CA 96001
916-225-2300

Shasta Wildlife Rescue
P.O. Box 757
Redding, CA 96099
916-246-9453

Humboldt Wildlife Care Center
P.O. Box 4141
Arcata, CA 95521
Raptor and Mammal Team
707-839-0295
Marine Mammal Team
707-839-4847
Songbird and Seabird Team
707-822-7552

Burney Falls Wildlife Rescue
P.O. Box 29,
20129 Mowry Lane
Burney, CA 96013
916-335-3855

REGION 2
DF&G
1701 Nimbus Road
Rancho Cordova, CA 95670
916-355-0978

UCD Raptor Center
c/o VMTH Business Office
Davis, CA 95616
916-752-6091

Davis Wildlife Care Assn.
P.O. Box 676
Davis, CA 95617

Sacramento Wildlife Care Assoc.
3615 Auburn Boulevard
Sacramento, CA 95821
916-383-7922

Bidwell Nature Center
Bidwell Park
Chico, CA 95926
916-342-3710

Lake Tahoe Wildlife Care, Inc.
P.O. Box 10557
South Lake Tahoe, CA 95705
916-577-2273

Amador County Wildlife Care Center
P.O. Box 362
Jackson, CA 95642
209-296-4218

San Joaquin County Zoological Society Wildlife Rehabilitation Group
11793 Mickey Grove Road
Lodi, CA 95240
209-462-3702

Five Mile Creek Raptor Center
1851 West Lincoln Road
Stockton, CA 95240
209-477-0602

Feather River Wildlife Care
P.O. Box 1964
Marysville, CA 95901
916-695-1788

Sacramento Science Center
3615 Auburn Boulevard
Sacramento, CA 95821
916-449-8255

REGION 3
DF&G
P.O. Box 47
Yountville, CA 94599
707-944-5500

Alexander Lindsey Junior Museum of Natural Science
1901 First Avenue
Walnut Creek, CA 94596
415-935-1978

Central Coast Rehabilitation Guild
P.O. Box 3257
San Luis Obispo, CA 93403

Napa Valley Naturalists
1791 G Street
Napa, CA 94558
707-224-9300

The Humane Society of Santa Clara Valley
2530 Lafayette Street
Santa Clara, CA 95050
408-727-3383

International Bird Rescue Research Center
2701 8th Street Aquatic Park Avenue
Berkeley, CA 94710
415-841-9086

Jan Peter
2552 Oakwood Court
Napa, CA 94559
707-255-8413

Marin Humane Society
171 Bel Marin Keys Blvd.
Novato, CA 94947
415-883-4621

Marin Wildlife Center
76 Albert Park Lane
San Rafael, CA 94915
415-454-6961

Monterey County SPCA
P.O. Box 3058
Monterey, CA 93940
408-373-2631

Native Animal Rescue of Santa Cruz
2200 Seventh Avenue
Santa Cruz, CA 95062
408-462-0726

San Benito County SPCA
P.O. Box 1074
Hollister, CA 95024
408-637-8635

San Francisco SPCA
2500 16th Street
San Francisco, CA 94103
414-554-3000

Suisun Marsh Natural History Association
1171 Kellogg Street
Suisun, CA 94585
707-425-4158

Wendy Harget
1703 Hunters Point Road
Upper Lake, CA 95485

Wildlife Rescue, Inc.
4000 Middlefield Road
Palo Alto, CA 94303
415-494-7283

Youth Science Institute
296 Garden Hill Drive
Los Gatos, CA 95030
408-356-4945

(continued)

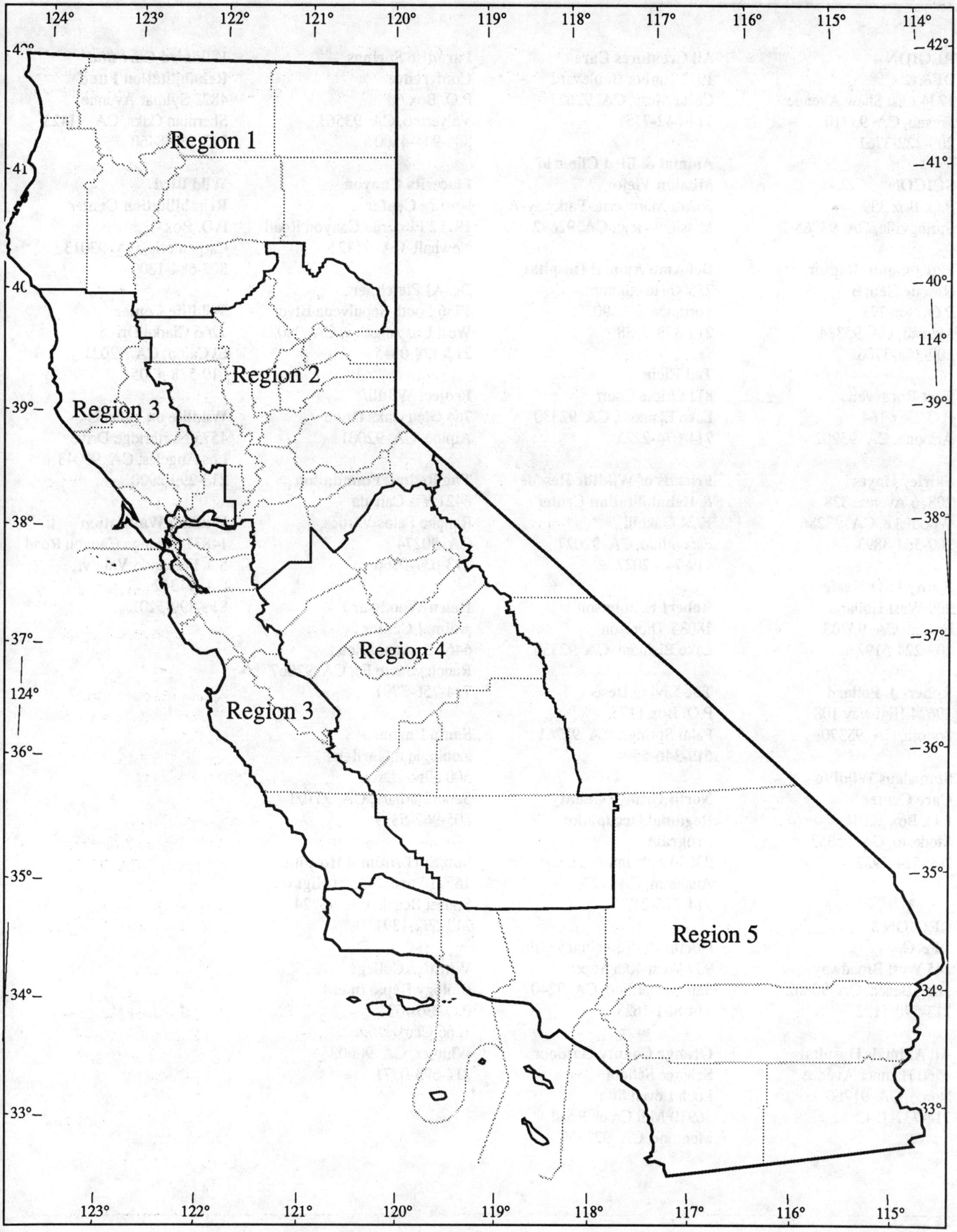

Region 1

Region 2

Region 3

Region 4

Region 3

Region 5

REGION 4
DF&G
1234 East Shaw Avenue
Fresno, CA 93710
209-222-3761

SCICON
P.O. Box 339
Springville, CA 93265

San Joaquin Raptor Rescue Center
P.O. Box 778
Merced, CA 95314
209-358-3706

Fort Roosevelt
P.O. Box 164
Armona, CA 93202

Shirley Hayes
20876 Avenue 328
Woodlake, CA 93286
209-564-3893

Kathy L. O'Keefe
209 West Holland
Fresno, CA 93705
209-224-5192

Robert J. Pollard
19624 Highway 108
Sonora, CA 95370

Stanislaus Wildlife Care Center
P.O. Box 1201
Modesto, CA 95353
209-524-7922

REGION 5
DF&G
245 West Broadway
Long Beach, CA 90802
213-590-5132

All Animals Hospital
1560 Hamner Avenue
Norco, CA 91760
714-737-1242

All Creatures Care
1912 Harbor Boulevard
Costa Mesa, CA 92627
714-642-7151

Animal & Bird Clinic of Mission Viejo
25290 Marguerite Parkway-A
Mission Viejo, CA 92692

Del Amo Animal Hospital
23500 Hawthorne
Torrance, CA 90505
213-378-9338

Ted Eich
811 Oriole Court
Lake Elsinore, CA 92330
714-674-2225

Friends of Wildlife Rescue & Rehabilitation Center
1724 Oakhill
Escondido, CA 92027
619-743-2027

Robert E. Johnson
18085 Thoreson
Lake Elsinore, CA 92330

The Living Desert
P.O. Box 1175
Palm Springs, CA 92261
619-346-5694

North Orange County Regional Occupation Program
2360 La Palma Avenue
Anaheim, CA 92801
714-776-2170

Northside Veterinary Clinic
939 West 40th Street
San Bernardino, CA 92407
714-881-1623

Orange County Outdoor Science School
Loch Leven Site
36910 Mill Creek Road
Mentone, CA 922359

Paradise Springs Conference
P.O. Box 68
Valyermo, CA 93563
805-944-4500

Placerita Canyon Nature Center
19152 Placerita Canyon Road
Newhall, CA 91321

Dr. Al Pletchner
1736 South Sepulveda Blvd.
West Los Angeles, CA 90025
213-478-0545

Project Wildlife
764 Glen Oaks Drive
Alpine, CA 92001

The Redtail Foundation
6421 Via Canada
Rancho Palos Verdes, CA 90274
213-637-9566

Helen Woodward Animal Center
6461 El Apajo Road
Rancho Sante Fe, CA 92067
714-756-3791

Santa Barbara Zoological Gardens
500 Ninos Drive
Santa Barbara, CA 93103
805-962-5339

Sun Surf Animal Hospital
16571 Pacific Coast Highway
Sunset Beach, CA 90724
213-592-1391

Whittier College Biology Department
P.O. Box 634
1266 Clark Drive
Whittier, CA 90608
213-693-0771

Wildbird Care and Rehabilitation Fund
4822 Sylmar Avenue
Sherman Oaks, CA 91423
213-789-8460

Wild Bird Rehabilitation Center
P.O. Box 4
Carpenteria, CA 93013
805-684-1300

Wildlife Center
1266 Clarke Drive
El Cajon, CA 92021
619-588-4705

Wildlife on Wheels
4575 Northridge Drive
Los Angeles, CA 90043
213-299-2900

Wildlife Waystation
14831 Tujunga Canyon Road
San Fernando Valley, CA 91342
818-899-5201

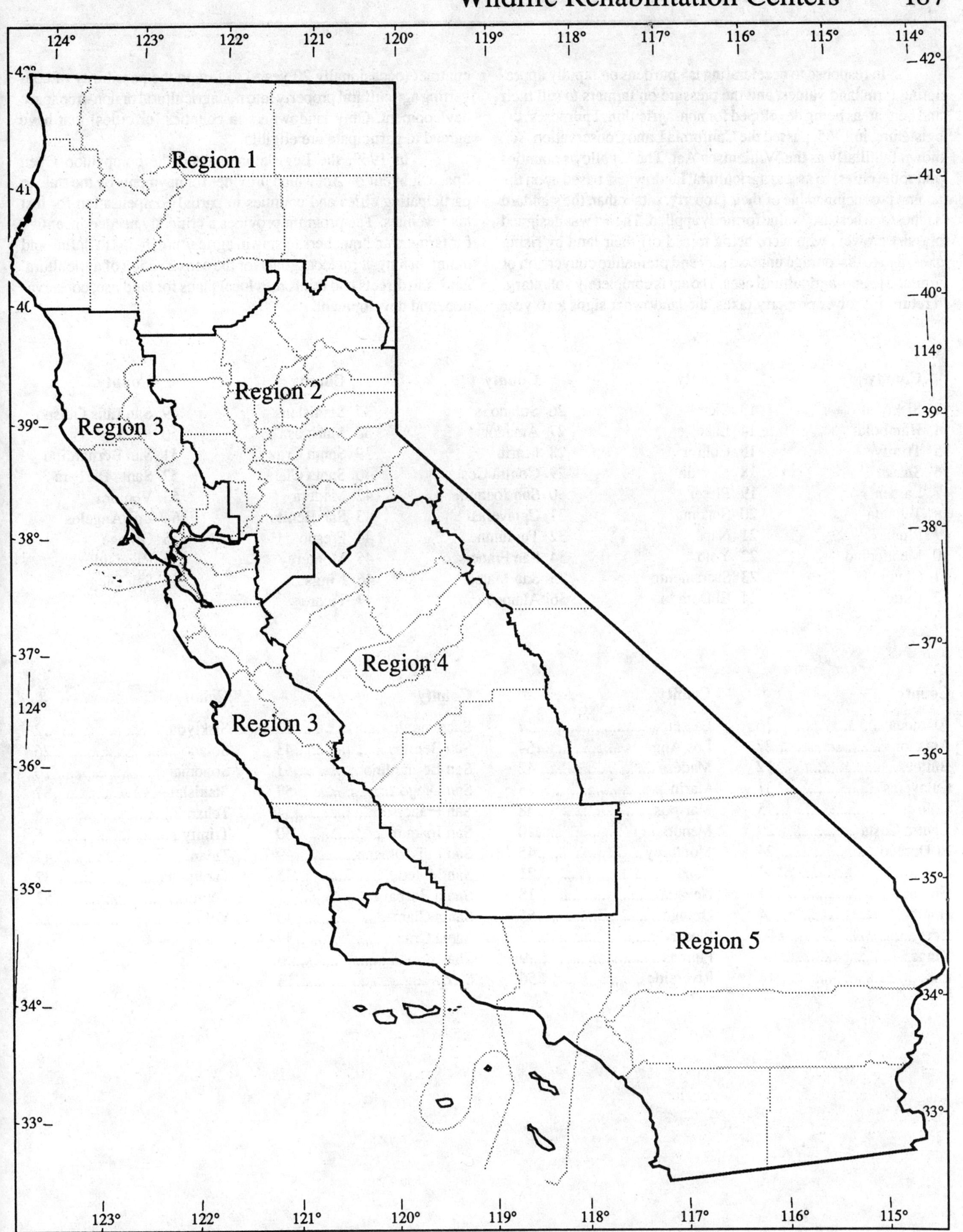
Region 1
Region 2
Region 3
Region 3
Region 4
Region 5
124°
123°
122°
121°
120°
119°
118°
117°
116°
115°
114°
42°
41°
40°
39°
38°
37°
36°
35°
34°
33°

Williamson Act Counties

In response to accelerating tax burdens on rapidly appreciating farm land values, and the pressure on farmers to sell their land near areas being developed for non-agricultural purposes, the legislature, in 1965, passed the California Land Conservation Act, known familiarly as the Williamson Act. The act allows counties (and some cities) to assess agricultural landowners based upon the income producing value of their property, rather than the standard "highest and best use" value formerly applied. The act was designed to assist growers who were being forced off their land by rising taxes, and to discourage unnecessary and premature conversion of farmland to non-agricultural uses. The act is completely voluntary. In return for lower property taxes, the landowner signs a 10 year contract (occasionally 20 years) to forego the possibility of converting agricultural property into nonagricultural or non-free space devleopment. Only landowners in counties (or cities) that have agreed to participate are eligible.

In 1969, the Legislature passed the companion Open Space Subvention Act which provides for payments by the state to participating cities and counties in partial compensation for lost tax revenues. The program provides a critically needed incentive for farmers and ranchers to stay in agriculture, helps to define and maintain long-term local goals for the conservation of agricultural land, and directs and reinforces local plans for land use, conservation, and development.

	County		County		County		County		County
2	Siskiyou	13	Sierra	26	Solano	37	Stanislaus	49	San Luis Obispo
4	Humboldt	14	Lake	27	Amador	38	Mariposa	50	Kern
5	Trinity	15	Colusa	28	Marin	39	Santa Cruz	51	San Bernardino
6	Shasta	18	Nevada	29	Contra Costa	40	Santa Clara	52	Santa Barbara
7	Lassen	19	Placer	30	San Joaquin	42	Madera	53	Ventura
8	Tehama	20	Sonoma	31	Calaveras	43	San Benito	54	Los Angeles
9	Plumas	21	Napa	32	Tuolumne	44	Fresno	55	Orange
10	Mendocino	22	Yolo	34	San Francisco	45	Monterey	56	Riverside
11	Glenn	23	Sacramento	35	San Mateo	46	Kings	57	San Diego
12	Butte	24	El Dorado	36	Alameda	47	Tulare		

County	#	County	#	County	#	County	#
Alameda	36	Lassen	7	Sacramento	23	Siskiyou	2
Amador	27	Los Angeles	54	San Benito	43	Solano	26
Butte	12	Madera	42	San Bernardino	51	Sonoma	20
Calaveras	31	Marin	28	San Diego	57	Stanislaus	37
Colusa	15	Mariposa	38	San Francisco	34	Tehama	8
Contra Costa	29	Mendocino	10	San Joaquin	30	Trinity	5
El Dorado	24	Monterey	45	San Luis Obispo	49	Tulare	47
Fresno	44	Napa	21	San Mateo	35	Tuolumne	32
Glenn	11	Nevada	18	Santa Barbara	52	Ventura	53
Humbolt	4	Orange	55	Santa Clara	40	Yolo	22
Kern	50	Placer	19	Santa Cruz	39		
Kings	46	Plumas	9	Shasta	6		
Lake	14	Riverside	56	Sierra	13		

The Z'Berg-Warren-Keene-Collier Forest Taxation Reform Act of 1976 derives from the Williamson Act. The Reform Act, which was amended in 1977, 1983, and 1984, provides for timberlands a type of tax incentive and protection enjoyed by agricultural land under the Williamson Act. Indeed, the Timberland Protection Zone Agreement is a 10-year restriction on the use of timberland, which replaced the agricultural preserve status timberlands held under the Williamson Act. These zones "are lands assessed by counties or the California State Board of Equalization for growing timber as the highest and best land use".

TPZ lands have a 10 year restriction on the use of the land to grow and harvest timber. Property taxes are generally assessed at a reduced rate based on these restrictions. Each county sets ordinances defining minimum TPZ parcel size and accepted compati-ble uses.

The intent of the law is spelled out in the California Timberland Productivity Act of 1982, which notes that it is the policy of the state to "discourage premature or unnecessary conversion of timberland to urban and other uses," to "discourage expansion of urban services into timberland," and to "encourage investment in timberlands based on a reasonable expectation of harvest."

Compatible uses were also provided in the 1982 Act which notes that such uses may include but are not limited to "management for watershed, management for fish and wildlife habitat or hunting and fishing" as well as for other measures related to timber harvest.

Twenty-eight counties hold 5,497,843 acres in Timberland Production Zones, estimated at 74 percent of the productive private forest lands of the state.

	County
1	Del Norte
2	Siskiyou
3	Modoc
4	Humboldt
5	Trinity
6	Shasta
7	Lassen
8	Tehama
9	Plumas
10	Mendocino
11	Glenn
12	Butte
13	Sierra
14	Lake
17	Yuba
18	Nevada
19	Placer
20	Sonoma
24	El Dorado
25	Alpine
27	Amador
31	Calaveras
32	Tuolumne
35	San Mateo
38	Mariposa
39	Santa Cruz
44	Fresno
47	Tulare

County	#
Alpine	25
Amador	27
Butte	12
Calaveras	31
Del Norte	1
El Dorado	24
Fresno	44
Glenn	11
Humboldt	4
Lake	14
Lassen	7
Mariposa	38
Mendocino	10
Modoc	3
Nevada	18
Placer	19
Plumas	9
San Mateo	35
Santa Cruz	39
Shasta	6
Sierra	13
Siskiyou	2
Sonoma	20
Tehama	8
Trinity	5
Tulare	47
Tuolumne	32
Yuba	17

The U.S. Department of Agriculture's Water Bank Program was conceived as a measure to aid prairie state land owners to maintain waterfowl nesting sites. The program provides financial incentives to farmers and ranchers who wish to maintain their wetlands against the threat of encroaching development or incompatible agricultural practices. In 1972, the program was applied broadly and ten states now participate: Arkansas, California, Louisiana, Minnesota, Mississippi, Montana, Nebraska, North Dakota, South Dakota, and Wisconsin.

A county must apply for eligibility. Individuals holding suitable wet areas within the county may then agree to ten year renewable contracts to maintain natural wetlands on their property and to provide wildlife habitat on land adjacent to the wetland. The owners or operators agree not to drain, burn, fill, or otherwise destroy the wetland character of areas placed under the agreement, nor to use such areas for agricultural purposes. The landowners also agree to carry out the wetland conservation plans for their lands in accordance with the terms of the agreements and not to adopt any practice that would tend to defeat the purpose of the agreements.

Land eligible to be placed under the water bank agreement must have a high priority type wetland and sufficient adjacent areas to make a viable agreement. The long term goals of the Water Bank Program include improvement of water quality and enhancement of the natural beauty of the landscape in addition to maintaining habitat for water fowl. As of the end of 1989, California landowners held 95 agreements in 10 counties protecting 6,713 acres of wetlands and 18,352 acres of adjacent lands for which they received $382,178 in annual fees.

See: Bibliography 142.

	County
2	Siskiyou
3	Modoc
7	Lassen
9	Plumas
12	Butte
13	Sierra
15	Colusa
16	Sutter
22	Yolo
41	Merced

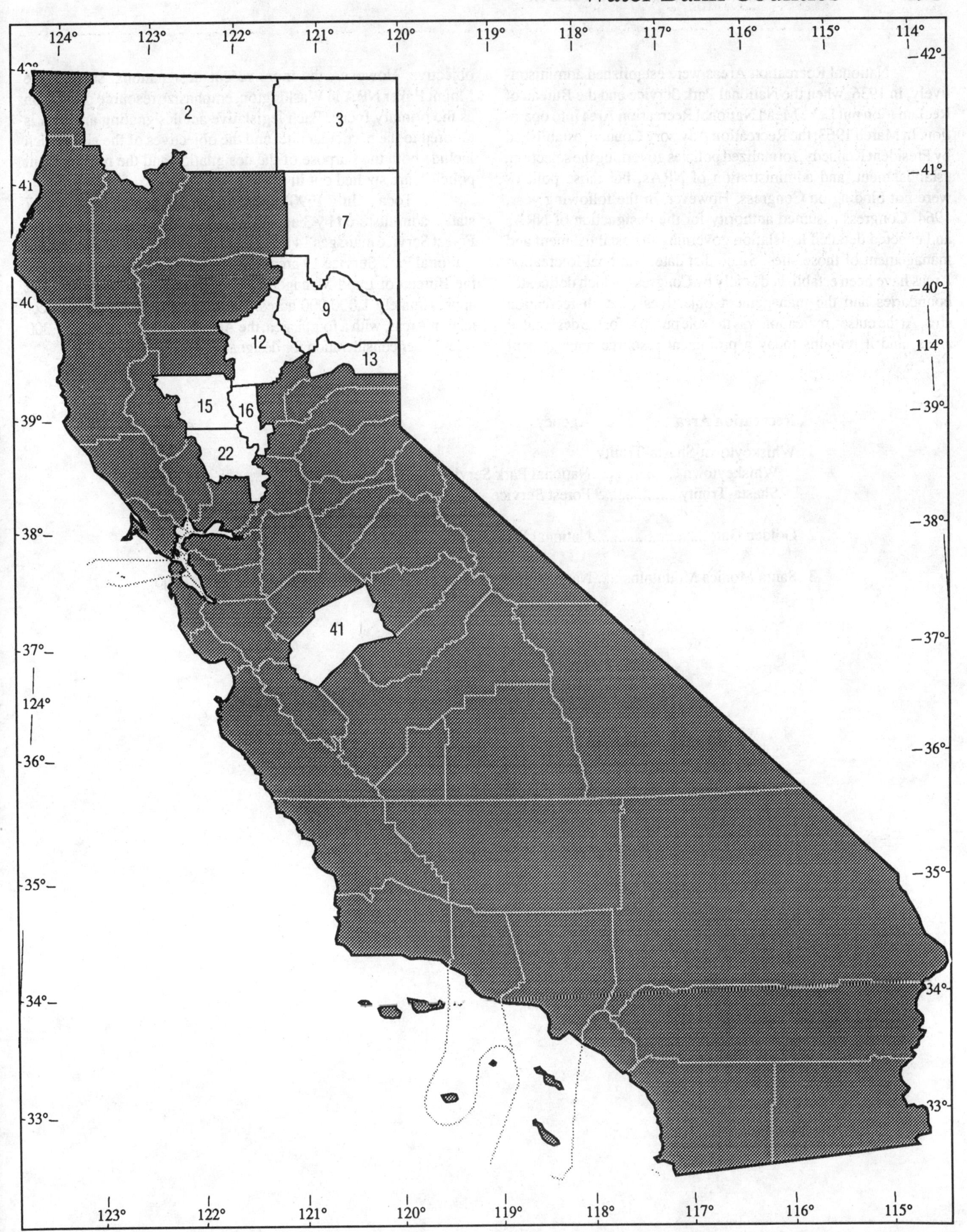
124°
123°
122°
121°
120°
119°
118°
117°
116°
115°
114°
42°
41°
40°
39°
38°
37°
36°
35°
34°
33°
2
3
7
9
12
13
15
16
22
41

National Recreation Areas were established administratively, in 1936, when the National Park Service and the Bureau of Reclamation put Lake Mead National Recreation Area into operation. In March 1963, the Recreation Advisory Council, established by President Kennedy, formalized policies governing the selection, establishment, and administration of NRAs, but those policies were not binding on Congress. However, in the following year, 1964, Congress assumed authority for the designation of NRAs and enacted detailed legislation governing the establishment and management of those sites. Since that date, National Recreation Areas have been established solely by Congress, which delineates boundaries and the management objectives of each recreation area. At the outset, recreation was the sole purpose for the designated areas, and it remains today a prominent resource management objective. However, the more recent designations, such as the Mount Baker NRA in Washington, emphasize resource protection as the primary focus. Each legislative act designating an NRA is tailored to the particular site, and the objectives of the site, which include both the purpose of the designation and the management policies, are spelled out in the legislative act.

Today (July 1990), there are 34 designated NRAs in 27 states, administered by 3 agencies of the federal government: the Forest Service manages 14 units with a total of 1,370,148 acres, the National Park Service 19 areas and a total of 3,772,708 acres, and the Bureau of Land Management 1 area in Alaska authorized at approximately 1,000,000 acres. California holds 3 National Recreation Areas, with a fourth area, the American River NRA of 81,300 acres under consideration for designation late in 1990.

	Recreation Area	Agency	County	Acres
1	Whiskeytown-Shasta-Trinity			
	Whiskeytown	National Park Service	Shasta	42,503
	Shasta-Trinity	Forest Service	Shasta	209,554
2	Golden Gate	National Park Service	Marin, San Francisco	72, 815
3	Santa Monica Mountains	National Park Service	Los Angeles, Ventura	150,000

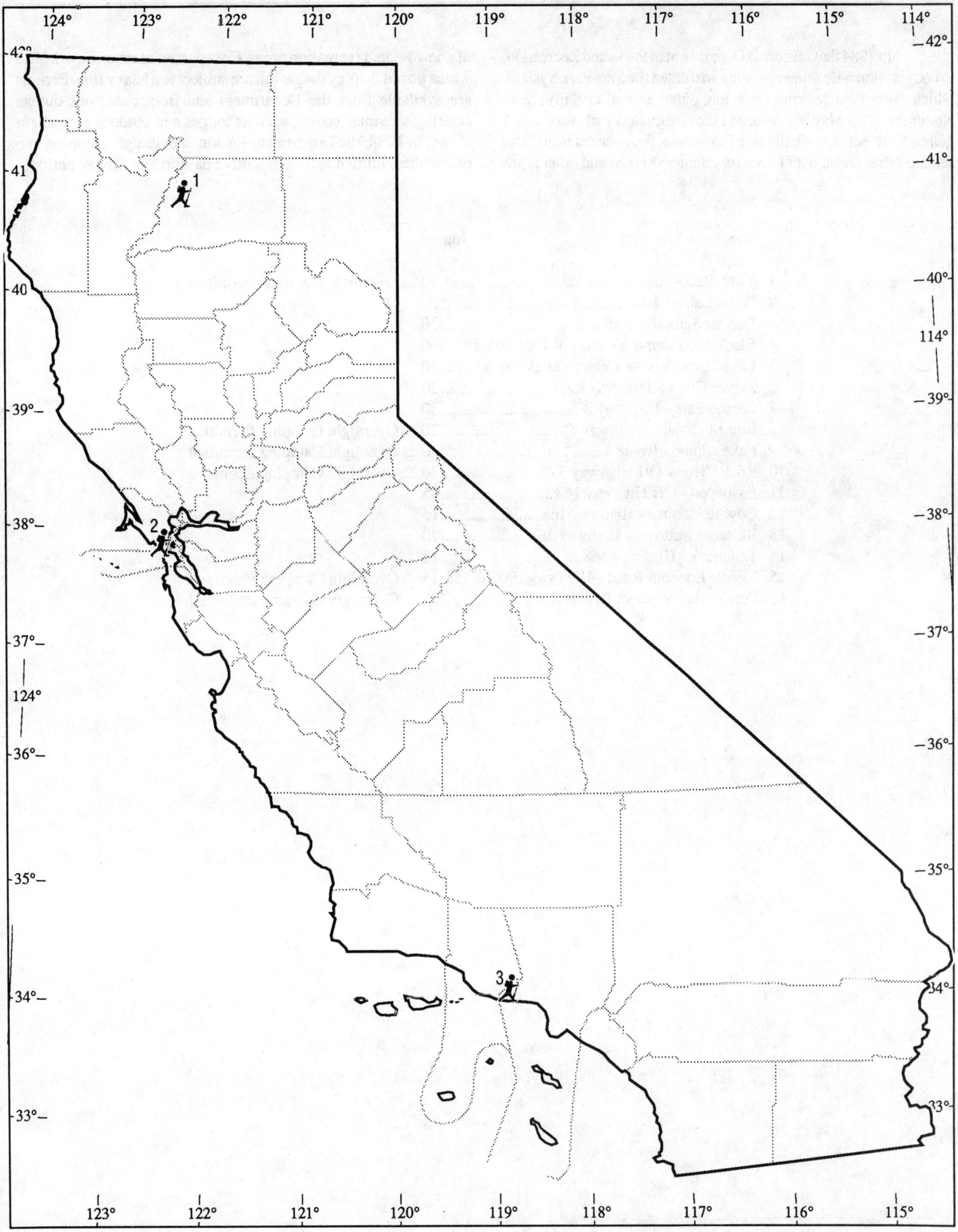
124°
123°
122°
121°
120°
119°
118°
117°
116°
115°
114°
42°
41°
40°
39°
38°
37°
36°
35°
34°
33°
1
2
3

In 1984 the California Department of Parks and Recreation, in cooperation with other agencies, instituted the Sno-Park System, which allows car parking for a fee, either annual or daily, from November 1 to May 30. Space at the designated park sites is not guaranteed, but is available on a first-come, first-served basis. The selected sites are adjacent to cross-country ski runs and other types of snow related recreation areas. Cars parked at the sno-park sites which do not display the permit are subject to a heavy fine. Permits are available from the Department and from numerous outlets catering to winter sports, such as lodges and outdoor equipment shops. In 1988, the Department instituted overnight camping at 6 of the sites, limited to 7 consecutive days in any 30 day period.

	Sno-Park Site	**Spaces**	
1	Yuba Pass - Highway 49	30	Overnight Camping Permitted
2	Eagle Lakes - I-80	30	
3	Donner Summit - I-80	30	
4	Blackwood Canyon - West of Highway 89	30	
5	Lake Tahoe Visitor Center - Highway 89	30	
6	Taylor Creek - Highway 89	30	
7	Carson Pass - Highway 88	30	
8	Iron Mountain - Highway 88	20	Overnight Camping Permitted
9	Lake Alpine - Route 4	30	Overnight Camping Permitted
10	Rock Creek - Off Highway 395	30	Overnight Camping Permitted
11	Eastwood - Off Highway 168	15	
12	Coyote Marmot - Highway 168	15	
13	Balsam Meadows - Highway 168	20	
14	Tamarack - Highway 168	15	
15	Needle Lookout Road - Highway 190	15	Overnight Camping Permitted
16	Ponderosa - Western Divide Road	15	Overnight Camping Permitted

124° 123° 122° 121° 120° 119° 118° 117° 116° 115° 114°

42° 41° 40° 39° 38° 37° 36° 35° 34° 33°

1 2 3 4 5 6 7 8 9 10 11 12 13 14 15 16

Overnight Camping Permitted

Day Use Only

123° 122° 121° 120° 119° 118° 117° 116° 115°

California's Off-Highway Motor Vehicle Recreation Program was initiated in 1971 with the Off-Highway Motor Vehicle Act. In 1982, the law was amended to establish a Division of Off-Highway Motor Vehicle Recreation within the Department of Parks and Recreation, but with a separate 7 member commission appointed jointly by the governor and the legislature. The amended law was titled the Off-Highway Motor Vehicle Recreation Act. Again in 1988, the law was amended primarily to expand off-highway vehicle facilities by subvention to local governments and joint undertakings with federal agencies, and to attempt some protection for the lands damaged by off-highway vehicles. The current law is operative until January 1, 1993. Prominent among the off-highway vehicle sites are the 7 State Vehicular Recreation Areas with a total 42,120 acres.

Areas Managed by the U.S. Forest Service

1 San Gabriel Canyon
2 Rowher Flat
3 Mill Canyon Areas
4 Little Rock
5 Corral Canyon
6 Wildomar
7 Mace Mill
8 Stumpy Meadows
9 Loon Lake
10 Barrett Lake
11 Wheeler Crest
12 Coyote Flat
13 Monache Meadows
14 Poleta
15 Scott Mountain
16 Medicine Lake Snowmobile Park
17 Deer Mountain Snowmobile Park
18 Ashpan Snowmobile Park
19 Morgan Summit Snowmobile Park
20 Bogard Snowmobile Area
21 Kings Beach
22 Blackwood Canyon
23 Alamo Mountain
24 Ballinger Canyon
25 Black Mountain (Pozo-LaPanza)
26 Santa Barbara Ranger District
27 Ortega Trail
28 Davis Flat
29 Lake Pillsbury
30 Elk Mountain Area
31 Gold Lake
32 Lake Davis
33 Lake Arrowhead Area
34 Big Bear Lake Area
35 Tule River Ranger District
36 Frog Meadow
37 Kennedy Meadows
38 Hayfork Area
39 Kings River
40 Shaver Lake Area
41 Hite Cove

U.S. Forest Service (continue)

42 Miami Creek
43 Tamarack Ridge Snowmobile Trailhead
44 Niagara Ridge Area
45 Date Flat Area
46 Foresthill OHV Area
47 Sierraville Ranger District
48 Nevada City Ranger District
49 Truckee Ranger District
50 Fordyce Jeep Trail
51 Little Truckee Summit
52 Prosser Hills Area
53 Big Bend Area
54 Downieville Ranger District
55 Monitor Pass
56 Hope Valley-Blue Lakes

Areas Managed by the U.S. Bureau of Land Management

1 Red Hills OHV Area
2 Clear Creek Management Area
3 Fort Sage OHV Area
4 Samoa Spit
5 Black Sands Beach
6 Shasta ORV Area
7 Cow Mountain Recreation Area
8 Knoxville Recreation Area
9 Olancha Dunes
10 Panamint Dry Lake
11 Spangler Hills
12 Dove Springs-Red Rock Canyon
13 Jawbone Canyon
14 Dumont Dunes
15 Silurian Dry Lake
16 Stoddard Valley
17 Johnson Valley
18 Ford Dry Lake
19 Glamis-Gecko
20 Plaster City
21 Buttercup Valley

BLM (continued)

22 Lark Canyon (McCain Valley)
23 Rasor
24 Mammoth Wash
25 Arroya Salada
26 El Mirage-Shadow Mountains
27 Rice Valley Dunes

Areas Managed by the California Department of Parks & Recreation

1 Carnegie
2 Clay Pit
3 Hollister Hills
4 Hungry Valley
5 Ocotillo Wells
6 Pismo Dunes
7 Prairie City

Areas Managed by Counties, Cities, or Other Jurisdictions

1 De Anza Cycle Park
2 Frank Raines-Deer Creek OHV Park
3 Glen Helen OHV Park
4 Heber Dunes County Park
5 Huron Cycle Park
6 La Grange ORV Park
7 Osborne County Park
8 Park Moabi
9 Porterville OHV Park
10 Riverfront Park ORV Area
11 County of Santa Clara Motorcycle Park
12 Bassetts Snowmobile Park
13 Bucks Lake Snowmobile Park
14 North Tahoe Public Utilities District
15 Laguna Seca
16 Black Butte Lake
17 Isabella Lake

- U.S. Forest Service
- Bureau of Land Management
- Department of Parks & Recreation
- Counties, Cities, or other Jurisdictions

Numbering of areas is taken from Department of Parks & Recreation publication 89 52776, published 6-89.

California's 16,000 miles of highways are among the busiest in the world. To assist drivers planning trips across these heavily used roads, CALTRANS provides telephone information on current road conditions on all highways in California and major highways in Nevada. Reports include notices of highway closures, one-way controlled traffic, construction work areas, and other disruptions of traffic. During the winter months, driving conditions are featured, sometimes in code: Chains R1 - chains required on all vehicles except autos and pick-up trucks with snow tires; Chains R2 - chains required on all vehicles except 4-wheel drive vehicles with snow tires on all 4 wheels; Chains R3 - chains required on all vehicles with no exceptions. Callers with touchtone phones using the San Francisco, Oakland, Sacramento, or Los Angeles numbers can also dial in specific route numbers for a report on those particular roadways. CALTRANS also provides a supplementary system, the Caltrans Highway Information Broadcasters' Network, CHIBN, which provides similar information via computer terminals to radio and television stations, automobile clubs, and public agencies. Finally, CALTRANS will supply telephone numbers to obtain highway condition information in neighboring states.

	CHIN Location	Phone Number	
1	Yreka	916-842-1217	
2	Alturas	916-233-5761	
3	Eureka	707-444-3077	
4	Redding	916-244-1500	
5	Susanville	916-257-5126	
6	Quincy	916-283-1045	
7	Chico	916-895-8111	
8	Oroville	916-534-7900	
9	Ukiah	707-462-1055	
10	Marysville	916-743-4681	
11	Grass Valley	916-272-2171	
12	Auburn	916-885-3786	
13	South Lake Tahoe	916-622-7355	
14	Santa Rosa	707-585-0326	
15	Sacramento	916-445-7623	*
16	Placerville	916-622-7355	
17	Jackson	209-223-4455	
18	Vallejo	707-643-8421	
19	Walnut Creek	415-938-1180	
20	Stockton	209-931-4848	
21	Angels Camp	209-736-4564	
22	San Francisco	415-557-3755	*
23	Oakland	415-654-9890	*
24	Sonora	209-532-0227	
25	Modesto	209-521-2240	
26	San Jose	408-436-1404	
27	Merced	209-383-4291	
28	Bishop	619-873-6366	
29	Salinas	408-757-2006	
30	Fresno	209-227-7264	
31	San Luis Obispo	805-543-1985	
32	Bakersfield	805-393-7350	
33	Ventura	805-653-1821	
34	Newhall	805-259-8081	
35	Los Angeles	213-626-7231	*
36	San Bernardino	714-788-7600	
37	Riverside	714-788-7600	
38	Santa Ana	714-972-9980	
39	San Diego	619-293-3484	

CHIN Location	Phone Number	#	
Alturas	916-233-5761	2	
Angels Camp	209-736-4564	21	
Auburn	916-885-3786	12	
Bakersfield,	805-393-7350	32	
Bishop	619-873-6366	28	
Chico	916-895-8111	7	
Eureka	707-444-3077	3	
Fresno	209-227-7264	30	
Grass Valley	916-272-2171	11	
Jackson	209-223-4455	17	
Los Angeles	213-626-7231	35	*
Marysville	916-743-4681	10	
Merced	209-383-4291	27	
Modesto,	209-521-2240	25	
Newhall	805-259-8081	34	
Oakland	415-654-9890	23	*
Oroville	916-534-7900	8	
Placerville	916-622-7355	16	
Quincy	916-283-1045	6	
Redding	916-244-1500	4	
Riverside	714-788-7600	37	
Sacramento	916-445-7623	15	*
Salinas	408-757-2006	29	
San Bernardino	714-788-7600	36	
San Diego	619-293-3484	39	
San Francisco	415-557-3755	22	*
San Jose	408-436-1404	26	
San Luis Obispo	805-543-1985	31	
Santa Ana	714-972-9980	38	
Santa Rosa	707-585-0326	14	
Sonora	209-532-0227	24	
South Lake Tahoe	916-622-7355	13	
Stockton	209-931-4848	20	
Susanville	916-257-5126	5	
Ukiah	707-462-1055	9	
Vallejo	707-643-8421	18	
Ventura	805-653-1821	33	
Walnut Creek	415-938-1180	19	
Yreka	916-842-1217	1	

*** Touchtone Features**

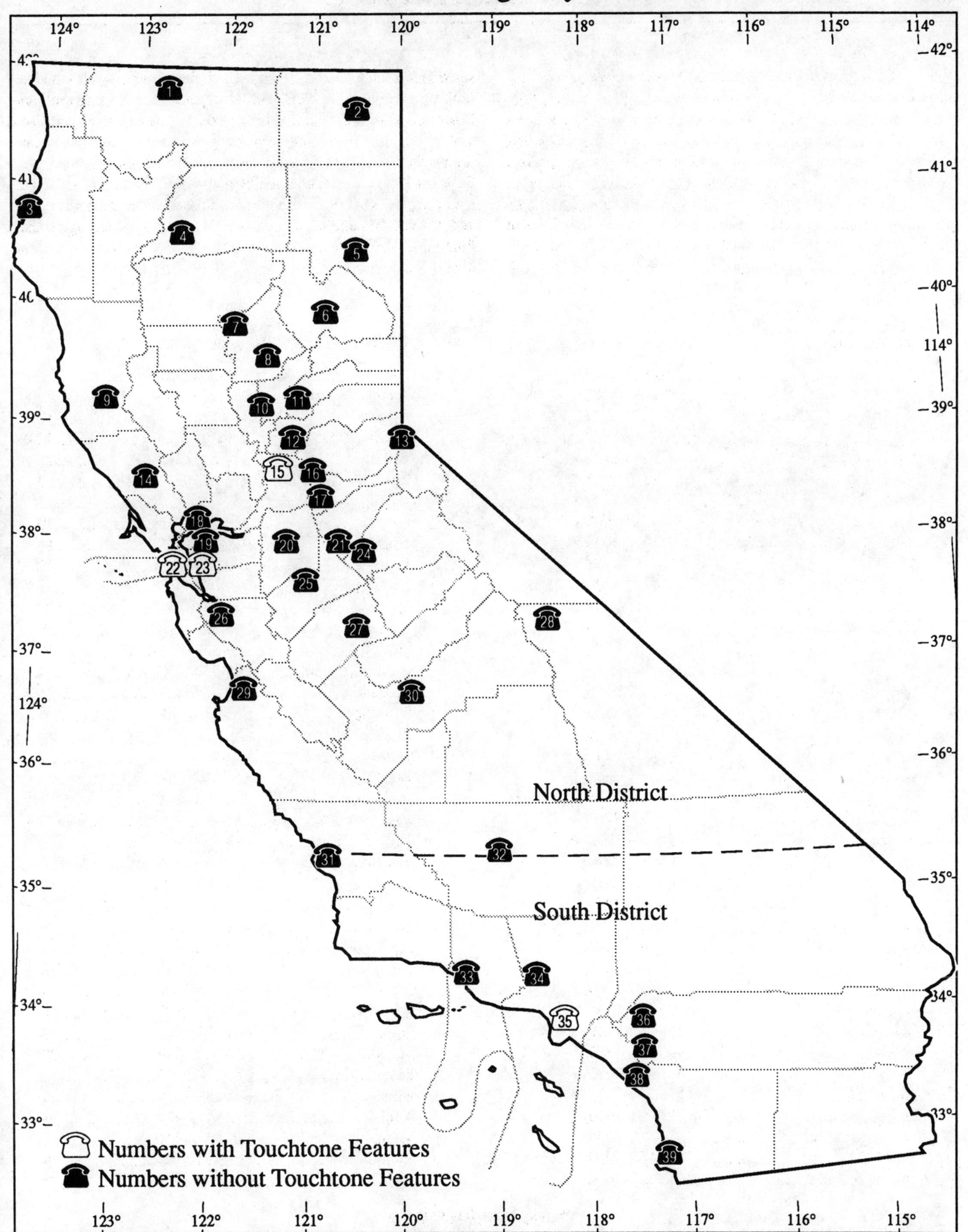

124°
123°
122°
121°
120°
119°
118°
117°
116°
115°
114°
42°
41°
40°
39°
38°
37°
36°
35°
34°
33°
North District
South District
1
2
3
4
5
6
7
8
9
10
11
12
13
14
15
16
17
18
19
20
21
22
23
24
25
26
27
28
29
30
31
32
33
34
35
36
37
38
39
Numbers with Touchtone Features
Numbers without Touchtone Features

California Guide

Directory

Agriculture, Department of
14th Street and Independence Ave. S.W.
Washington, D.C. 20230
202-447-2791

Forest Service, Dept. of Agriculture
P.O. Box 2417
Washington, D.C. 20013
202-447-3760

Pacific Southwest Regional Office
630 Sansome Street-Room 559
San Francisco, CA 94111
415-556-0123

Pacific Southwest Forest and Range Experiment Station
1960 Addison Street
Berkeley, CA 94704
415-486-3382

Fresno Forestry Sciences Laboratory
2081 East Sierra Avenue
Fresno, CA 93710
209-487-5794

Redding Science Laboratory
2400 Washington Avenue
Redding, CA 96001
916-246-5445

Redwood Science Laboratory
1700 Bayview Drive
Arcata, CA 95521
707-822-3691

Riverside Forest Fire Laboratory
4955 Canyon Crest Drive
Riverside, CA 92507
714-351-6523

National Forests, Headquarters and Ranger Districts

Angeles National Forest
701 North Santa Anita Avenue
Arcadia, CA 91006
818-574-5200

Saugus Ranger District
30800 Bouquet Canyon Road
Saugus, CA 91350
805-252-9710

Arroyo Seco Ranger District
Oak Grove Park
La Canada Flintridge, CA 91011
818-790-1151

Tujunga Ranger District
12371 Little Tujunga Road
San Fernando, CA 91342
818-790-1151

Mt. Baldy Ranger District
110 North Wabash Avenue
Glendora, CA 91740
818-335-1251

Valyermo Ranger District
34146 Longview Road
Pearblossom, CA 93553
805-944-2187

Cleveland National Forest
880 Front Street-Room 5N14
San Diego, CA 92188
619-293-5050

Trabuco Ranger District
1147 East Sixth Street
Corona, CA 91720
714-736-1811

Descanso Ranger District
3348 Alpine Boulevard
Alpine, CA 92001
619-445-6235

Palomar Ranger District
332 South Juniper
Excondido, CA 92025
619-745-2421

Eldorado National Forest
100 Forni Road
Placerville, CA 95667
916-622-5061

Eldorado Information Center
3070 Camino Heights Drive
Camino, CA 95709
916-644-6048

Amador Ranger District
26820 Silver Drive & Highway 88
Star Route 3
Pioneer, CA 9566
209-295-4241

Placerville Ranger District
3491 Carson Court
Placerville, CA 95667
916-644-2324

Georgetown Ranger District
Georgetown, CA 95634
916-333-4312

Pacific Ranger District
Pollock Pines, CA 95726
916-644-2349

Placerville Nursery
2375 Fruitridge Road
Camino, CA 95709
916-622-9600

Inyo National Forest
873 North Main Street
Bishop, CA 93514
619-873-5841

Mt. Whitney Ranger District
P.O. Box 8
Lone Pine, CA 93545
619-876-5542

Mammoth Ranger District
P.O. Box 148
Mammoth lakes, CA 93546
619-934-2505

White Mountai Ranger District
798 North Main Street
Bishop, CA 93514
619-873-4207

Mono Lake Ranger District
P.O. Box 10
Lee Vining, CA 93541
619-647-6525

Klamath National Forest
1312 Fairland Road
Yreka, CA 96097
916-842-6131

Oak Knoll Ranger District
22541 Highway 96
Klamath River, CA 96050
916-465-2241

Ukonom Ranger District
P.O. Drawer 410
Orleans, CA 95556
916-627-3291

Goosenest Ranger District
37805 Highway 97
Macdoel, CA 96058
916-398-4391

Salmon River Ranger District
Sawyers Bar, CA 96027
(ask Fort Jones operator for Sawyers Bar 4600)
-or-
P.O. Box 280
Etna, CA 96027
916-467-5757

Happy Camp Ranger District
P.O. Box 377
Happy Camp, CA 96039
916-493-2243

Scott River Ranger District
11263 South Highway 3
Fort Jones, CA 96032
916-468-5351

Lake Tahoe Basin Management Unit
P.O. Box 8465
870 Emerald Bay Road
South Lake Tahoe, CA 95731
916-573-2600

Tahoe Visitor Center
916-541-0209
(summers only)

William Kent Information Station
William Kent Campground - West Shore
916-583-3642
(open summers only)

Lassen National Forest
55 South Sacramento St.
Susanville, CA 96130
916-257-2151

Hat Creek Ranger District
P.O. Box 220
Fall River Mills, CA 96028
916-336-5521

Almanor Ranger District
P.O. Box 767
Chester, CA 96020
916-258-2141

Eagle Lake Ranger District
472-013 Johnstonville Road
Susanville, CA 96130
916-257-2595

Los Padres National Forest
6144 Calle Real
Goleta, CA 93117
805-683-6711

Ojai Ranger District
1190 East Ojai Avenue
Ojai, CA 93023
805-646-4348

Monterey Ranger District
406 South Mildred
King City, CA 93930
408-385-5434

Santa Lucia Ranger District
1616 North Carlotti Drive
Santa Maria, CA 93454
805-925-9538

Mt. Pinos Ranger District
Star Route, Box 400
Frazier Park, CA 93225
805-245-3731

Santa Barbara Ranger District
Star Route, Los Prietos
Santa Barbara, CA 93105
805-967-3481

Mendocino National Forest
420 East Laurel Street
Willows, CA 95988
916-934-3316

Stonyford Ranger District
Stites Ladoga Road
Stonyford, CA 95979
916-963-3128

Corning Ranger District
22000 Corning Road
P.O. Box 1019
Corning, CA 96021
916-824-5196

Upper Lake Ranger District
Middlecreek Road
P.O. Box 96
Upper Lake, CA 95485
707-275-2361

Covelo Ranger District
Route 1, Box 62-C
Covelo, CA 95428
707-983-6118

Chico Tree Improvement Center
2741 Cramer Lane
Chico, CA 95926
916-895-1176

Modoc National Forest
441 North Main Street
Alturas, CA 96101
916-233-5811

Doublehead Ranger District
P.O. Box 818
Tulelake, CA 96134
916-667-2247

Big Valley Ranger District
P.O. Box 885
Adin, CA 96006
916-299-3215

Warner Mountain Ranger District
P.O. Box 220
Cedarville, CA 96104
916-279-6116

Devil's Garden Ranger District
P.O. Box 5
Canby, CA 96015
916-233-4611

Plumas National Forest
P.O. Box 1500
159 Lawrence St.
Quincy, CA 95971
916-283-2050

La Porte Ranger District
Challenge Ranger Station
P.O. Box F
Challenge, CA 95925
916-675-2462

Beckwourth Ranger District
Mohawk Ranger Station
P.O. Box 7
Blairsden, CA 96013
916-836-2575

Milford Ranger District
Laufman Ranger Station
Milford, CA 96121
916-253-2223

Quincy Ranger District
1400 East Main Street
P.O. Box 69
Quincy, CA 95971
916-283-0555

Greenville Ranger District
P.O. Box 329
Greenville, CA 95947
916-284-7126

Oroville Ranger District
875 Mitchell Ave.
Oroville, CA 95965
916-534-6500

San Bernardino National Forest
1824 Commercenter Circle
San Bernardino, CA 92408
714-383-5588

Cajon Ranger District
Lytle Creek Ranger Station Star Route
Fontana, CA 92335
714-887-2576

San Gorgonio Ranger District
Mill Creek Station
34701 Mill Creek Road
Mentone, CA 92359
714-794-1123

Big Bear Ranger District
P.O. Box 290
Fawnskin, CA 92333
714-866-3437

San Jacinto Ranger District
Idyllwild Ranger Station
P.O. Box 518
Idyllwild, CA 92349
714-659-2117

Sequoia National Forest
900 West Grand Avenue
Porterville, CA 93257
209-784-1500

Hot Springs Ranger District
Route 4, Box 548
California Hot Springs, CA 93207
805-548-6503

Cannell Meadow Ranger District
P.O. Box 6
Kernville, CA 93238
619-376-3781

Hume Lake Ranger District
36273 East Kings Canyon Road
Dunlap, CA 93621
209-338-2251

Greenhorn Ranger District
Federal Building-Room 322
Bakersfield, CA 93301
805-861-4212

Tule Ranger District
32588 Highway 190
Porterville, CA 93257

Shasta-Trinity National Forests
2400 Washington Avenue
Redding, CA 96001
209-539-2607

McCloud Ranger District
Drawer 1
McCloud, CA 96057
916-964-2184

Weaverville Ranger District
P.O. Box T
Weaverville, CA 96093
916-623-2131

Big Bar Ranger District
Star Route, Box 10
Big Bar, CA 96010
916-623-6106

Mt. Shasta Ranger District
204 West Alma
Mt. Shasta, CA 96067
916-926-4511

Yolla Bolly Ranger District
Platina, CA 96076
916-352-4211

Hayfork Ranger District
P.O. Box 159
Hayfork, CA 96041
916-628-5227

Shasta Lake Ranger District
6543 Holiday Drive
Redding, CA 96003
916-275-1587

Northern California Service Regional
Fire Control Center
6106 Airport Road
Redding, CA 96002
916-246-5285

Sierra National Forest
1130 O Street
Fresno, CA 93721
209-487-5155

Mariposa Ranger District
P.O. Box 747
Mariposa, CA 95338
209-966-3638

Bass Lake Ranger District
41969 Highway 41
Oakhurst, CA 93644
209-683-4665

Minarets Ranger District
North Fork, CA 93643
209-877-2218

Kings River Ranger District
Trimmer Route
Sanger, CA 93657
209-855-8321

Dinkey Ranger Station
Dinkey Route
Shaver Lake, CA 93664
209-841-3404
(summers only)

Pineridge Ranger District
P.O. Box 300
Shaver Lake, CA 93664
209-841-3311

Six Rivers National Forest
507 F Street
Eureka, CA 95501
707-442-1721

Mad River Ranger District
Star Route, Box 300
Bridgeville, CA 95526
707-574-6233

Gasquet Ranger District
P.O. Box 228
Gasquet, CA 95543
707-457-3131

Orleans Ranger District
P.O. Drawer B
Orleans, CA 95556
916-627-3291

Lower Trinity Ranger District
P.O. Box 668
Willow Creek, CA 95573
916-629-2118

Humboldt Nursery
4886 Cottage Grove
McKinleyville, CA 95521
707-839-3256

Stanislaus National Forest
19777 Greenley Road
Sonora, CA 95370
209-532-3671

Mi-Wok Ranger District
P.O. Box 100
Mi-Wuk Village, CA 95346
209-586-3234

Calaveras Ranger District
P.O. Box 500
Hathaway Pines, CA 95233
209-795-1381

Summit Ranger District
Highway 108-E & Pinecrest
Star Route, Box 1295
Sonora, CA 95370
209-965-3434

Groveland Ranger District
Star Route, Box 75-G
Groveland, CA 95321
209-962-7825

Tahoe National Forest
Highway 49 and Coyote Street
Nevada City, CA 95959
916-265-4531

Nevada City Ranger District
Highway 49 and Coyote Street
Nevada City, CA 95959
916-265-4538

Downieville Ranger District North
Yuba Ranger Station
Star Route, Box 1
Camptonville, CA 95922
916-288-3231

Sierraville Ranger District
Highway 89, P.O. Box 95
Sierraville, CA 96126
916-994-3401

Foresthill Ranger District
22830 Auburn-Foresthill Road
Foresthill, CA 95631
916-367-2224

Truckee Ranger District
P.O. Box 399
Truckee, CA 95734
916-587-3558

Toiyabe National Forest
1200 Franklin Way
Sparks, NV 89431
702-355-5301

Bridgeport Ranger District
P.O. Box 595
Bridgeport, CA 93517

Carson Ranger District
1536 South Carson Street
Carson City, NV 89701

Soil Conservation Service
P.O. Box 2890
Washington, D.C. 20015
202-447-4543

Soil Conservation Service
2828 Chiles Road
Davis, CA 95616
916-440-2848

Commerce, Department of
14th Street, Constitution Avenue & E Street N.W.
Washington, D.C. 20230
202-377-2000

National Oceanic and Atmospheric Administration, Dept. of Commerce
14th Street, Constitution Avenue & E Street N.W.
Washington, D.C. 20230
202-377-2985

National Weather Service - Los Angeles
10445 South Sepulveda Boulevard
Los Angeles, CA 90045
213-215-2338
213-209-7211
recording 213-554-1212
recording (marine) 213-477-1463

National Weather Service - San Francisco
660 Price Avenue
Redwood City, CA 94063
415-364-7974

National Marine Fisheries Service, NOAA
Southwest Regional Office
300 S. Ferry Street
Terminal Island, CA 90731

Tuna-Porpoise Management Branch
1520 State Street-Suite 200
San Diego, CA 92101
619-293-6540

Northern Area Office
777 Sonoma Avenue-Room 325
Santa Rosa, CA 95404
707-525-4315

Habitat Conservation Branch
3150 Paradise Drive
Tiburon, CA 94920
415-435-3149

National Estuarine Reserve Research Program, NOAA
1825 Connecticut Avenue N.W.
Washington, D.C. 20235
202-673-5126

Elkhorn Slough National
Estuarine Research Reserve
1700 Elkhorn Road
Watsonville, CA 95076
408-728-0560

Tijuana River
National Estuarine Research Reserve
3990 Old Town Avenue-Suite 300C
San Diego, CA 92110
619-237-6766

Pacific Estuaries Research Laboratory
Biology Department
San Diego State University
San Diego, CA 92182
619-265-5809

National Marine Sanctuaries Program, NOAA
1825 Connecticut Avenue N.W.
Washington, D.C. 20235
202-673-5126

Channel Islands National
Marine Sanctuary
735 State Street
Santa Barbara, CA 93101
805-966-7107

Cordell Banks National Marine Sanctuary
(Temporary) Fort Mason, Building #201
San Francisco, CA 94123
415-556-3509

Gulf of the Farallones National Marine
Sanctuary
Fort Mason-Building #201
San Francisco, CA 94123
415-556-3509

National Environmental Satellite, Data, and Information Service, NOAA
Satelline Field Services Station
660 Price Avenue
Redwood City, CA 94063
415-876-9122

Energy, Department of
1000 Independence Avenue S.W.
Washington, D.C. 20585
202-252-5000

333 Broadway
Oakland, CA 94612
415-273-7829

Federal Energy Regulatory Commission, Dept. of Energy
333 Market Street 6th Floor
San Francisco, CA 94105
415-974-7150

Interior, Department of
C Street, 18th & 19th Streets N.W.
Washington, D.C. 20240
202-343-3171

Regional Environmental Office
450 Golden Gate Avenue-Room 14444
San Francisco, CA 94102
415-556-8200

United States Fish and Wildlife Service, Department of Interior
Regional Office
500 Northeast Multnomah St.,Suite 1692
Portland, OR 97232
503-231-6121

Laguna Niguel Field Office
24000 Avila Road
Laguna Niguel, CA 92677
714-643-4270

Sacramento Field Office
2800 Cottage Way-Room E-1803
Sacramento, CA 95825
916-978-4613

Division of Endangered Species
Sacramento Field Station
1230 N Street-14th Floor
Sacramento, CA 95814
916-978-4866

Office of Sea Otter Coordination
2800 Cottage Way-Room E-1818
Sacramento, CA 95825
916-978-4873

Arcata Fisheries Assistance Office
125 16th Street-Room 209
Arcata, CA 95521
707-822-7201

Coleman National Fish Hatchery
Route 1, Box 2105
Anderson, CA 96007
916-365-8622

Red Bluff Fisheries Assistance Office
P.O. Box 667
Red Bluff, CA 96080
916-527-3043

Stockton Fisheries Assistance Office
401 North Wilson Way
Stockton, CA 95205
209-466-4421

Tehama-Colusa Fish Facilities
P.O. Box 1050
Red Bluff, CA 96080
916-527-7440

Division of Law Enforcement
1290 Howard Avenue-Room 325
Burlingame, CA 94010
414-344-5900

P.O. Box 4401
Chico, CA 95927
916-342-8724

P.O. Box 5377
Fresno, CA 93755
209-487-5733

400 Oceangate-Suite 1000
Long Beach, CA 90802
213-436-1183

5758 Century Blvd., Room E-211
Los Angeles, CA 90045
213-215-2033

2800 Cottage Way, Room E-1924
Sacramento, CA 95825
916-978-4861

880 Frong Street, Room 4S23
San Diego, CA 92188
619-293-5063

425 Henrietta Avenue
Los Osos, CA 93402
805-528-7980

Kern-Pixley National Wildlife Refuge
Complex
P.O. Box 670
Delano, CA 93216
805-725-2767
Satellite Refuges: Hopper Mountain; Seal Beach; Blue Ridge; Bitter Creek

Klamath Basin National Wildlife Refuges
Route 1, Box 74
Tulelake, CA 96134
916-667-2231
Satellite Refuges: Clear Lake; Klamath Forest; Lower Klamath; Tule Lake.

Modoc National Wildlife Refuge
P.O. Box 1610
Alturas, CA 96101
916-233-3572

Sacramento National Wildlife Refuge Complex
Route 1, Box 311
Willows, CA 95988
916-934-2801
Satellite Refuges: Colusa; Delevan; Sutter; Butte Sink; Willow Creek-Lurline

Salton Sea National Wildlife Refuge
P.O. Box 120
Calipatria, CA 92233
619-348-5278
Satellite Refuges: Tijuana Slough; Coachella Valley

San Francisco Bay National Wildlife Refuge Complex
P.O. Box 524
Newark, CA 94560
415-792-0222
Satellite Refuges: Humboldt Bay; Farallon; Salinas Wildlife Management Area; Ellicott Slough; San Pablo Bay; Antioch Dunes; Castle Rock

San Luis National Wildlife Refuge
P.O. Box 2176
Los Banos, CA 93635
209-826-3508
Satellite Refuges: Kesterson; Merced; Grasslands

San Joaquin Valley Drainage Program
2800 Cottage Way-Room W-2143
Sacramento, CA 95825

Cooperative Fishery Research Unit
Humboldt State University
Arcata, CA 95521
707-826-3268

Condor Research Center
2291-A Portola Road
Ventura, CA 93001
805-644-1766

National Park Service, Dept. of Interior
Western Regional Office
450 Golden Gate Avenue, Box 36063
San Francisco, CA 94102
415-556-4122

Cabrillo National Monument
P.O. Box 6670
San Diego, CA 92106
619-254-5450

Channel Islands National Park
1699 Anchors Way Drive
Ventura, CA 93003
805-644-8157

Death Valley National Monument
Death Valley, CA 92328
619-786-2331

Devils Postpile National Monument
c/o Sequoia & Kings Canyon Natl. Parks
Three Rivers, CA 93271
209-565-3341

Eugene O'Neill National Historic Site
Danville, CA
c/o 4202 Alhambra Avenue
Martinez, CA 94553
415-820-1818

Golden Gate National Recreation Area
Fort Mason
San Francisco, CA 94123
415-556-4122

Fort Point National Historic Site
P.O. Box 29333
Presidio of San Francisco, CA 94129
415-556-1693

Muir Woods National Monument
Mill Valley, CA 94941
415-388-2595

John Muir National Historic Site
4202 Alhambra Avenue
Martinez, CA 94553
415-228-8860

Joshua Tree National Monument
74485 National Monument Drive
Twentynine Palms, CA 92277
714-367-7511

Lassen Volcanic National Park
Mineral, CA 96063
916-595-4444

Lava Beds National Monument
P.O. Box 867
Tulelake, CA 96134
916-667-2282

Pinnacles National Monument
Paicines, CA 95043
408-389-4578

Point Reyes National Seashore
Point Reyes, CA 94956
415-663-8522

Redwood National Park
1111 Second Street
Crescent City, CA 95531
707-464-6101

Santa Monica Mountains National Recreation Area
22900 Ventura Boulevard-Room 140
Woodland Hills, CA 91306
818-888-3340

Sequoia & Kings Canyon National Parks
Three Rivers, CA 93271
209-565-3341

Whiskeytown-Shasta-Trinity National Recreation Area
P.O. Box 188
Whiskeytown, CA 96095
916-241-6584

Yosemite National Park
P.O. Box 577
Yosemite National Park, CA 95389
209-372-0200

Geological Survey, Dept. of Interior
12201 Sunrise Valley Drive
Reston, Virginia 22092
703-860-7444

Western Regional Headquarters
Menlo Park, CA 94025
415-323-8111

Public Inquiries Offices for
Geological Survey
555 Battery Street-Room 504
San Francisco, CA 94111
415-556-5627

Public Inquiries Offices for
Geological Survey
300 North Los Angeles Street-Room 7638
Los Angeles, CA 90012
213-894-2850

Seismic Engineering for
Geological Survey
P.O. Box 6113
Lawndale, CA 90261
213-536-6672

Water Resources Division
California District Office
2800 Cottage Way-Room W2234
Sacramento, CA 95825
916-978-4633

Bureau of Indian Affairs,
Department of Interior
C Street, 18th & 19th Streets N.W.
Washington, D.C. 20240

Bureau of Land Management,
Department of Interior
C Street, 18th & 19th Streets N.W.
Washington, D.C. 20240
202-343-4662

Bureau of Land Management
2800 Cottage Way-Room E2841
Sacramento, CA 95825
916-484-4465

Bakersfield District
800 Truxtun Avenue-Room 302
Bakersfield, CA 93301
805-861-4191

Caliente Resource Area
520 Butte Street
Bakersfield, CA 93305
805-861-4191

Folsom Resource Area
63 Natoma Street
Folsom, CA 95630
916-985-4474

Hollister Resource Area
P.O. Box 365, Parkhill
Hollister, CA 95023
408-637-8183

California Desert District
1695 Spruce Street
Riverside, CA 92507
714-351-6383

Barstow Resource Area
150 Coolwater Lane
Barstow, CA 92311
714-256-3591

El Centro Resource Area
333 South Waterman Avenue
El Centro, CA 92243
619-352-5842

Indio Resource Area
1900 Tahquitz-McCallum Way
Palm Springs, CA 92262
619-323-3896

Needles Resource Area
101 West Spices Road
Needles, CA 92363
619-326-3896

Ridgecrest Resource Area
112-East Dolphin Street
Ridgecrest, CA 93555
619-375-7125

Southern California Metropolitan
Project Area
1695 Spruce Street
Riverside, CA 92507
714-351-6379

Susanville District
705 Hall Street, P.O. Box 1090
Susanville, CA 96130
916-257-5381

Alturas Resource Area
120 Main Street, P.O. Box 771
Alturas, CA 96101
916-233-4666

Surprise Resource Area
602 Cressler Street, P.O. Box 460
Cedarville, CA 95104
916-279-6101

Eagle Lake Resource Area
2545 Riverside Drive, P.O. Box 1090
Susanville, CA 96130
916-257-5381

Ukiah District
555 Leslie Street
Ukiah, CA 95482
707-462-3873

Clear Lake Resource Area
555 Leslie Street
Ukiah, CA 95482
707-462-3873

Arcata Resource Area
1125 16th Street, P.O. Box 1112
Arcata, CA 95521
707-822-7648

Redding Resource Area
355 Hemsted Drive
Redding, CA 96001
916-246-5325

Bureau of Reclamation,
Department of Interior
C Street, 18th & 19th Streets N.W.
Washington, D.C. 20240
202-343-4662

Mid-Pacific Region
2800 Cottage Way
Sacramento, CA 95825
916-978-4919

Fresno Central Valley Project
Construction Office
1130 O Street-Room 2201
Fresno, CA 93721
209-487-5116

Shasta Office
Shasta Dam
Redding, CA 96003
916-275-1554

Lower Colorado River Region
Office of the Regional Director
P.O. Box 427
Boulder City, Nevada 89005
702-293-8419

Minerals Management Service
Pacific Outer Continental Shelf
Regional Office
1340 West Sixth Street
Los Angeles, CA 90017
213-894-2050

Minerals Management Service
Ventura District Office
145 North Brent Street-Room 202
Ventura, CA 93003
805-648-5131

Office of Strategic and
International Minerals
11 Golden Shore-Suite 260
Long Beach, CA 90802
213-514-6140

Enviromental Protection Agency
401 M Street S.W.
Washington, D.C. 20460
202-382-2080

Environmental Protection Agency
215 Fremont Street
San Francisco, CA 94105
415-974-8153

San Francisco Bay Estuary Project
101 8th Street
Oakland, CA 94607
415-464-7990

Federal Information Sources for California

Agriculture, Dept. of - Forest Service
630 Sansome Street
San Francisco, CA 94111
415-556-0122

Commerce, Department of - National Oceanic and Atmospheric Admin.
7600 Sand Point Way N.E.
Seattle, Washington 98115
206-526-6026

Energy, Department of - Federal Energy Regulatory Commission
901 Market Street
San Francisco, CA 94103
415-974-7145

Interior, Department of - Bureau of Land Management
2800 Cottage Way
Sacramento, CA 95825
916-978-4746

National Park Service
450 Golden Gate Avenue
San Francisco, CA 94102
415-556-5186

U.S. Fish and Wildlife Service
500 Northeast Multnomah Street
Portland, OR 97232
503-231-6118

U.S. Geological Survey
345 Middlefield Road
Menlo Park, CA 94025
415-853-8300

Labor Department - Occupational Safety and Health Administration
71 Stevenson Street
San Francisco, CA 94105
415-995-5672

Environmental Protection Agency
215 Fremont Street
San Francisco, CA 94105
415-974-8153

Nuclear Regulatory Commission
1450 Maria Lane
Walnut Creek, CA 94596
414-943-3700

General Services Administration - Federal Information Center
880 Front Street
San Diego, CA 92188
619-293-6030
714-836-2386 (Orange County)

300 North Los Angeles Street
Los Angeles, CA 90012
213-894-3800

1825 Bell Street-Room 108
Sacramento, CA 95825-1080
916-978-4010

450 Golden Gate Avenue
P.O. Box 36082
San Francisco, CA 94102
415-556-6600

Board on Geographic Names
U.S. Geological Survey National Center
12201 Sunrise Valley Drive
Reston, VA 22092
703-648-4513

Air Resources Board
P.O. Box 2815
Sacramento, CA 95812
916-322-2990
FAX 916-322-9612

Boating and Waterways, Department of
1629 S Street
Sacramento, CA 95814
916-445-2615

Coastal Commission, California
631 Howard Street-4th Floor
San Francisco, CA 94105
415-543-8555

North Coast District
350 E Street
Eureka, CA 95501
707-443-1623

North Central District
631 Howard Street-4th Floor
San Francisco, CA 94105
415-543-8555

Central Coast District
701 Ocean Street
Santa Cruz, CA 95060
408-426-7390

South Central District
925 De La Vina
Santa Barbara, CA 93101
805-963-6871

South Coast District
245 Broadway-Suite 380
Long Beach, CA 90802
213-590-5071

San Diego District
6154 Mission Gorge Road-Suite 220
San Diego, CA 92120
619-280-6992

Legislative Office
921 11th Street-Room 1200
Sacramento, CA 95814
916-445-6067

Coastal Conservancy, California
1330 Broadway-Room 1100
Oakland, CA 94612
415-464-1015
FAX 415-464-0470

Coastal Conservancy, California
915 Capitol Mall-Room 216
Sacramento, CA 95814
916-323-4688
FAX 916-445-0766

Conservation, Department of
1416 9th Street 13th Floor
Sacramento, CA 95814
916-322-7683

Land Conservation Office
1516 9th Street-Room 400
Sacramento, CA 95814
916-324-0859

Mines and Geology, Division of
1416 9th Street-Room 1341
Sacramento, CA 95814
916-445-1825

Publications Office
1516 9th Street-Room 407
Sacramento, CA 95814
916-445-5716

Surface Mining and Reclamation, Office of
6503 Bercut Drive
Sacramento, CA 95814
916-323-8567

Surface Mining and Reclamation, Office of
107 South Broadway
Los Angeles, CA 90012
213-620-3560
FAX 213-620-3961

Oil and Gas, Division of
1416 9th Street-Room 1310
Sacramento, CA 95814
916-445-9686
FAX 916-323-0424

Geothermal Section
1516 9th Street-Room 402
Sacramento, CA 95814
916-323-1788

Recycling, Division of
819 9th Street
Sacramento, CA 95814
916-323-3508

Energy Resources, Conservation and Development Commission
1516 9th Street
Sacramento, CA 95814
916-324-3000

Environmental Affairs Agency
1162 Q Street
P.O. Box 2815
Sacramento, CA 95812
916-322-4203

Fish and Game Commission
1416 9th Street-12th Floor
Sacramento, CA 95814
916-445-5708

Fish and Game, Department of
1416 9th Street-12th Floor
Sacramento, CA 95814
916-445-3531
FAX 916-324-8553

Environmental Services Divison
916-445-1383
Inland Fisheries Division
916-445-1383
Marine Resources Division
916-445-8386
& Long Beach
(213) 590-5189
Wildlife Management Divison
916-445-5561
Wildlife Protection Division
916-324-7243
Hazardous Materials
916-322-9210

Wildlife Conservaton Board
1416 9th Street
Sacramento, CA 95814
916-445-8448

Bay Delta Fisheries Project
4001 North Wilson Way
Stockton, CA 95205

Nongame Heritage Program
1220 S Street
Sacramento, CA 95814
916-322-2493

Natural Diversity Data Base
1220 S Street
Sacramento, CA 95814
916-322-2495
Animals
916-322-249
Plants
916-323-8970
Aquatic Communities
916-322-2495
Plant Communities
916-324-6857
Lands and Natural Areas Project
916-322-6469
Endangered Plant Project
916-324-3814

Fish and Game - Region I
601 Locust Street
Redding, CA 96001
916-225-2300
FAX 916-225-2381

Fish and Game - Region II
1701 Nimbus Road
Rancho Cordova, CA 95670
916-355-0978
FAX 916-355-7162

Fish and Game - Region III
7329 Silverado Trail
Napa, CA 94558
P.O. Box 47
Yountville, CA 94599
707-944-5500
FAX 707-944-5563

Fish and Game - Region IV
1234 East Shaw Street
Fresno, CA 93710
209-222-3761
FAX 209-445-6426

Fish and Game - Region V
330 Golden Shore-Room D
Long Beach, CA 90802
213-590-5132

Forestry, Board of
1416 9th Street
Sacramento, CA 95814
916-445-2753

Forestry and Fire Protection, Department of
1416 9th Street-15th Floor
Sacramento, CA 95814
916-445-3976

North Coast Region
135 Ridgeway
Santa Rosa, CA 95402
707-576-2275

Sierra Cascade Region
6105 Airport Road
Redding, CA 96002
916-225-2445

South Sierra Region
1234 East Shaw Avenue
Fresno, CA 93710
209-222-3714

Central Coast Region
2221 Garden Road
Monterey, CA 93940
408-649-2801

Southern California Region
2524 Mulberry Street
Riverside, CA 92501
714-782-4140

Demonstration State Forest Program
1416 9th Street
Sacramento, CA 94244
916-322-0169

Boggs Mountain State Forest
P.O. Box 839
Cobb, CA 95426
707-928-4375

Ellen Pickett State Forest
(contact Demonstration State Forest Program)

Jackson State Forest
802 North Main Street
P.O. Box 1185
Fort Bragg, CA 95437
707-964-5674

Los Posades State Forest
(contact Demonstraton State Forest Program)

Latour State Forest
6105 Airport Road
Redding, CA 96002
916-225-2495

Mount Zion State Forest
(contact Demonstration State Forest Program)

Mountain Home State Forest
P.O. Box 517
Springvale, CA 93265
209-539-2321 - summer
209-539-2855 - winter

Heritage Preservation Commission, California
1020 O Street-Room 130
Sacramento, CA 95814
916-445-4293

Native American Heritage Commission
915 Capitol Mall-Room 288
Sacramento, CA 95814
916-322-7791

Parks and Recreation, Department of
1416 9th Street
P.O. Box 942896
Sacramento, CA 94296
916-445-6477

Off-Highway Motor Vehicle Recreation Commission
916-445-0305
Parks and Recreation Commission, State
916-324-6976
Recreational Trails Commission, Calif.
916-445-0835
Off Highway Motor Vehicle Recreation Division
916-323-9897
Resource Protection Division
916-322-5636

Northern Region
3033 Cleveland Avenue-Room 110
Santa Rosa, CA 95401
707-576-2185
FAX 707-576-2558

Southern Region
1333 Camino Del Rio South-Room 200
San Diego, CA 92108
619-237-7411
FAX 619-298-6241

Resources Agency
1416 9th Street-Room 1311
Sacramento, CA 95814
916-445-3758

San Francisco Bay Conservation And Development Commission
30 Van Ness Avenue-Room 2011
San Francisco, CA 94102
415-557-3686

Santa Monica Mountains Conservancy
107 South Broadway-Room 7117
Los Angeles, CA 90012
213-620-2021

Seismic Safety Commission
1900 K Street-Room 100
Sacramento, CA 95814
916-322-4917

State, Department of
Toxic Substances Control Division
714 P Street
Sacramento, CA
916-324-1826

State Lands Commission
1807 13th Street
Sacramento, CA 95814
916-322-7777

Tahoe Regional Planning Agency, California
3053 Harrison Avenue
P.O. Box 14467
South Lake Tahoe, CA 95702
916-541-0249

Transportation Commission, California
1120 N Street
Sacramento, CA 95814
916-445-1690

Transportation, Department of
1120 N Street
Sacramento, CA 95814
916-445-4616
FAX 916-445-8353

Waste Management Board, California
1020 9th Street-Room 300
Sacramento, CA 95814
916-322-3330

Water Resources, Department of
1416 9th Street
P.O. Box 942836
Sacramento, CA 94236
916-445-9248

Water Commission, California
916-445-8750
Reclamation Board
916-445-9454
Flood Management, Division of
916-324-4784

Safety of Dams, Divison of
2201 X Street
Sacramento, CA 95818
916-445-7606

Northern District
2440 Main Street
P.O. Box 607
Red Bluff, CA 96080
916-526-6530
FAX 916-527-2222

Central District
3251 S Street
P.O. Box 160088
Sacramento, CA 95816
916-445-6831
FAX 916-322-7184

San Joaquin District
3374 East Shields Avenue
Fresno, CA 93726
209-445-5443
FAX 209-445-5370

Southern District
849 South Broadway-Room 500
P.O. Box 6598
Los Angeles, CA 90055
213-620-4107
FAX 213-620-5645

Water Resources Control Board, State
901 P Street
Sacramento, CA 95814
P.O. Box 100
Sacramento, CA 95801
916-322-3132

Water Quality, Division of
916-445-9552
Water Rights, Division of
916-322-4503

North Coast - Region I
1440 Guerneville Road
Santa Rosa, CA 95403
707-576-2220
FAX 707-523-0135

Bay - Region 2
1115 Jackson Street-Room 6040
San Francisco, CA 94607
415-464-0516

Central Coast - Region 3
1102A Laurel Lane
San Luis Obispo, CA 93401
805-549-3147

Los Angeles - Region 4
107 South Broadway-Room 4027
Los Angeles, CA 90012
213-620-4460

Central Valley - Region 5
3443 Routier Road
Sacramento, CA 95827
916-361-5600

Lahontan - Region 6
2092 Lake Tahoe Boulevard
South Lake Tahoe, CA 95731
916-544-3481
FAX 916-544-2271

Colorado River Basin - Region 7
73-271 Highway 111-Room 21
Palm Desert, CA 92260
619-946-7491

Santa Ana - Region 8
6809 Indiana Avenue-Room 200
Riverside, CA 92506
714-782-4130

San Diego - Region 9
9771 Claremont Mesa Blvd., Room B
San Diego, CA 92124
619-265-5114

Wildlife Conservation Board
1416 9th Street
Sacramento, CA 95814
916-445-8448

California Association of Councils of Governments
106 K Street-Room 200
P.O. Box 808
Sacramento, CA 95814
916-441-5930

Association of Bay Area Governments (ABAG)
Metro Center
101 Eigth Street
Oakland, CA 94607
415-464-7900
mailing address:
P.O. Box 2050
Oakland, CA 94604
Member Jurisdictions: Alameda, Contra Costa, Marin, Napa, San Francisco, San Mateo, Santa Clara, Solano, and Sonoma counties, and 91 cities

Association of Monterey Bay Area Governments (AMBAG)
Carmel Hill Professional Center
977 Pacific Street
Monterey, CA 93940
408-373-6116
mailing address:
P.O. Box 190
Monterey, CA 93942
Member Jurisdictions: Monterey and Santa Cruz counties and 15 cities

Butte County Association of Governments
7 County Center Drive
Oroville, CA 95965
916-538-7601
Member Jurisdictions: Butte County and 5 cities

Central Sierra Planning Council and Economic Development District
83 South Stewart Street
Sonora, CA 95370
209-532-8768
Member Jurisdictions: Alpine, Amador, Calaveras, and Tuolumne counties and 7 cities

Fresno Council of Local Governments (FCLG)
2100 Tulare Street-Suite 619
Fresno, CA 93721
209-233-4148
Member Jurisdictions: Fresno County and 15 cities

Humboldt County Association of Governments (HCAG)
531 K Street-#309
Eureka, CA 95501
707-444-8208
mailing address:
P.O. Box 156
Eureka, CA 95502
Member Jurisdictions: Humboldt County and 7 cities

Inyo-Mono Association of Governmental Entities (IMAGE)
P.O. Drawer L
Independence, CA 93526
619-878-2411, extension 2263
Member Jurisdictions: Inyo and Mono counties

Kern County Council of Governments (Kern COG)
Kress Building-2nd Floor
1401 19th Street
Bakersfield, CA 93301
805-861-2191
Member Jurisdictions: Kern County and 11 cities

Kings County Regional Planning Agency (KCRPA)
Government Center
1400 West Lacey Boulevard
Hanford, CA 93230
209-582-3211, extension 2670
Member Jurisdictions: Kings County and 4 cities

Lake County-City Areawide Planning Council
160 Fifth Street
Lakeport, CA 95453
707-263-5441
Member Jurisdictions: Lake County and 2 cities

Mendocino Council of Governments (MCOG)
Courthouse-Room 110
Ukiah, CA 95482
707-463-4470
Member Jurisdictions: Mendocino County and 4 cities

Merced County Association of Governments (MCAG)
3339 M Street
Merced, CA 95348
209-723-3153
Member Jurisdictions: Merced County and 6 cities

Sacramento Area Council of Governments (SACOG)
106 K Street-Suite 200
Sacramento, CA 95814
916-441-5930
mailing address:
P.O. Box 808
Sacramento, CA 95804
Member Jurisdictions: Placer (part), Sacramento, Sutter, Yolo, and Yuba counties and 14 cities

San Benito County Council of Governments
3220 Southside Road
Hollister, CA 95023
408-637-3725
Member Jurisdictions: San Benito County and 2 cities

San Diego Association of Governments (SANDAG)
Security Pacific Plaza
1200 Third Avenue-Suite 524
San Diego, CA 92101
619-236-5300
Member Jurisdictions: San Diego County and 16 cities

San Joaquin County Council of Governments (SJCCOG)
1860 East Hazelton Avenue
Stockton, CA 95205
209-944-2233
Member Jurisdictions: San Joaquin County and 6 cities

San Luis Obispo Area Council of Governments (SLOACOG)
County Government Center
San Luis Obispo, CA 93408
805-549-5714
Member Jurisdictions: San Luis Obispo County and 7 cities

Santa Barbara County-Cities Area Planning Council
922 Laguna Street
Santa Barbara, CA 93101
805-963-7193
Member Jurisdictions: Santa Barbara County and 5 cities

Sierra Planning Organization and Economic Development District
1230 High Street-Suite 224
Auburn, CA 95603
916-823-4703
Member Jurisdictions: El Dorado (part), Nevada, Placer (part), and Sierra counties and 10 cities

Siskiyou Association of Governmental Entities (SAGE)
County Courthouse Annex
311 Fourth Street
Yreka, CA 96097
916-842-3531, extension 245
mailing address:
P.O. Box 1085
Yreka, CA 96097
Member Jurisdictions: Siskiyou County and 9 cities

Southern California Association of Governments (SCAG)
600 S. Commonwealth Ave., Suite 1000
Los Angeles, CA 90005
213-385-1000
Member Jurisdictions: Imperial, Los Angeles, Orange, Riverside, San Bernardino, and Ventura counties and 125 cities

Subregional Councils:
Coachella Valley Association of Governments (CVAG)
74-133 El Paseo-Suite 4
Palm Desert, CA 92260
619-346-1127
Member Jurisdictions: Riverside County and 8 cities

Imperial Valley Association of Governments (IVAG)
940 West Main Street
El Centro, CA 92243
619-339-4290
Member Jurisdictions: Imperial County and 7 cities

San Bernardino Association of Governments (SANBAG)
444 North Arrowhead Avenue #101
San Bernardino, CA 92401
714-884-8276
Member Jurisdictions: San Bernardino County and 17 cities

Ventura County Association of Governments
950 County Square Drive #100
Ventura, CA 93003
805-654-2882
Member Jurisdictions: Ventura County and 10 cities

Stanislaus Area Association of Goverments (SAAG)
1315 I Street
Modesto, CA 95354
209-571-6200
Member Jurisdictions: Stanislaus County and 9 cities

Tahoe Regional Planning Agency (TRPA)
195 Highway 50
Roundhill, NV
702-588-4547
mailing address:
P.O. Box 1038
Zephyr Cove, NV 89448
Member Jurisdictions: Carson City (part), Douglas (part), El Dorado (basin portion), Placer (basin portion), and Washoe (part) counties, Lake Tahoe Watershed Basin, and 1 city

Tri-County Area Planning Council
c/o Tehama County Planning Department
Courthouse Annex-Room I
Red Bluff, CA 96080
916-527-2200
Member Jurisdictions: Colusa, Glenn, and Tehama counties, and 7 cities

Tulare County Association of Governments (TCAG)
County Court House-Room 111
Visalia, CA 93291
209-733-6303
Member Jurisdictions: Tulare County and 8 cities

Local Agency Formation Cmmissions

California Association of Local Agency Formation Commissions
Erwin Meier Administration Center
625 Court Street-Room 202
Woodland, CA 95695
916-666-8048

Alameda
1221 Oak Street-Room 555
Oakland, CA 94612
415-874-7863

Alpine
P.O. Box 158
Markleeville, CA 96120
916-694-2281

Amador
108 Court Street
Jackson, CA 95642
209-223-6380

Butte
7 County Center Drive
Oroville, CA 95965-3397
916-534-4784

Calaveras
County Government Center
891 Mountain Ranch Road
San Andreas, CA 95249
209-754-3841

Colusa
Colusa County Planning & Building
1217 Market Street
Colusa, CA 95932
916-458-8877

Contra Costa
County Administration Building
651 Pine Street-8th Floor
Martinez, CA 94553
415-372-4090

Del Norte
700 Fifth Street
Crescent City, CA 95531
707-464-7253

El Dorado
360 Fair Lane
Placerville, CA 95667
916-626-2438

Fresno
2220 Tulare Street-Suite 119
Fresno, CA 93721
209-488-1688

Glenn
125 South Murdock
Willows, CA 95988
916-934-3388

Humboldt
3015 H Street
Eureka, CA 95501
707-445-7508

Imperial
c/o Planning Department, Courthouse
El Centro, CA 92243
619-339-4236

Inyo
Inyo County Administrator's Office
P.O. Drawer L
Independence, CA 93526
619-878-2411, extension 2263

Kern
1430 Truxtun Avenue-Suite 802
Bakersfield, CA 93301
805-861-2343

Kings
Government Center
Hanford, CA 93230
209-582-3211, extension 2670

Lake
255 North Forbes Street
Lakeport, CA 95453
707-263-2383

Lassen
Courthouse Annex-Room 103
South Lassen Street
Susanville, CA 96130
916-257-8311, extension 267

Los Angeles
383 Hall of Administration
500 West Temple Street
Los Angeles, CA 90012
213-974-1448

Madera
209 West Yosemite Avenue
Madera, CA 93637
209-675-7703

Marin
Marin Civic Center-Suite 401A
San Rafael, CA 94903
415-499-7395

Mariposa
Planning Department
P.O. Box 2039
Mariposa, CA 95338
209-966-3222

Mendocino
Courthouse
Ukiah, CA 95482
707-463-4470

Merced
c/o Planning Department
2222 M Street
Merced, CA 95340
209-385-7654

Modoc
202 West Fourth Street
Alturas, CA 96101
916-233-3939

Mono
P.O. Box 8060
Mammoth Lakes, CA 93546
619-934-7504

Monterey
P.O. Box 180
Salinas, CA 93902
408-757-2561
408-424-8611, extension 496

Napa
1195 Third Street-Room 210
Napa, CA 94559
707-253-4416

Nevada
Cranmer Engineering, Inc.
P.O. Box 1240
Grass Valley, CA 95945
916-273-7284

Orange
1200 North Main Street-Suite 215
Santa Ana, CA 92701
714-834-2239

Placer
175 Fulweiler Avenue
Auburn, CA 95603
916-823-4644

Plumas
P.O. Box 10437
Quincy, CA 95971
916-283-2000

Riverside
Robert T. Andersen
Administrative Center
4080 Lemon Street-12th Floor
Riverside, CA 92501-3651
714-787-2786

Sacramento
700 H Street-Room 7625
Sacramento, CA 95814
916-440-6458

San Benito
3220 Southside Road
Hollister, CA 95023
408-637-5313

San Bernardino
175 West Fifth Street-2nd Floor
San Bernardino, CA 92415-0490
714-387-5866

San Diego
County Administration Center
1600 Pacific Highway-Room 452
San Diego, CA 92101
619-531-5400

San Joaquin
1810 East Hazelton Avenue
Stockton, CA 95205
209-944-2196

San Luis Obispo
Administrative Office
County Government Center
San Luis Obispo, CA 93408
805-549-5011

San Mateo
County Government Center
Redwood City, CA 94063
415-363-4224

Santa Barbara
922 Laguna Street
Santa Barbara, CA 93101
805-966-1611, extension 7006

Santa Clara
County Government Center
70 West Hedding Street-East Wing
San Jose, CA 95110
408-299-4908, Assistant Executive
408-299-4321, Secretary

Santa Cruz
Governmental Center
701 Ocean Street-Room 318-D
Santa Cruz, CA 95060
408-425-2694

Shasta
1443 West Street
Redding, CA 96001
916-225-5333

Sierra
P.O. Box 530
Downieville, CA 95936
916-289-3251

Siskiyou
P.O. Box 1085
Yreka, CA 96097
916-842-3531, extensions 242, 245

Solano
Environmental Management
601 Texas Street
Fairfield, CA 94533
707-429-6599

Sonoma
575 Administration Drive-Room 104A
Santa Rosa, CA 95401
707-527-2577

Stanislaus
1100 H Street
Modesto, CA 95354
209-571-6203

Sutter
P.O. Box 1555
Yuba City, CA 95992
916-673-7932

Tehama
RMI Courthouse Annex
Red Bluff, CA 96080
916-527-2200

Trinity
P.O. Box 936
Weaverville, CA 96093
916-623-1351

Tulare
County Civic Center
Courthouse-Room 111
Visalia, CA 93291
209-733-6284

Tuolumne
2 South Green Street
Sonora, CA 95370
209-533-5611

Ventura
County Government Center
800 South Victoria Avenue
Ventura, CA 93009
805-654-2576

Yolo
Erwin Meier Administration Center
625 Court Street-Room 202
Woodland, CA 95695
916-666-8048

Yuba
Courthouse-Third Floor
215 Fifth Street
Marysville, CA 95901
916-741-6464

Planning Agencies

Alameda County Planning Department
399 Elmhurst Street-Room 136
Hayward, CA 94544
415-670-5400

Alpine County Planning Department
Route 1, Box 37
Markleeville, CA 96120
916-694-2255

Amador County Planning Department
108 Court Street
Jackson, CA 95642
209-223-6380

Butte County Planning Department
7 County Center Drive
Oroville, CA 95965
916-538-7601

Calaveras County Planning Department
891 Mountain Ranch Road
Government Center
San Andreas, CA 95249
209-754-3841

Colusa County Planning
& Building Department
220 12th Street
Colusa, CA 95932
916-458-8877

Contra Costa County Community
Development Department
651 Pine Street-4th Floor-North Wing
Martinez, CA 94539
415-646-2026

Del Norte County Planning Department
700 Fifth Street
Crescent City, CA 95531
707-464-7253

El Dorado County Planning Division
360 Fair Lane
Placerville, CA 95667
916-626-2438

Fresno County Public Works
& Development Services Department
2220 Tulare Street-6th Floor
Fresno, CA 93721
209-453-5010

Glenn County Planning Department
125 South Murdock Street
Willows, CA 95988
916-934-3388

Humboldt County Plannig
& Building Department
3015 H Street
Eureka, CA 95501
707-445-7541

Imperial County Planning Department
939 Main Street
El Centro, CA 9243
619-339-4236

Inyo County Planning Department
P.O. Drawer L
Independence, CA 93526
619-878-2411

Kern County Planning & Development
Services Department
1415 Truxtun Avenue
Bakersfield, CA 93301
805-861-2615

Kings County Planning Agency
Government Center
1400 West Lacey Boulevard-Bldg. 6
Government Center
Hanford, CA 93230
209-582-3211

Lake County Planning Department
255 North Forbes Street
Lakeport, CA 95453
707-263-2221

Lassen County Planning Department
Courthouse Annex-Room 103
Susanville, CA 96130
916-257-8311

Los Angeles County Regional
Planning Department
320 West Temple Street
Los Angeles, CA 90012
21-974-6401

Madera County Planning Department
135 West Yosemite Avenue
Madera, CA 93637
209-675-7821

Marin County Planning Department
Civic Center-Room 308
San Rafael, CA 94903
415-499-6269

Maiposa County Planning
& Building Services Department
589 Low Gap Road
Courthouse
Ukiah, CA 95482
707-463-4281

Merced County Planning Department
2222 M Street
Merced, CA 95340
209-385-7654

Modoc County Planning Department
202 West Fourth Street
Alturas, CA 96101
916-233-3939

Mono County Planning Department
P.O. Box 8
Mammoth Lakes, CA 93546
619-932-7911

Monterey County Planning Department
P.O. Box 1208
Salinas, CA 93901
408-422-9018

Napa County Conservation, Development
& Planning Department
1195 Third Street-Room 210
Napa, CA 94559
707-253-4416

Nevada County Planning Department
950 Maidu Avenue
Nevada City, CA 95959
916-265-1440

Orange County Environmental
Management Agency
P.O. Box 4048
Santa Ana, CA 92702
714-834-5380

Placer County Planning Department
11414 B Avenue
Auburn, CA 95603
916-823-4721

Plumas County Planning Department
P.O. Box 10437
Quincy, CA 95971
916-283-2000

Riverside County Planning Department
4080 Lemon Street-9th Floor
Riverside, CA 92376
714-787-6181

Sacramento County Planning & Community Development Department
827 Seventh Street-Room 230
Sacramento, CA 95814
916-440-6141

San Benito County Planning Department
3220 Southside Road
Hollister, CA 95023
408-637-5313

San Bernardino County Planning Department
385 North Arrowhead Avenue
San Bernardino, CA 92415
714-387-4141

San Diego County Planning & Land Use Department
5201 Ruffin Road-Suite B
San Diego, CA 92123
619-565-3001

San Francisco City/County Planning
450 McAllister-#401
San Francisco, CA 94102
415-558-6264

San Joaquin County Planning & Building Inspection Department
1810 East Hazelton Avenue
Stockton, CA 95205
209-944-3722

San Luis Obispo County Planning & Building Department
County Government Center
San Luis Obispo, CA 93408
805-549-5600

San Mateo County Planning & Development Division
590 Hamilton Avenue
Redwood City, CA 94063
415-363-4161

Santa Barbara County Resource Management Department
123 East Anapamu Street
Santa Barbara, CA 93101
805-963-7135

Santa Clara County Planning Department
70 West Hedding Street
San Jose, CA 95110
408-299-2521

Santa Cruz County Planning Department
701 Ocean Street-#400
Santa Cruz, CA 95060
408-425-2828

Shasta County Planning Department
1855 Placer Street-Room 102
Redding, CA 916-225-5332

Sierra County Planning Department
P.O. Box 530
Downieville, cA 95936
916-289-3251

Siskiyou County Planning Department
P.O. Box 1085
Yreka, CA 96097
916-842-3531

Solano Environmental Management Department
County Courthouse
601 Texas Street
Fairfield, CA 94533
707-429-6561

Sonoma County Planning Department
575 Administration Drive-Room 105A
Santa Rosa, CA 95401
707-527-2412

Stanislaus County Planning & Community Development Department
1100 H Street
Modesto, CA 95354
209-571-6330

Sutter County Planning Department
P.O. Box 1555
Yuba City, CA 95992
916-741-7400

Tehama County Planning Department
Courthouse Annex-Room I
Red Bluff, CA 96080
916-527-2200

Trinity County Planning Department
P.O. Box 936
Weaverville, CA 96073
916-623-1351

Tulare County Planning & Building Department
County Civic Center-Room 111
Visalia, CA 93291
209-733-6254

Tuolumne County Planning Department
2 South Green Street
Sonora, CA 95370
209-533-5611

Ventura County Planning Division
800 South Victoria Avenue
Ventura, CA 93004
805-654-2488

Yolo County Community Development Agency
292 West Beamer Street
Woodland, CA 95695
916-666-8020

Yuba County Planning & Building Services Department
938 14th Street
Marysville, CA 95901
916-741-6266

Fish and Game Commissions

Alpine County Fish and Game Commission
c/o Board of Supervisors
P.O. Box 158
Markleeville, CA 96120
916-624-2281

Butte County Fish and Game Commission
726 West Eleventh Avenue
Chico, CA 95926
916-342-7470

Calaveras County Fish and Game Commission
Government Center
San Andreas, CA 95249
209-754-3312

El Dorado County Fish and Game Commission
Flying C Road
Cameron Park, CA 95619
916-677-1197

Fresno County Recreation and Wildlife Commission
c/o Fresno County Parks Division
2220 Tulare Street, Suite 800
Fresno, CA 93721
209-488-3004

Glenn County Fish and Game Commission
P.O. Box 1202
Willows, CA 95988
916-934-8150

Humboldt county Fish and Game Advisory Committee
c\o Board of Supervisors
825 Fifth Street
Eureka, CA 95501
707-445-7471

Imperial County Fish and Game Commission
P.O. Box, 1407
Brawley, CA 92227
619-344-3050

Inyo-Mono Counties Fish and Game Advisory Commission
407 West Line Street
Bishop, CA 94514
619-873-6719

Kern County Wildlife Resources Commission
Kern County Parks and Recreation Department
1110 Golden Gate Avenue
Bakersfield, CA 93301
805-861-2345

Lake County Fish and Wildlife Advisory Committee
883 Lakeport Boulevard
Lakeport, CA 95453
707-263-2271

Lassen County Fish and Game Commission
830 Arnold Street
Susanville, CA 96130
916-257-2475

Los Angeles County Fish and Game Commission
Parks and Recreation Department
433 South Vermont Avenue
Los Angeles, CA 90015
213-738-2959

Madera County Fish and Game Commission
128 Madera Avenue
Madera, CA 93637
209-675-7879

Marin County Wildlife and Fisheries Advisory Committee
Civic Center, Room 422
San Rafael, CA 94903
415-499-6352

Mendocino County Fish and Game Advisory Committee
Courthouse
Ukiah, CA 95482
707-468-4281

Mono County, see Inyo-Mono Counties

Monterey County Fish and Game Commission
21 WEst Alisal, Suite 122
Salinas, CA 93901
408-757-8309

Napa County Wildlife Conservation Commission
c/o Napa County Conservation, Development, and Planning Department
1195 Third Street, Room210
Napa, CA 94559
707-253-4416

Nevada County Fish and Wildlife Commission
P.O. Box 1437
Nevada City, CA 95959
916-265-6911

Orange County Fish and Game Commission
P.O. Box 4048
Santa Ana, CA 92702
714-492-4718

Placer County Fish and Game Commission
9405 Craterville Road
Auburn, CA 95603
916-885-6209

Plumas County Fish and Game Commission
P.O. Box 1386
Portola, CA 96122
916-832-4380

Riverside County Fish and Game Commission
P.O. Box 3507
Riverside, CA 92519
714-787-2551

Sacramento County Fish and Game Advisory Committee
c/o Department of Parks and Recreation
3701 Branch Center Road, Room 106
Sacramento, CA 95827
916-366=2061

San Benito County Fish and Game Commission
P.O. Box 1
San Juan Bautista, CA 95045
408-623-4904

San Bernardino County Fish and Game Commission
157 West Fifth Street, 2nd Floor
San Bernardino, CA 92415
714-387-5940

San Diego County Fish and Wildlife Advisory Committee
c/o Department of Planning and Land Use
5201 Ruffin Road, Suite B2
San Diego, CA 92123
619-565-3031

San Luis Obispo County Fish and Game Committee
P.O. Box 406
Morro Bay, CA 93442
805-772-1241

Santa Barbara County Fish and Game Committee
933 Roble Lane
Santa Barbara, CA 93103
805-965-6636

Santa Clara County Fish and Game Commission
70 West Hedding Street
San Jose, CA 95110
408-299-4321

Santa Cruz County Fish and Game Advisory Commission
Government Center
701 Ocean Street
Santa Cruz, CA 95060
408-425-2861

Sierra County Fish and Game Commission
P.O. Box 252
Downieville, CA 95936
916-289-3295

Siskiyou County Fish and Game Commission
P.O. Box 396
Tulelake, CA 96134
916-667-2789

Solano County Fish and Game Advisory Committee
320 Benson Street
Vallejo, CA 94590
707-642-5170

Sonoma County Fish and Wildlife Advisory Board
2604 Ventura Avenue, Room 101P
Santa Rosa, CA 95401
707-527-2371

Stanislaus County Fish and Wildlife Committee
1716 Morgan Road
Modesto, CA 95353
209-525-6339

Sutter County Fish and Game Advisory Committee
463 Second Street
Yuba City, CA 95991
916-673-6650

Tulare County Fish and Game Commission
Agriculture Building, Co. Civic Center
Main and Woodland Drive
Visalia, CA 95291
209-733-6391

Tuolumne County Sportsmen, Inc.
P.O. Box 656
Sonora CA, 95370
209-532-7033

Ventura County Fish and Game Commission
3900 Pelican Way
Oxnard, CA 93030
805-487-7711, extension 4291

Yolo County Fish and Game Commission
373 North College Street
Woodland, CA 95695
916-666-8265

United Nations Environment Programme
P.O. Box 30552
Nairobi, Kenya
333930

UNEP Liaison Office
2 United Nations Plaza
UNDC 2 Building-Room 0803
New York, NY 10017
212-754-8138

UNEP Liaison Office
1889 F Street N.W.
Washington, D.C. 20006
202-289-8456

Council of the Programme on Man and the Biosphere
UNESCO
7 Place de Fontenoy, F 75007
Paris, France
45-68-10-00

Man and the Biosphere
Program Coordinator
National Park Service
P.O. Box 37127
Washington, D.C. 20013-7127
202-343-8122

World Heritage Committee
UNESCO - Cultural Heritage Division
7 Place de Fontenoy, F 75007
Paris, France
45-68-10-00

World Heritage List Program
National Park Service
P.O. Box 37127
Washington, D.C. 20013-7127
202-343-8122

California Conservation Organizations

Abalone Alliance
2940 16th Street #310
San Francisco, CA 94103
415-861-0592

American Farmland Trust
512 2nd Street
San Francisco, CA 94107
415-543-2098

California League of Conservation Voters
965 Mission Street-#705
San Francisco, CA 94103
415-397-7780

California Native Plant Society
909 12th Street
Sacramento, CA 95814
916-447-2677
415-841-5575

California Public Interest Research Group
46 Shattuck Square #11
Berkeley, CA 94704
415-642-9952

California Wilderness Coalition
2655 Portage Bay East-Suite 5
Davis, CA 95616
916-758-0380

Californians Against Waste
P.O. Box 289
Sacramento, CA 95802
916-443-5422

Citizens for a Better Environment
942 Market Street #505
San Francisco, CA 94102
415-788-0690

Ecology Center
1403 Addison Street
Berkeley, CA 94702
415-548-2220

Ecology Center of Southern California
P.O. Box 35473
Los Angeles, CA 90035
213-559-9160

Environmental Defense Fund
5655 College Avenue-Room 304
Oakland, CA 94618
415-658-8008

Friends of the River
Fort Mason-Building C
San Francisco, CA 94123
415-771-0400
and
909 12th Street #207
Sacramento, CA 95814
916-442-3155

Friends of the Sea Otter
P.O. Box 221220
Carmel, CA 93922
408-625-3290

The Fund for Animals
Fort Mason, Building C-Room 262
San Francisco, CA 94123
415-474-4020

Greenpeace
Fort Mason-Building E
San Francisco, CA 94123
415-474-6767

League for Coastal Protection
P.O. Box 421698
San Francisco, CA 94142
415-777-0221

League of Women Voters of California
926 J Street #1000
Sacramento, CA 9814
916-442-7215

League to Save Lake Tahoe
P.O. Box 10110
South Lake Tahoe, CA 95731
916-541-5388/5389

Mono Lake Committee
P.O. Box 29
Lee Vining, CA 93541
619-647-6386/6596

National Audubon Society
555 Audubon Place
Sacramento, CA 95825
916-481-5332

Natural Resources Defense Council
90 New Montgomery Street-Room 620
San Francisco, CA 94105
415-777-0220

The Nature Conservancy
785 Market Street
San Francisco, CA 94103
415-777-0487

Northcoast Environmental Center
879 9th Street
Arcata, CA 95521
707-822-6918

Oceanic Society
Fort Mason, Building E
San Francisco, CA 94123
414-441-5970

Peninsula Conservation Center Foundation
2253 Park Boulevard
Palo Alto, CA 94306
415-328-5313

People for Open Space
116 New Montgomery Street
San Francisco, CA 94105
415-543-4291

Planning and Conservation League
909 12th Street #203
Sacramento, CA 95814
916-444-8726

San Diego Ecology Center
2270 5th Avenue
San Diego, CA
619-238-1984

Save San Francisco Bay Association
P.O. Box 925
Berkeley, CA 94701
415-849-3053

Save the Redwoods League
114 Sansome Street-#605
San Francisco, CA 94104
415-362-2352

Sierra Club
See page 228

The Trust For Public Land
116 New Montgomery Street
San Francisco, CA 94105
415-495-4014

Whale Center
3929 Piedomont Avenue
Oakland, CA 94611
415-654-6621

Zero Population Growth
1025 9th Street #217
Sacramento, CA 95814
916-446-1033

National Conservation Organizations

American Farmland Trust
1717 Massachusetts Avenue N.W.
Washington, D.C. 20036
202-332-0764

American Littoral Society
Sandy Hook
Highlands, NJ 07732
201-291-0055

American Lung Association
1740 Broadway
New York, NY 10019
212-245-8000

American Rivers Conservation Council
322 4th Street N.E.
Washington, D.C. 20002
202-547-6900

Citizens for a Better Environment
942 Market Street #505
San Francisco, CA 94102
415-788-0690

Clean Water Action Project
733 15th Street N.W.-Rom 1110
Washington, D.C. 20005
202-638-1196

The Conservation Foundation
1255 23rd Street N.W.
Washington, D.C. 20037
202-797-4300

Defenders of Wildlife
1244 19th Street N.W.
Washington, D.C. 20036
202-659-9510

Earth First!
P.O. Box 5871
Tucson, AZ 85703

Enviromental Action
1525 New Hampshire Avenue N.W.
Washington, D.C. 20036
202-745-4870

Environmental Defense Fund
444 Park Avenue South
New York, NY 10016
212-686-4191

Environmental Policy Institute
218 D Street S.E.
Washington, D.C. 20003
202-544-2600

Environmental Task Force
1012 14th Street N.W.
Washington, D.C. 20005
202-842-2222

Friends of the Earth
530 7th Street S.E.
Washington, D.C. 20003
202-543-4312

Global Tomorrow Coalition
1325 G Street N.W.-Room 1003
Washington, D.C. 20005
202-879-3040

Greenpeace U.S.A.
1611 Connecticut Avenue N.W.
Washington, D.D. 20009
202-462-1177

International Union for the Conservation of Nature & Natural Resources
Avenue du Mont Blanc, DH-1196
Gland, Switzerland
022-64.71.81

The Izaak Walton League of America
1701 North Fort Meyer Drive #110
Arlington, VA 22209
703-528-1818

League of Conservation Voters
320 4th Street N.E.
Washington, D.C. 20002
202-547-7200

League of Women Voters
1730 M Street N.W.
Washington, D.C. 20036
202-429-1965

National Audubon Society
950 Third Avenue
New York, NY 10022
212-832-3200

National Parks & Conservation Association
1701 18th Street N.W.
Washington, D.C. 20009
202-265-2717

National Wildlife Federation
1412 16th Street N.W.
Washington, D.C. 20036
202-797-6800

Natural Areas Association
320 South 3rd Street
Rockford, IL 61108
815-964-6660

Natural Resources Defense Council
122 East 42nd Street
New York, NY 10168
212-949-0049

The Nature Conservancy
1800 North Kent Street #800
Arlington, VA 22209
703-841-5300

Oceanic Society
Stanford Marine Center
Mayee Avenue
Clamford, CT 06902
203-327-9786

Solar Lobby
1001 Connecticut Avenue #638
Washington, D.C. 20036
202-466-6350

Sierra Club
See page 228

The Trust for Public Land
116 New Montgomery Street
San Francisco, CA 94105
415-495-4014

Union of Concerned Scientists
26 Church Street
Cambridge, MA 02238
617-547-5552

Whale Center
3929 Piedmont Avenue
Oakland, CA 94611
415-654-6621

The Wilderness Society
1400 I Street N.W.
Washington, D.C. 20005
202-842-3400

Worldwatch Institute
1776 Massachusetts Avenue N.W.
Washington, D.C. 20036
202-452-1999

World Wildlife Fund
1255 23rd Street N.W.-Room 200
Washington, D.C. 20037
202-293-4800

Zero Population Growth
1346 Connecticut Avenue N.W.
Washington, D.C. 20036
202-332-2200

Sierra Club
National Headquarters
730 Polk Street
San Francisco, CA 94109
415-776-2211

Washington, D.C. Office
408 C Street N.E.
Washington, D.C. 20002
202-547-1141

Northern California/Nevada Office
5428 College Avenue
Oakland, CA 94618
415-654-7847

Southern California Office
3550 West 6th Street #323
Los Angeles, CA 90020
213-387-6528

Sacramento Office
1014 9th Street #201
Sacramento, CA 95814
916-444-6906

Legal Defense Fund
2044 Fillmore Street
San Francisco, CA 94115
415-567-6100

Sierra Club Foundation
730 Polk Street
San Francisco, CA 94109
415-923-5640

Angeles Chapter
3550 West 6th Street #321
Los Angeles, CA 90020
213-387-4287

Kern-Kaweah Chapter
1619 West Monte Vista
Visalia, CA 93277
209-625-3205

Loma Prieta Chapter
2448 Watson Court
Palo Alto, CA 94303
415-494-9901

Los Padres Chapter
P.O. Box 90924
Santa Barbara, CA 93190
(805) 965-8709

Mother Lode Chapter
P.O. Box 1335
Sacramento, CA 95812
916-444-2180

Redwood Chapter
P.O. Box 466
Santa Rosa, CA 95402
707-544-7651

San Diego Chapter
3820 Ray Street
San Diego, CA 92104
619-299-1744

San Francisco Bay Chapter
6014 College Avenue
Oakland, CA 94618
415-653-6127

San Gorgonio Chapter
568 North Mountain View Avenue #130
San Bernardino, CA 92401
714-381-5015

Santa Lucia Chapter
P.O. Box 15755
San Luis Obispo, CA 93406
805-544-1777

Tehipite Chapter
P.O. Box 5396
Fresno, CA 93755
209-233-1820

Toiyabe Chapter
P.O. Box 8096
Reno, NV 89507
702-323-3162

Ventana Chapter
P.O. Box 5667
Carmel, CA 93921

Acronyms and Initialisms

Bibliography

ACEC Area of Critical Environmental Concern (Interior, Department of)
ABAG "AYBAG" Association of Bay Area Governments
ACP Agricultural Conservation Program (Agriculture, Department of)
AIDS Agency for International Development (Federal)
AMBAG "AMBAG" Association of Monterey Bay Area Governments
APCD Air Pollution Control District
ARB Air Resources Board (State)
BCAG "BECAG" Butte County Associaton of Governments
BIA Bureau of Indian Affairs (Interior, Department of)
BLM Bureau of Land Management (Interior, Department of)
BOR Bureau of Reclamation (Interior, Department of)
CAL-OSHA "CAL-OHSHA" California Occupational Safety and Health Administration
CACOG "CACOG" California Association of Councils of Governments
CALAFCO "CALAFFCOE" California Association of Local Agency Formation Commissions
CALTRANS "CALTRANS" California Department of Transportation
CCC California Conservation Corps
CDFFP California Department of Forestry and Fire Protection
CDPL California Desert Protection League
CEQ Council on Environmental Quality (Executive Office)
CEQA "SEEKWAH" California Environmental Quality Act (1970)
CNACC California Natural Areas Coordinating Council
CNDDB California Natural Diversity Data Base
CNPS California Native Plant Society
COG "COG" Council of Governments
CTC California Transportation Commission
CVAG "SEEVAG" Coachella Valley Association of Governments
CWC California Wilderness Coalition
CUWARFA California Urban Waterfront Area Restoration Financing Authority
DFG Department of Fish and Game (State)
DMG Division of Mines and Geology (State)
DOF Department of Forestry and Fire Protection (State)
DOW Defenders of Wildlife
DPR Department of Parks and Recreation (State)
DWR Department of Water Resources (State)
EIA Energy Information Administration (Energy, Department of)
EIR Environmental Impact Report (State)
EIS Environmental Impact Statement (Federal)
EPIC "EPIC" Energy Conservation Program Guide for Industry and Commerce (Energy, Department of)
ERMIE Environmental Resource Management Element
ESC Endangered Species Committee (Interior, Department of)
FERC "FERK" Federal Energy Regulatory Commission
FIC Federal Information Center
FLPMA "FLIPMAH" Federal Land Policy and Management Act (1970)
FS Forest Service (Agriculture, Department of)
FWQA Federal Water Quality Act (1987)
FWS Fish and Wildlife Service (Interior, Department of)
GEMS "GEMS" Global Environmental Monitoring System (United Nations)
GRID "GRID" Global Resource Information Database (United Nations)
HCAG "AITCHCAG" Humboldt County Association of Governments
IMAGE "IMAGE" Inyo Mono Association of Government Agencies
IRPTC International Register of Potentially Toxic Chemicals (United Nations)
IUCN International Union for Conservation of Nature and Natural Resources
IVAG "EYEVAG" Imperial Valley Association of Governments
KCRPA Kings County Regional Planning Agency
LAFCP "LAFFCOE" Local Agency Formation Commission
LCP Local Coastal Program
MAB "MAB" Man and the Biosphere Programme (UNESCO)
MBC Migratory Bird Commission (Federal)
MCAG "EMCAG" Merced County Association of Governments
MCOG "EMCOG" Mendocino Council of Governments
MEMD Marine and Estuarine Management Division (Commerce, Department of)
MMC Marine Mammal Commission (Federal)
MMS Minerals Management Service (Interior, Department of)
MPO Metropolitan Planning Organization
NAP Non-Attainment Area Plan (air quality)
NEPA "NEEPAH" National Environmental Policy Act
NFS National Forest System
NOAA "NOAH" National Oceanic and Atmospheric Administration(Commerce, Department of)
NOS National Ocean Survey
NPS National Park Service (Interior, Department of)

NRA National Recreation Area
NRC Nuclear Regulatory Commission (Federal)
NRS Natural Reserve System (University of California)
NTS National Trails System
NWPS National Wilderness Preservation System (Federal)
OCS Outer Continental Shelf
OHV Off-Highway Vehicle
ONA Outstanding Natural Area (Interior, Department of)
OPR Office of Planning and Research (State)
OSHA "OHSHA" Occupational Safety and Health Administration (Federal)
PCL Planning and Conservation League
PIRG "PERG" Public Interest Research Group
PNA Protected Natural Area (Interior, Department of)
PSTIP Proposed State Transportation Improvement Program (Caltrans)
PUC Public Utilities Commission
PUC Public Utilities Code
PUNA Public Use Natural Area (Interior, Department of)
RARE II "RARE TWO" Roadless Area Review and Evalution-II (second evaluation, 1979)
RNA Research Natural Area (Federal)
RTIP Regional Transportation Improvement Program
RTP Regional Transportaion Plan
RTPA Regional Transportation Planning Agency
SAAG "SAG" Stanislaus Area Association of Governments
SACOG "SAYCOG" Sacramento Area Council of Governments
SAGE "SAGE" Siskiyou Association of Government Entities
SANBAG "SANBAG" San Bernardino Association of Governments
SANDAG "SANDAG" San Diego Association of Governments
SCAG "SKAG" Southern California Association of Governments
SCC State Coastal Conservancy
SERC Smithsonian Environmental Research Center (Federal)
SFEP San Francisco Estuary Project (Environmental Protection Agency)
SIA Special Interest Area (Agriculture, Department of)
SIP State Implementation Plan (air quality)
SJCCOG San Joaquin County Council of Governments
SLOACOG "SLOWCOG" San Luis Obispo Area Council of Governments
SHBRP Santa Monica Bay Restoration Project (Environmental Protection Agency)
STA State Transit Assistance Fund
STIP State Transportation Improvement Program
SVRATS State Vehicular Recreation Area and Trail System
TCAG "TEECAG" Tulare County Association of Governments
TPL Trust for Public Land
TPZ Timberland Production Zone
TRPA Tahoe Regional Planning Agency (Bi-State)
UMTA Urban Mass Transportation Administration
UNEP "YOUNEP" United Nations Environment Programme
UNESCO "YOUNESCOE" United Nations Educational, Scientific, and Cultural Organization
USACE United States Army Corps of Engineers
USFS United States Forest Service (Agriculture, Department of)
USFWS United States Fish & Wildlife Service (Interior, Department of)
USGS United States Geological Survey (Interior, Department of)
WAPA "WAHPAH" Western Area Power Administration (Energy, Department of)
WBP Water Bank Program (Agriculture, Department of)
WCB Wildlife Conservation Board (State)
WHR California Wildlife-Habitat Relationships System
ZPG Zero Population Growth

Books

1 Airola, Daniel A. *Guide to the California Wildlife Habitat Relationship System.* Sacramento: Department of Fish and Game, 1988

2 Allen, Thomas B. *Guardian of the Wild: The Story of the National Wildlife Federation, 1936-1986.* Bloomington: Indiana University Press, 1987.

3 *Areas of Critical Concern and Proposals For Their Protection.* Revised ed. Sacramento: Sacramento Audobon Society, 1989

4 *Atlas of California Coastal Marine Resources.* Sacramento: Department of Fish & Game, 1980.

5 Ayensu, Edward and De Filippo, Robert A. *Endangered and Threatened Plants of the United States.* Washington, D.C.: Smithsonian, 1978.

6 Bakker, Elna. *An Island Called California.* 2nd ed. Berkeley: University of California Press, 1984.

7 Barbour, Michael G. and Major, Jack. *Terrestrial Vegetation of California.* New ed. Davis, California: California Native Plant Society, 1988.

8 Beck, Warren A. and Haase, Ynez D. *Historical Atlas of California.* Norman: University of Oklahoma Press, 1974.

9 Bender, Gordon L. *Reference Handbook On the Deserts of North America.* Westport, Connecticut: Greewood Press, 1982.

10 *Boating Trails for California Rivers.* Sacramento: Resources Agency, 1978.

11 Brainard, John C. and McGrath, Roger N. *The Directory of National Environmental Organizations.* 2nd ed. St. Paul, Minnesota: United States Environmental Directories, Inc., 1986.

12 Browning, Bruce M. *Natural Resources of Elkhorn Slough: Their Present and Future Use.* Sacramento: Department of Fish and Game, 1972.

13 Burke, Mary T. et al. *Natural Landmarks of the Sierra Nevada.* Davis, California: Department of Botany, University of California, Davis, 1982.

14 Bushwick, Nancy and Heimstra, Hal D. *National Directory of Farmland Protection Agencies.* Washington, D.C.: The Farmland Project, 1983.

15 *California Atlas and Gazetteer.* Vol. I Northern California, Vol. II Southern California. Freeport, Maine: DeLorme Publishing Co., 1986.

16 California Coastal Commission. *California Coastal Access Guide.* 3rd ed. Berkeley: University of California Press, 1982.

17 ---------------. *California Coastal Resource Guide.* Berkeley: University of California Press, 1987.

18 *California's Coastal Wetlands.* La Jolla, California: California Sea Grant College Program, 1978.

19 *California County Fact Book.* Sacramento: County Supervisors Association of California, 1988-89 (annual).

20 *California Gazetteer.* Wilmington, Delaware: American Historical Publications Inc., 1985.

21 *The California Planner's Book of Lists.* Sacramento: Office of Planning and Research, 1989 (annual).

22 *California Statistical Abstract.s* Sacramento: Department of Finance, 1988. (annual.)

23 *California Wind Atlas.* Sacramento: California Energy Commission, 1985.

24 *California's Important Fish and Wildlife Habitat: An Inventory.* Portland, Oregon: U.S. Fish and Wildlife Service, 1980.

25 Callaham, Robert Z. and Stangenberger, Alan G. *Directory to Expertise and Facilities Related to Wildlands.* Berkeley: Wildlands Resource Center, University of California, 1985.

26 Carr, Patrick ed. *The Sierra Club: A Guide.* San Francisco: The Sierra Club, 1989.

27 *Channel Islands National Marine Sanctuary Management Plan.* Washington, D.C.: Oceanic and Atmospheric Administration, 1983.

28 Cheatham, Norden H. (Dan) and Haller, J. Robert. *An Annotated List of California Habitat Types.* Berkeley: University of California Land & Water Reserves System, 1975.

29 *Citizens' Report on the Diked Historic Baylands of San Francisco Bay.* Sausalito, California: Bay Institute of San Francisco, 1987

30 Clark, John, Lok, Debra D. and Caudill, James D. *California Coastal Catalog.* Sandy Hook, New Jersey: American Littoral Society, 1980.

31 Clements, John. *California Facts.* Dallas: Clements Research Inc., 1985.

32 *Climatic Atlas of the United States.* Reprint ed. Asheville, North Carolina: National Oceanic and Atmospheric Administration, 1983 (1968).

33 Cohen, Michael P. *The History of the Sierra Club 1892-1970.* San Francisco: Sierra Club Books, 1988

34 Comp, T. Allan. *Blueprint For the Environment: A Plan for Federal Action.* Salt Lake City: Howe Brothers, 1989

35 Conradson, Diane R. *Exploring Our Baylands.* Point Reyes, California: Coastal Parks Association Inc., 1982.

36 *Conservation Directory: A List of Organizations, Agencies, and Officials Concerned with Natural Resources Use and Management.* Washington, D.C.: National Wildlife Federation, 1989 (annual).

37 Cooperrider, Allen Y., Boyd, Raymond J. and Stuart, Hanson R. *Inventory and Monitoring of Wildlife Habitat.* Denver, Colorado: Bureau of Land Management, 1986.

38 Crippen, J. *An Inventory of Large Lakes in California.* Menlo Park: United States Geologic Survey, 1969.

39 *Dams Within the Jurisdiction of the State of California.* Bulletin 17-88. Sacramento: Department of Water Resources, 1988.

40 Denney, Richard J. Jr.; Monahan, Michael A.; and Hickok, Michael L. *California Environmental Law Handbook.* 3rd ed. Rockville, Maryland: Government Institutes, 1989.

41 Donley, Michael W. et al. *Atlas of California.* Culver City, California: Pacific Book Center, 1979.

42 Doron, William D. *Legislating For the Wilderness: RARE II and the California National Forests.* Millwood, New York: Associated Faculty Press Inc., 1986.

43 Douglas, Edward M. *California Mountain Passes.* Washington, D.C.: Federal Board of Surveys and Maps, 1929.

44 ---------------. *Gazetteer of the Mountains of California.* Preliminary (incomplete) edition. Washington, D.C.: Federal Board of Surveys and Maps, 1929.

45 Durrenberger, Robert W. and Johnson, Robert B. *California: Patterns On the Land.* 5th ed. Palo Alto: Mayfield Publishing Company, 1976.

46 *Earthquakes and Earthquake Faults of California.* Van Nuys, California: Varna Enterprises, 1980.

47 *Energy Resources in California.* Sacramento: California Energy Commission, 1981.

48 Engbeck, Joseph H. *State Parks of California: From 1864 to the Present.* Portland, Oregon: Charles H. Belding, 1980.

49 Estes, Carol and Sessions, Keith W., eds. *Controlled Wildlife: A Three Volume Guide to U.S. Wildlife Laws and Permit Procedures.* Lawrence, Kansas: Association of Seystematics, 1983-1985.

50 Eu, March Fong. *Roster: California State, County, City, and Township Officials. State Officials of the United States.* Sacramento: California General Services Department, 1989.

51 Fairfax, Salley K. and Yale, Carolyn F. *Federal Lands: A Guide to Planning, Management, and State Reserves.* Washington, D.C.: Island Press, 1987

52 Fay, James S.; Fay, Stephanie W., and Boehm, Ronald J. *California Almanac.* 3rd ed. Novato, California: Pacific Data Resources, 1987.

53 *Fish and Wildlife 2000: A Plan for the Future.* Washington, D.C.: Bureau of Land Management. 1988.

54 Fisher, Michael S. and Schiel, David R. *The Ecology of Giant Kelp Forests in California: A Community Profile.* Slidell, Louisiana: U.S. Fish and Wildlife Service, 1985.

55 Foster, Lynne. *Adventuring in the California Desert: The Sierra Club Travel Guide to the Great Basin, Mojave, and Colorado Desert Regions of California.* San Francisco: Sierra Club Books, 1987.

56 Frayer, W. E.; Peters, Dennis R.; and Pywell, H. Ross. *Wetlands of the California Central Valley: Status and Trends, 1939 to Mid-1980's.* Portland, Oregon: U.S. Fish and Wildlife Service, 1989.

57 Frome, Michael. *The Forest Service.* 2nd Ed. Revised. Boulder, Colorado: Westview Press, 1983

58 Gilpin, Alan. *Dictionary of Environmental Terms.* St. Lucia, Australia: University of Queensland Press, 1976.

59 Ginsberg, Jonathan; Mintz, Robert; and Walter, William S. *The Fragile Balance: Environmental Problems of the California Desert.* Palo Alto: Stanford Environmental Law Society, 1976.

60 Goodwin, Richard H. and Niering, William. *Inland Wetlands of the United States Evaluated as Potential National Natural Landmarks.* Washington, D.C.: Government Printing Office, 1971.

61 Griffin, Paul E. and Young, Robert N. *Atlas of California: A Living Geography.* San Francisco: Fearon Publishers, 1956.

62 *A Guide to Our Federal Lands.* Washington, D.C.: National Geographic Society, 1984.

63 Halliday, William R. *Caves of California: A Special Report of the Western Speleological Society in Cooperation With the National Speleological Society.* Seattle, Washington: Halliday, 1962.

64 *The Harbinger File: A Directory of Citizen Groups, Government Agencies and Environmental Education Programs Concerned with California Environmental Issues.* Santa Cruz, California: Harbinger Communications, 1988-89 (biannual).

65 Harris, Stephen L. *Fire Mountains of the West: The Cascade and Mono Lake Volcanoes.* Missoula, Montana: Mountain Press Publishing Company, 1988.

66 Hillinger, Charles. *The California Islands.* Los Angeles: Academy Publishers, 1985.

67 Holing, Dwight. *California Wild Lands: A Guide to the Nature Conservancy Preserves.* San Francisco: Chronicle Books, 1988.

68 Hood, Leslie, ed. *Inventory of California Natural Areas.* 14 volumes. Sonoma, California: California Natural Areas Coordinatiing Council, 1975.

69 Hoose, Phillip M. *Building an Ark: Tools for the Preservation of Natural Diversity Through Land Protection.* Covelo, California: Island Press, 1981.

70 Horan, Andrew. *California: National Forests.* Helena, Montana: Falcon Press, 1988.

71 Hornbeck, David and Kane, Phillip. *California Patterns: A Geographical and Historical Atlas.* Palo Alto: Mayfield Publishing Company, 1983.

72 Hubler, Kathryn and Henderson, Timothy R. *Directory of State Environmental Agencies.* Washington, D.C.: Environmental Law Institute, 1982.

73 Iacopi, Robert. *Earthquake Country.* Menlo Park: Lane Books, 1964.

74 Jaeger, Edmond C. *The California Deserts.* Stanford: The Stanford University Press, 1955.

75 Jessup, Deborah Hitchock. *Guide to State Environmental Programs.* Washington, D.C.: Bureau of National Affairs, 1988

76 Josselyn, Michael (ed.) *Wetland Restoration and Enhancement in California.* La Jolla, California: California Sea Grant College Program, 1982.

77 Kahrl, William L. et al. *California Water Atlas.* Sacramento: State of California, 1978.

78 Kitazono, Oki. *Directory of Nature Centers and Related Environmental Education Facilities.* New York: National Audubon Society, 1979.

79 Lantis, David W., Steiner, Rodney, and Karinen, Arthur F. *California: Land of Contrast.* Revised 3rd ed. Dubuque, Iowa: Kendall/Hunt Publishing Co., 1981.

80 Larson, Peggy. *The Deserts of the Southwest.* San Francisco: Sierra Club Books, 1977.

81 Lewis, Robin D. and McKee, Kinberly K. *A Guide to the Artificial Reefs of Southern California.* Sacramento: Department of Fish and Game, 1989

82 *Liason Conservation Directory For Endangered and Threatened Species.* Washington, D.C.: U.S. Fish and Wildlife Service, 1983.

83 Loam, Jayson, and Sohler, Gary. *Hot Springs and Hot Pools of the Southwest. Revised Ed.* Berkeley: Wilderness Press, 1985.

84 Lollock, Don, et al. *California Wetlands: An Element of the California Outdoor Recreation Planning Program.* Sacramento: Department of Parks and Recreation, 1988.

85 Mayer, Kenneth E. and Laudenslayer, William F. Jr. *A Guide to Wildlife Habitats of California.* Sacramento: Department of Forestry and Fire Protection, 1988.

86 McDonnell, Lawrence R. *Rivers of California.* San Francisco: Pacific Gas and Electric Co., 1962.

87 Mohlenbrock, Robert H. *Field Guide to the U.S. National Forests.* New York: Congdon and Weed, 1984.

88 Muhn, James and Stuart, Hanson R. *Opportunity and Challenge: The Story of BLM.* Washington, D.C. Bureau of Land Management, 1988.

89 *National Estuarine Inventory: Data Atlas. Volume 1: Physical and Hydrologic Characteristics.* Rockville, Maryland: National Oceanic and Atmospheric Administration, 1985.

90 *National Gazetteer of the United States of America,. California* Preliminary Edition. Washington, D.C.: U.S. Geological Survey, 1987.

91 *The National Parks: Index 1987.* Washington, D.C.: National Park Service, 1987.

92 Norris, Robert M. and Webb, Robert W. *Geology of California.* New York: Wiley, 1976.

93 *The Natural Resources of Elkhorn Slough: Their Present and Future Use.* Sacramento: California Department of Fish and Game, January 1972.

94 Oakeshott, Gordon B. *California's Changing Landscapes: A Guide to the Geology of the State.* 2nd ed. New York: McGraw Hill, 1978.

95 Osborne, Michael. *Granite, Water and Light: Waterfalls of Yosemite Valley.* Yosemite, California: Yosemite Natural History Association, 1983.

96 Philbrick, R. N., ed. *Proceedings of the Symposium on the Biology of the California Islands.* Santa Barbara: Santa Barbara Botanic Gardens, 1967.

97 Power, Dennis M., ed. *The California Islands: Proceedings of a Multidisciplinary Symposium.* Santa Barbara: Santa Barbara Museum of Natural History, 1980.

98 *Public Land Statistics 1988.* Washington, D.C.: Bureau of Land Management, 1989. (annual).

99 Riley, Laura and Riley, William. *Guide to the National Wildlife Refuges.* Garden City, NY: Anchor Press, 1979.

100 Ringold, Paul L. and Clark, John. *The Coastal Almanac: For 1980—The Year of the Coast.* San Francisco: W. H. Freeman and Company, 1980.

101 *Sacramento—San Joaquin Delta Atlas.* Sacramento: Department of Water Resources, 1987.

102 Sarnoff, Paul. *The New York Times Encyclopedic Dictionary of the Environment.* New York: Quadrangle Books, 1971.

103 Simkin, Tom et al. *Volcanoes of the World.* Washington, D.C.: Smithsonian Institution, 1981.

104 Smith, Emil J. Jr. and Johnson, Thom H. *The Marine Life Refuges and Reserves of California.* Revised ed. Sacramento: Department of Fish and Game, 1989.

105 Todd, David Keith. *The Water Encyclopedia.* Fort Washington, NY: Water Information Center, 1970.

106 Trzyna, Thaddeus and Gotelli, Ilze. *California Environmental Directory: A Guide to Organizations and Resources.* 4th ed. Claremont, California: California Institute of Public Affairs, 1988.

107 Trzyna, Thaddeus C. *The California Handbook: A Congressional Guide to Sources of Information and Actions With Selected Background Material.* 5th ed. Claremont, California: Institute of Public Affairs, 1988.

108 *Tule Elk in California: A Report to the Legislature.* Sacramento: Department of Fish and Game, 1989.

109 *United Nations List of National Parks and Protected Areas.* Gland, Switzerland: International Union for Conservation of Nature and Natural Resources, 1985.

110 *United States Earthquakes.* Washington, D.C.: National Earthquake Information Center, 1989. (annual).

111 *The United States Goverment Manual 1988/1989.* Washington, D.C.: The Federal Register, 1988 (annual).

112 *U.S. Government Offices in California: A Directory.* 3rd ed. Claremont, California: California Institute of Public Affairs, 1987.

113 Wallace, Joe C. *An Inventory of Medium-Sized Lakes in California.* Menlo Park: U.S. Geological Survey, 1970.

114 Warner, Richard E. and Hendrix, Kathleen M. *California's Riparian Systems.* Berkeley: University of California Press, 1984.

115 Wengert, Norman; Dyer, N. A. and Deutsch, Henry A. *The Purposes of the National Forests: A Historical Reinterpretation of Policy Development.* Fort Collins, Colorado: U.S. Forest Service, 1979.

116 *Wetlands: Their Use and Regulation.* Washington, D.C.: U.S. Congress Office of Technology Assessment, 1984.

117 *Where The Wildlife Roam: A Report on Wildlife Habitat on Public Lands in California Administered by the U.S. Bureau of Land Management.* Washington, D.C.: Bureau of Land Management, 1989.

118 Wilkinson, Charles F. and Anderson, Michael H. *Land and Resource Planning in the National Forests.* Washington, D.C.: Island Press, 1987.

119 *The World's Greatest Natural Areas: An Indicative Inventory of Natural Sites of World Heritage Quality.* Gland, Switzerland: IUCN, Commission on National Parks and Protected Areas, 1982.

120 Wrona, June ed. *Recreation 2000: A Strategic Plan.* Washington, D.C.: Bureau of Land Management 1988.

121 Zedler, Joy B. and Nordby, Christopher S. *The Ecology of Tijuana Estuary, California: An Estuarine Profile.* Slidell, Louisiana: U.S. Fish and Wildlife Service, 1985.

122 Zeiner, David C., Laudenslayer, William F. Jr. and Mayer, Kenneth E. *California's Wildlife: Vol I Amphibians and Reptiles, Vol II Birds, Vol III Mammals.* Sacramento: Department of Fish and Game, 1988+.

Journal Articles

123 Arnold, John. "California Underwater Parks!," *Parklands,* 2 No. 3, Summer 1988: 14-15.

124 Bogue, Gary L. "The Price You Pay For a Wild Pet," *Outdoor California,* November-December 1977: 39-40; January-February 1978: 25-26.

125 Browning, Bruce M. "Coastal Wetlands—Two-Thirds Gone," *Outdoor California,* January-February 1980: 24, 27.

126 Burnett, John L. "Glacier Trails of California," *Mineral Information Service* (later: *California Geology*) 17 No. 3, March 1964: 33-34, 44-51.

127 "California Desert 1988," *Sunset,* March 1988: 97-111.

128 "Central Valley Wetlands: Winter Home for 10 to 12 Million Water Fowl," *Outdoor California.* January-February 1980: 30-32.

129 Chesterman, Charles W. "Volcanism in California," *California Geology,* August 1971: 139-147.

130 "California State Coastal Conservancy, Annual Report," 1976.

131 Clark, Jeanne. "California's Wild Places," *Outdoor California,* January-February 1980: 19-20, 21-54.

132 Cohen;, Russell D. "The National Natural Landmark Program: A Natural Areas Protection Technique for the 1980's and Beyond," *UCLA Journal of Environmental Law and Policy.* Fall 1982. 119-163.

133 Corker, Mike. "California Wetlands for California Waterfowl," *Outdoor California,* March-April 1986: 20-22.

134 Delisle, Glenn E. "Environmental Interests vs. Water Use Interests: They Can Conflict," *Outdoor California,* July-August 1979: 15-17.

135 Dick, Dave. "Relocating Tule Elk," *Outdoor California,* January-February 1978: 1-3, 10.

136 Edgar Blake. "Suffering Symbol of the Mojave," *Pacific Discovery,* Winter 1990: 17-21

137 Erlich, Paul R. "Habitats in Crisis," *Wilderness,* Spring 1987: 12-15

138 Fell, George "The Natural Areas Movement in the United States, Its Past and Its Future", *Natural Areas Journal,* 3 No. 4, 1983: 47-55.

139 Fisher, Charles K. "No Guarantees for Minimum Water Flows for Fish and Wildlife," *Outdoor California,* July-August 1979: 10-11.

140 Ford, Lawrence D. and Norris, Kenneth S. "The University of California Natural Reserve System," *Bioscience,* July-August 1988: 463-470.

141 Fox, Stephen. "We Want No Straddlers," *Wilderness,* Winter 1984: 5-19

142 Gray, Randall L. "Water Bank Program Protects Wetlands," *Outdoor California,* March-April 1979: 9.

143 Gregg, William P. Jr. and McGean, Betsy Ann. "Biosphere Reserves: Their History and Their Promise," *Orion Nature Quarterly,* Summer 1985: 40-51.

144 Hansen, Richard J. "The DFG's Fish and Wildlife Water Pollution Control Lab," *Outdoor California,* July-August 1979: 24-26.

145 Hart, Earl W. "Zoning for Surface Fault Hazards in California: The New Special Studies Zone Maps," *California Geology,* October 1974: 227-231.

146 Hashagen, Ken. "California's Hatchery System," *Outdoor California,* March-April 1988: 5-8.

147 Hill, Mary. "Living Glaciers of California," *California Geology,* August 1975: 171-177.

148 --------------. "Glaciers of Mt. Shasta," *California Geology,* April 1977: 75-80.

149 Hone, Elizabeth. "Tales of the Sea Otter," *Pacific Discovery,* July-September 1988: 10-15.

150 Hoopaugh, Dave. "What Happened to the Salmon", *Outdoor California,* July-August 1979: 12-14.

151 Hooten, Mark M. "Caring for Wildlife Babies," *Outdoor California*, January-February 1988: 10.
152 "An Introduction to Wetlands—What are they?" *Outdoor California*, January-February 1980: 20.
153 Jarvis, Elena. ""Rambo's Racers," *Defenders*, January-February 1990: 32-34.
154 Jarvis, Michael. "The Beleaguered Guardians of California's Coast," *California Journal*, October 1989: 403-406.
155 Jones, Bruce E. "Once Considered Wastelands," *Outdoor California*, January-February 1980: 21-23.
156 Jones, Bruce E. and Sands, Ann. "Watershed Management," *Outdoor California*, January-February 1980: 25-27.
157 Juday, Glenn Patrick. "The Outcome of Research Natural Areas in National Forest Planning," 1986, *Natural Areas Journal*, 6, No. 1, 1986: 43-53.
158 Killian, Ronald. "Selected Natural Diversity Bibliography with Annotations" *Natural Areas Journal*, 2 No. 4, 1982: 12-27.
159 Loft, Eric R. "A Success Story for California," *Outdoor California*, July-August 1989: 1-4.
160 Mallette, Bob. "Wildlife Areas of California," *Outdoor California*, November-December 1983: 21-23.
161 Moreno Tracey. "Help for Injured, Orphaned, Wildlife," *Outdoor California.*, January-February 1988: 11-14.
162 Mott, William Penn. "Looking Beyond National Park Boundaries," *Natural Areas Journal*, 8 No. 2, 1988: 80-82.
163 Nagano, Christopher D. and Sakai, Walter H. "Making the World Safe For Monarchs," *Outdoor California*, January-February 1988: 5-9.
164 Nash, Roderick. "Path to Preservation," *Wilderness*, Summer 1984: 5-11
165 Nathenson, Si. "Elephant Seals Win Battle for Survival," *Outdoor California*, May-June 1980:7-8.
166 "Programme on Man and the Biosphere, (MAB) Task Force on Criteria and Guidelines for the Choice and Establishment of Biosphere Reserves," *MAB Report Series*, No. 22, May 1974: 20-24.
167 Rae, Stephen P. "Vernal Pools: A Late Spring Phenomenon," *Outdoor California*, January-February 1980: 29.
168 Raub, W.B. et al. "Perennial Ice Masses of the Sierra Nevada, California," *Proceedings of the Association of Hydrological Science*, No. 126, 1980: 33-34.
169 Reffalt, William C. "Wetlands in Extremis: A Nationwide Survey," *Wilderness*, Winter 1985: 28-41.
170 Reisner, Marc. "A Decision for the Desert," *Wilderness*, Winter 1986: 33-38, 48-53.
171 Rutsch, Alvin G. "Forty Years of Wildlife Preservation," *Outdoor California*, May-June 1988: 1-9.
172 Scott, Douglas W. "The National Preservation System-Its Place in Natural Area Protection," *Natural Areas Journal*, 4 No. 4, 1984: 6-19
173 Sheehan, Joe. "State Mandated Program to Restore and Enhance California Wetlands," *Outdoor California*, January-February 1980: 31-32.
174 Schmieder, Robert. "Cordell Bank: An Oceanic Marvel," *Defenders*, May-June 1988: 25-29.
175 Slack, Gordy. "The Politics of Preservation," *Pacific Discovery*, Winter 1990: 22-31.
176 Smith, Kent A. "Food and Cover for Wildlife Depends on Water," *Outdoor California*, July-August 1980: 21-23.
177 Speth, John. "New Approach to Save Remaining Wetlands in Bay-Delta," *Outdoor California*, January-February 1980: 28-29.
178 Spicer, Richard C. "Selecting Geological Sites for National Natural Landmark Designation," *Natural Areas Journal*, 7 No.4, 1987: 157-173.
179 Stankey, George H. "Wilderness Preservation Activity at the State Level: A National Review," *Natural Areas Journal*, 4 No. 4, 1984: 20-28.
180 Stebbins, Robert C. "A Desert at the Crossroads," *Pacific Discovery*, Winter 1990: 3-16, 32-36.
181 Steele, Jim. "San Nicolas Island," *Outdoor California*, September-October 1979: 29-30.
182 Steinhart, Peter. "Desert Folly, Desert Hopes," *Defenders*, January-February 1990: 10-31.
183 Stevens, Donald E. "Since 1970 Water Projects Have Caused A Young Bass Survival Reduction To An Average of About 30 Percent," *Outdoor California*, July-August 1979: 18-20.
184 Tharratt, Bob and Lollock, Don. "Habitat: The Key to California's Fish and Wildlife Future," *Outdoor California*, January-February 1980: 65-69.
185 Trent, D.D. "California's Ice Age Lost: The Palisade Glacier, Inyo County," *California Geology*, December 1983: 264-269.
186 Trumbly, James M. and Gray, Kenneth L. "The California Wilderness Preservation System," *Natural Areas Journal*, 4 No. 4, 1984: 29-35.
187 Unkel, Chris. "California's Wildlands Program," *Outdoor California*, January-February 1989: 5-9.
188 ---------------. "Ecological Reserves Save Wildlife," *Outdoor California*, January-February 1986: 22-26.
189 Unkel, Chris and Weisner, Pete. "The Change in DFG Funding," *Outdoor California*, January-February 1989: 10-11.
190 Unkel, William C. "Natural Diversity and National Forest Planning," *Natural Areas Journal*, 5 No. 4 1985: 8-13.
191 Vandie, Vince. "Water As an Ecological Balance Wheel," *Outdoor California*, July-August 1979: 9.
192 Watkins, T. H. "The Conundrum of the Forest," *Wilderness* Spring 1986: 12-24, 34-49.

193 Weber, Michael, "Marine Sanctuaries: A Forgotten Cause?" *Defenders*, May-June 1988: 17-23.

194 "What Is a Glacier," *California Geology*, February 1974: 23-32.

195 Wilburn, Jack. "Beach and Dune," *Outdoor California*, July-August 1986: 13-17.

196 ---------------. "Coastal Salt Marshes," *Outdoor California*, July-August 1984: 26-28.

197 ---------------. Redwood Forest," *Outdoor California*, January-February 1985: 13-16.

198 ---------------. "Riparian Woodland," *Outdoor California*, May-June 1986: 20-23.

199 "Wilderness America," *Wilderness*, Spring, 1989: 3-15.

200 Williams, David C. and Campbell, Faith. "How the Bureau of Land Management Designates and Protects Areas of Critical Environmental Concern: A Status Report With a Critical Review By The Natural Resources Defense Council," *Natural Areas Journal*, 8 No. 4, 1988: 231-237.

201 Wolfe, Gary and Nagano, Christopher D. "Migratory Monarch Butterflies: Wintering On California's Coast," *California Scenic*, February 1988: 20-22.

202 Wright, Gene."The National Estuarine Reserve Research System: A Review," *Natural Areas Journal*, 7 No. 2, 1987: 75-78.

Index

Abalone Cove Ecological Reserve, 122, 124
Marine Life Refuge, 114
Abbot, Mount - see Mount Abbot
Abbott Lake Ecological Reserve, 122, 124
Aberdeen Volcanic Area, 40
Administrative Sites, U.S. Fish and Wildlife Service, 132
Admiral William H. Standley Redwood Park, 104
Afton Canyon
Area of Critical Environmental Concern, 156
Important Habitat Area, 78
Riparian Habitat Demonstration Area, 152
Wetland, 62
Agua Hedionda Wetland, 60
Agua Tibia Wilderness, 88
Air Resources Board, 10
Alameda County, 2
Williamson Act County, 188
Alameda South Shore Wildlife Refuge, 58
Alamo Mountain, 198
Alamo River, 44, 46
Albany Mudflats Ecological Reserve, 122, 124
Marine Life Refuge, 114
Albion River, 46
Wetland, 60
Alkali Sink Ecological Reserve, 122, 124
Allensworth Ecological Reserve, 122, 124
Alligator Rock Area of Critical Environmental Concern, 156
Almanor, Lake - see Lake Almanor
Alpine County, 2
Alpine Habitats, 80
Alton Avenue Bike Trail, 178
Alturas, County Seat, 2
Amador County, 2
Williamson Act County, 188
Amargosa Canyon Area of Critical Environmental Concern, 156
Amargosa River, 44, 46
California Nature Conservancy Preserve, 172, 174
Amboy Crater
National Natural Landmark, 142
Volcanic Area, 40
American River, 44, 46
Hatchery, 116
National Recreation Area, 194
Wild and Scenic, 48
American River Bluffs National Natural Landmark, 142
Anacapa Island, 66, 162
Area of Special Biological Significance, 158
Ecological Reserve, 114, 122, 124
Offshore Land, 112
Anaheim Bay Wetland, 60
Ancient Bristlecone Pine Forest Special Interest Area, 146
Anderson Island Natural Preserve, 100
Anderson Marsh
Historic Park, 94
Natural Preserve, 100
Andrew Molera Redwood Park, 104
Angeles Chapter, Sierra Club, 180
Angeles National Forest, 84
Animals Extinct in California, 81
Año Nuevo Island, 64
Habitat Site, 74, 76
Offshore Land, 112
University of California Natural Reserve System, 166
Año Nuevo Point and Island
Area of Special Biological Significance, 158
National Natural Landmark, 142
Año Nuevo State Reserve, 98
Marine Life Refuge, 114
Ansel Adams National Wilderness, 88
Antelope Valley
State Reserve, 98
Wildlife Area, 106
Antioch Dunes National Wildlife Refuge, 132
Antioch Fault, 36
Antone Meadows Natural Preserve, 100
Anza-Borrego Desert State Park, 20, 70
Biosphere Reserve, 128
National Natural Landmark, 142
State Wilderness, 102
Anza-Borrego Sky Trail, 86
Appalachian Trail, 86
Apricum Hill Ecological Reserve, 122, 124
Aquatic Habitats, 80
Arcata Bay Wetland, 60
Areas of Critical Environmental Concern-Bureau of Land Management, 154, 156
Areas of Special Biological Significance, 60
Water Resources Control Board, 158
Arguello Fault, 38
Armstrong Redwoods
Redwood Park, 104
State Reserve, 98
Arrowhead Landmark Special Interest Area, 146
Arroyo Salada OHV Area, 198
Arroyo Seco, 44, 46
Artificial Reefs-Department of Fish and Game, 126, 127
Arubay Volcanic Area, 40
Ash Creek, 152
Wildlands Program Area, 110
Wildlife Area, 106
Ash Valley
Area of Critical Environmental Concern, 154, 156
Research Natural Area, 136
Ashpan OHV Area, 198
Association of Bay Area Governments, 4
Association of Monterey Bay Area Governments, 4
Atascadero Artificial Reef, 126, 127
Atascadero Beach Pismo Clam Preserve Marine Life Refuge, 114
Atascadero Creek Marsh Wetland, 62
Auburn, County Seat, 2
Audubon Canyon Ranch
National Natural Landmark, 140
Sanctuary, 170
Audubon Society Sanctuaries, 170
Azalea State Reserve, 98
Backbone Creek Research Natural Area, 134
Bacon Island, 56
Bad Water, 28
Bair Island Ecological Reserve, 122, 124
Bakersfield, County Seat, 2
Baldwin Hills Dam, 54
Baldwin Lake California Nature Conservancy Preserve, 174
Ballinger Canyon OHV Area, 198
Barker Slough, 56
Pumping Plant, 56
Barrett Lake, 198
Barstow Woolly Sunflower Area of Critical Environmental Concern, 156
Bartlett Spring Habitat Site, 74
Basin Ranges, 26, 28, 30
Bass Hill Wildlife Area, 106, 108
Bassetts Snowmobile Park OHV Area, 198
Batiquitos Lagoon
Ecological Reserve, 122, 124
Wetland, 60
Battle Creek Wildlife Area, 106, 108
Baxter, Mount - see Mount Baxter
Bay Park Refuge, San Leandro Bay, 58
Bayside Park Lagoon, Burlingame, 58
Bear Creek Catch and Release Stream, 120
Bear Gulch Cave, 34
Bear River, 44, 46
Bedrock Spring Area of Critical Environmental Concern, 154
Begg Rock Area of Special Biological Significance, 158
Belmont Slough, 58
Benbow Lake Redwood Park, 104
Benecia State Recreation Area, 58
Bennett Juniper California Nature Conservancy Preserve, 172, 174
Bennett Mountain Lake Wetland, 62
Berryessa, Lake - see Lake Berryessa
Bethel Tract, 56
Big Basin Redwood(s) Park, 20, 104
Big Bear Lake OHV Area, 198
Big Bear Valley California Nature Conservancy Preserve, 174
Big Bend OHV Area, 198
Big Hot Warm Spring, 42
Big Lagoon
Forest Natural Preserve, 100
Wetland, 60, 62
Wildlife Area, 106, 108
Big Morongo Canyon
Area of Critical Environmental Concern, 156
California Nature Conservancy Preserve, 174
Riparian Habitat Demonstration Area, 152
Big Pine Fault, 38
Big River, 44, 46
Estuary Important Habitat Area, 78
Wetland, 60
Big Sand Springs Area of Critical Environmental Concern, 154
Big Sandy Wildlife Area, 106, 108
Big Sur River, 46
Big Trees Trail, 86
Bigelow Cholla Outstanding Natural Area, 148
Bighorn Sheep, 74, 76
Biogeographical Provinces, 128
Biosphere Reserves-UNESCO, 128

Bird Rock Area of Special Biological
    Significance, 158
Biscar Wildlife Area, 106
Bishop Pine California Nature Conservancy
    Preserve, 172, 174
Bishop Tract, 56
Bitter Creek National Wildlife Refuge, 132
Bizz Johnson Trail, 86, 178
Black Butte Lake OHV Area, 198
Black Butte River, 46
Black Buttes Volcanic Area, 40
Black Chasm Cave National Natural Landmark, 142
Black Mountain
    Area of Critical Environmental Concern, 156
    OHV Area, 198
    Special Interest Area, 146
Black Point Trail, 86
Black Rock Rearing Ponds, 116
Black Sands Beach OHV Area, 198
Blacks Mountain
    Experimental Forest, 144
    Research Natural Area, 134
Blackwood Canyon OHV Area, 198
Blayney Hot Springs, 42
Blue Lake Trail, 86
Blue Lakes OHV Area, 198
Blue Ridge
    Area of Critical Environmental
        Concern, 154, 156
    Ecological Reserve, 122, 124
    National Wildlife Refuge, 132
Board of Forestry, 14
Bobelaine
    Ecological Reserve, 122, 124
    Sanctuary, 170
Bodega Bay, 58
    Wetland, 60, 62
Bodega Head Marsh Wetland, 62
Bodega Marine Life Refuge, 114
    Area of Special Biological Significance, 158
    University of California Natural Reserve
        System, 166
Bodfish Piute Cypress Special Interest Area, 146
Bogard OHV Area, 198
Boggs Lake California Nature Conservancy
    Preserve, 172, 174
Boggs Mountain State Forest, 82
Bogs and Marshes, 80
Bolinas Lagoon, 58
    Wetland, 60
Bolsa Bay Wetland, 60
Bolsa Chica
    Artificial Reef, 126, 127
    Ecological Reserve, 122, 124
    Marine Life Refuge, 114
Bolsa Chica Marsh Important Habitat Area, 78
Boney Mountain State Wilderness, 102
Bonny Doon California Nature Conservancy
    Preserve, 172, 174
Border Field State Park, 160
Bothin Marsh, 58
Bouldin Island, 56
Boundary-Kangaroo Trail, 86
Bouverie Sanctuary, 170
Bow Willow Palms Research Natural Area, 136
Box Springs University of California Natural
    Reserve System, 166
Boyden Caves, 34
Brack Tract, 56
Bradford Island, 56
Brady Creek Waterfall, 50
Brannan Island, 56
Brea-Olinda Wilderness Important Habitat Area, 78
Bridalveil Waterfall, 50
Bridgeport, County Seat, 2
Brimmer Lake Creek Wild and Scenic River, 48
Bristlecone Pine, 28
Browns Island, 56
Brushy Mountain Habitat Site, 74, 76
Buckeye Hot Spring, 42
Bucks Lake
    National Wilderness, 88
    Snowmobile Park OHV Area, 198
Bud Turner Trail, 178
Buena Vista Lagoon Ecological Reserve, 122, 124
Buena Vista Lake, 52
Buena Vista Wetland, 60
Bufano Peace Statue, 20
Bulam Glacier, 32
Bunnell Vernal Pools California Nature
    Conservancy Preserve, 172, 174
Bureau of Indian Affairs, 8
Bureau of Land Management, 8, 88, 112, 148,
    150, 194
    Areas of Critical Environmental
        Concern, 154, 156
Bureau of Reclamation, 8, 116
Burney Falls National Natural Landmark, 142
Burns Pinon Ridge University of California Natural
    Reserve System, 166
Burnt Lava Flow Special Interest Area, 146
Burton Creek Natural Preserve, 100
Butano Redwood Park, 104
Butte Basin Wetland, 62
Butte County, 2
    Association of Governments, 4
    Water Bank County, 192
    Williamson Act County, 188
Butte Creek Canyon Ecological Reserve, 122, 124
Butte Creek House Ecological Reserve, 122, 124
Butte Sink National Wildlife Refuge, 132
Butte Valley Wildlife Area, 106
Buttercup Valley OHV Area, 198
Butterfly Valley Special Interest Area, 146
Buttonwillow Habitat Site, 76
By Day Creek Ecological Reserve, 122, 124
Byron Tract, 56
Cabrillo National Monument National Park, 92
Cache Creek
    Habitat Site, 76
    Wildlife Area, 106, 108
Cache Creek Corridor Area of Critical
    Environmental Concern, 154, 156
Cahuilla, Lake - see Lake Cahuilla
Cahuilla Mountain Research Natural Area, 134
Calaveras
    Fault, 36, 38
    River, 44, 46
Calaveras Big Trees
    National Forest, 84
    Redwood Park, 104
    Special Interest Area, 146
Calaveras County, 2
    Williamson Act County, 188
Calaveras South Grove Natural Preserve, 100
Calico Early Man Site Area of Critical
    Environmental Concern, 156
California Acclimatization Society, 116
California Aqueduct, 56
California Association of Councils
    of Government, 4
California Association of Local Agency Formation
    Commissions, 6
California Caverns, 34
California Chapparal Research Natural Area, 136
California Coast Ranges Biosphere Reserve, 128
California Condor, 74, 76
California Desert Conservation Area, 150
California Desert Conservation Area Plan, 154
California Fish Commission, 116
California Highway Information Network-California
    Department of Transportation, 200
California Interagency Wildlife Task Group, 80
California Islands Wildlife Sanctuary, 112
California Land Conservation Act, 188
California Native Plant Society, 176
California Nature Conservancy Preserves and
    Easements, 150, 172, 176
California Poppy, State Symbol, 7
California Quail, State Symbol, 7
California Recreational Trails and Hostel Plan, 86
California Redwood, State Symbol, 7
California Sea Otter Game Refuge Marine Life
    Refuge, 114
California State Board of Equalization, 190
California Tahoe Regional Planning Agency, 4
California Timberland Productivity Act, 190
California Trails Act, 86
California Water Commission, 24
California Wild and Scenic Rivers Act, 48
California Wilderness Act, 150
California Wildlands Program, 106, 110
California Wildlife Habitat Relationships
    System, 74, 80
Camanche Dam, 54
Camanche Reservoir, 52
Camp Cady Wildlife Area, 106
Camp Creek Trail, 86
Camp Irwin Area of Critical Environmental
    Concern, 156
Camp Roberts Habitat Site, 76
Camp Rock Spring Area of Critical Environmental
    Concern, 156
Canal Ranch Tract, 56
Cannell Meadow Trail, 86
Caribou National Wilderness, 88
Carl Inn Research Natural Area, 138
Carlsbad Artificial Reef, 126, 127
Carmel Bay
    Area of Special Biological Significance, 158
    Ecological Reserve, 122, 124
    Marine Life Refuge, 114

Carmel River, 44, 46
Catch and Release Stream, 120
Lagoon and Wetland Natural Preserve, 100
Carnegie OHV Area, 198
Carpenteria
Special Interest Area, 146
Wetland, 60
Carpenteria Salt Marsh University of California Natural Reserve System, 166
Carquinez Strait, 58
Carrizo Canyon Ecolgical Reserve, 122, 124
Carrizo Creek Wetland, 62
Carrizo Natural Heritage Reserve, 150
Carrizo Plain California Nature Conservancy Preserve, 172, 174
Carson-Iceberg National Wilderness, 88
Carson River, 44, 46
Wild and Scenic, 48
Caruthers Canyon Outstanding Natural Area, 148
Cascade Range, 26, 28, 30
The Cascades, 50
Caspar Headlands State Reserve, 98
Castaic Dam, 54
Castle Crags National Wilderness, 88
Castle Rock(s)
National Wildlife Refuge, 132
Offshore Land, 112
Research Natural Area, 138
Cat Rock Offshore Land, 112
Catalina Marine Science Center Marine Life Refuge, 114
Cataract Gulch Cave, 34
Catch and Release Act, 120
Catch and Release Waters-Department of Fish and Game, 120
Caverns of Mystery, 34
Caves, 34
Cedar Creek
Riparian Habitat Demonstration Area, 152
Riparian Showcase Area, 152
Cedar Roughs
Area of Critical Environmental Concern, 154, 156
Research Natural Area, 136
Central California Coast Biosphere Reserve, 128
Central San Francisco Bay, 58
Central Sierra Planning Council and Economic Development District, 4
Central Valley Hatchery, 116
Cerro Gordo Area of Critical Environmental Concern, 154, 156
Chagoopa Waterfall, 50
Challenge Experimental Forest, 144
Chanchelulla National Wilderness, 88
Channel Islands, 66
Biosphere Reserve, 128
Marine Life Refuge, 114
National Marine Sanctuary, 162
National Monument, 66
National Park, 92
Channel Islands Harbor Artificial Reef, 126, 127
Chaos Crags Volcanic Area, 40
Charter Counties, 2
Chemise Mountain Primitive Area, 150
Chickering American River University of California Natural Reserve System, 166
Chilnualna Waterfall, 50
China Camp State Park, 58
China Point Ecological Reserve, 122, 124
China Ranch Important Habitat Area, 78
Chipps Island, 56
Chowchilla River, 44, 46
Christmas Canyon Area of Critical Environmental Concern, 154, 156
Chuckwalla Bench
Area of Critical Environmental Concern, 156
California Nature Conservancy Preserve, 174
Chuckwalla Valley Dune Thicket Area of Critical Environmental Concern, 156
Chuckwalla Valley Outstanding Natural Area, 148
Chumash Indians, 66
Chumash Painted Cave Historic Park, 94
Cibola National Wildlife Refuge, 132
Cima Volcanic Area, 40
Cinder Cone(s) Natural Area
National Natural Landmark, 140
Outstanding Natural Area, 148
Volcanic Area, 40
Cinder Flats Wildlife Area, 106, 108
Clair Engle Lake, 52
Clam Reserves, 112
Clark Canyon, 152
Clark McNary Act, 84
Clark Mountain
Area of Critical Environmental Concern, 156
Outstanding Natural Area, 148
Clavey River, 44, 46
Clay Pit OHV Area, 198
Clean Water Act, 62
Clear Creek Serpentine Area of Critical Environmental Concern, 154, 156
Clear Creek Trail, 86
Clear Lake, 52
Cascade, 50
Dam, 54
OHV Area, 198
National Wildlife Refuge, 132
Reservoir, 52
Volcanic Area, 40
Cleveland National Forest, 84
Clifton Court Forebay, 56
Coachella Valley
California Nature Conservancy Preserve, 174
Ecolgical Reserve, 122, 124
National Wildlife Refuge, 132
Coachella Valley Fringe-Toed Lizard Site Important Habitat Area, 78
Coal Oil Point University of California Natural Reserve System, 166
Coast Ranges, 26
Coast Redwood, 104
Coastal Act, 60
Coastal and Shoreline Habitats, 80
Coastal Commission, 60, 64, 164
Coastal Wetland Losses, 81
Coastal Zone Conservation Act, 64, 164
Coastal Zone Management Act, 160
Cold Creek Canyon California Nature Conservancy Preserve, 174
Coldwater Canyon Ecological Reserve, 122, 124
Coleman National Fish Hatchery, 116
Collins Eddy Wildlife Area, 106
Colorado Desert, 26, 70
Colorado River, 44, 46
Boating Trail, 86
Columns of the Giants Trail, 86
Colusa County, 2
County Seat, 2
Water Bank County, 192
Williamson Act County, 188
Colusa National Wildlife Refuge, 132
Committee on Ecological Reserves, 136
Concord Habitat Site, 76
Cone Creek Gradient Research Natural Area, 134
Cone Peak Western Slope Biosphere Reserve, 128
Coney Island, 56
Conness, Mount - see Mount Conness
Contra Costa Canal, 56
Contra Costa County, 2
Williamson Act County, 188
Contra Loma Reservoir, 56
Convention for the Protection of the World Cultural and Natural Heritage, 130
Coon Hollow Wildlife Area, 106
Coordination Areas, 132
Copco Lake Volcanic Area, 40
Cordell Bank, 162
Marine Life Refuge, 114
National Marine Sanctuary, 162
Corn Spring Area of Critical Environmental Concern, 156
Corral Canyon OHV Area, 198
Corral Hollow Ecological Reserve, 122, 124
Corte Madera Creek, 58
Corte Madera Marsh Ecological Reserve, 122, 124
Marine Life Refuge, 114
Cortese-Knox Local Government Act, 6
Coso Mountains Volcanic Area, 40
Cosumnes River, 44, 46
California Nature Conservancy Preserve, 172, 174
Riparian Woodlands National Natural Landmark, 140
Cottonwood Creek
Catch and Release Stream, 120
Wildlife Area, 106
Council of Fresno County Governments, 4
Counties and County Seats, 2
County Supervisors Association, 4
Courtright Intrusive Contact Zone Special Interest Area, 146
Cow Mountain OHV Area, 198
Coyote Creek Fault, 38
Coyote Flat OHV Area, 198
Coyote Hills Regional Park, 58
Coyote Mountains Fossil Site Area of Critical Environmental Concern, 156
Coyote Point, San Mateo, 58
Creative Act, 84

Creighton Ranch
California Nature Conservancy Preserve, 172, 174
Important Habitat Area, 78
Crescent City, County Seat, 2
Crescent City Marsh Wildlife Area, 106
Cronese Basin Area of Critical Environmental Concern, 156
Crowley Lake Catch and Release Lake, 120
Crystal Cave, 34
Crystal Cove Underwater Park, 96
Crystal Creek Waterfall, 50
Crystal Lake Hatchery, 116
Cub Creek Research Natural Area, 134
Cucamonga National Wilderness, 88
Cuckoo Island Corridor Important Habitat Area, 78
Cuesta Ridge Special Interest Area, 146
Cultural Preserves, 20
Cuyama River, 44, 46
Cuyamaca Mountain State Wilderness, 102
Dale Lake Area of Critical Environmental Concern, 156
Dams and Reservoirs-Division of Safety of Dams, 54
Dana, Mount - see Mount Dana
Dana Point Marine Life Refuge, 114
Darlingtonia Swampy Area Wetland, 62
Darrah Springs Hatchery, 116
Darwin Falls, 50
Area of Critical Environmental Concern, 154
Outstanding Natural Area, 148
Date Flat OHV Area, 198
Davis Flat OHV Area, 198
Davis Raptor Center, 110
Dawn Redwood, 104
Dawson Los Monos Canyon University of California Natural Reserve System, 166
De Anza Cycle Park OHV Area, 198
Dead Mountains Area of Critical Environmental Concern, 156
Death Valley National Monument, 70
Biosphere Reserve, 128
National Park, 92
World Heritage Site, potential, 130
Decker Island, 56
Wildlife Area, 106
Deep Canyon Wetland, 62
Deep Creek
Catch and Release Stream, 120
Hot Springs, 42
Deep Springs Marsh
National Natural Landmark, 140
Wetland, 62
Deer Creek
Important Habitat Area, 78
OHV Area, 198
Deer Mountain OHV Area, 198
Defense, Department of, 8
Del Loma Cave, 34
Del Mar Landing Ecological Reserve, 122, 124
Area of Special Biological Significance, 158
Marine Life Refuge, 114
Del Norte Coast Redwoods, 104
Biosphere Reserve, 128
World Heritage Site, 20, 130
Del Norte County, 2
Delevan National Wildlife Refuge, 132
Delonegha Hot Springs, 42
The Delta, 56
Delta Cross Channel, 56
Delta Mendota Canal, 56
Denning Spring Area of Critical Environmental Concern, 156
Department of Defense, 8
Department of Energy, 8
Department of Fish and Game, 8, 12, 112, 116, 118, 120, 122, 126, 176
Department of Forestry and Fire Protection, 8, 14, 82
Department of Parks and Recreation, 8, 16, 18, 20, 64, 94, 96, 98, 100, 102
Department of Water Resources, 24, 52, 116
Desert Tortoise Natural Area
Area of Critical Environmental Concern, 154, 156
California Nature Conservancy Preserve, 172, 174
Important Habitat Area, 78
Research Natural Area, 136
Desert Tortoise, State Symbol, 7
Deserts, 70
Desolation National Wilderness, 88
Developed Habitats, 80
Devil's Garden Research Natural Area, 134
Devils Postpile National Monument, 92
Diablo, Mount - see Mount Diablo
Dick Smith National Wilderness, 88
Dinkey Lakes National Wilderness, 88
Dinosaur Trackway Area of Critical Environmental Concern, 156
Dirty Sock Hot Spring, 42
Discovery Bay, 56
Discovery Trail, 86
Dixon Vernal Pools National Natural Landmark, 140
Doane Valley Natural Preserve, 100
Dog-Face Butterfly, State Symbol, 7
Doheny State Beach
Marine Life Refuge, 114
Underwater Park, 96
Dome Land National Wilderness, 88
Dominie Creek Wild and Scenic River, 48
Don Pedro Dam, 54
Donner Camp Trail, 86
Dos Palmos Oasis California Nature Conservancy Preserve, 174
Double Point Area of Special Biological Significance, 158
Dove Springs OHV Area, 198
Downie River, 46
Downieville
County Seat, 2
OHV Area, 198
Doyle Wildlife Area, 106
Dozier Grassland and Vernal Pools Important Habitat Area, 78
Drake's Bay Wetland, 60
Dry Lagoon Wetland, 60, 62
Duarte Bike Trail, 178
Dumont Dunes OHV Area, 198
Duncan Mills Marsh Wetland, 62
Dune Habitats, 80
Dune Lakes Wetland, 60
Duxbury Reef Reserve and Extension
Area of Special Biological Significance, 158
Marine Life Refuge, 114
Dyerville Giant, 104
Eagle Lake, 52
University of California Natural Reserve System, 166
Volcanic Area, 40
Waterfall, 50
Eagle Peak, 28
Earl, Lake - see Lake Earl
Earthquakes, 36
East Carson River Hot Spring, 42
East Mojave National Scenic Area, 70, 150
East Walker River Catch and Release Stream, 120
Ecological Reserve Act, 122, 124
Ecological Reserves-Department of Fish and Game, 112, 122, 124
Edom Hill-Willow Hole Area of Critical Environmental Concern, 156
Edwin L. Z'berg Natural Preserve, 100
Eel River, 44, 46
Catch and Release Stream, 120
Wetland, 60
Wild and Scenic, 48
Wildlife Area, 106
El Capitan, 28
El Capitan (Horsetail) Waterfall, 50
El Centro, County Seat, 2
El Dorado County, 2
Williamson Act County, 188
El Dorado Manzanita Area of Critical Environmental Concern, 154
El Estero-Carpenteria Wetland, 60
El Mirage OHV Area, 198
El Segundo Dunes Important Habitat Area, 78
Elder Creek National Natural Landmark, 140
Eldorado National Forest, 84
Elephant Seal, 74, 76
Elk Creek Habitat Site, 74
Elk Mountain OHV Area, 198
Elk River, 46
Elkhorn Plains
Area of Critical Environmental Concern, 154, 156
Ecolgical Reserve, 122, 124
Elkhorn Slough
California Nature Conservancy Preserve, 172, 174
Ecological Reserve, 122, 124
Marine Life Refuge, 114
National Estuarine Reserve Research System, 160
San Francisco Bay Access Area, 58
Wetland, 60
Wildlands Program Area, 110

Ellen Browning Scripps Natural Preserve, 100
Ellen Pickett State Forest, 82
Ellicott Slough National Wildlife Refuge, 132
Elliot Chaparral University of California Natural
  Reserve System, 166
Elsinore Fault, 38
Emerald Bay National Natural Landmark, 140
Emeryville Crescent
  Important Habitat Area, 78
  San Francisco Bay Access Area, 58
Emigrant National Wilderness, 88
Emigrant Summit Trail, 86
Empire Tract, 56
Endangered Species Act, 134
Energy, Department of, 8
Estero Americano Wetland, 60
Estero de Limantour
  Marine Life Refuge, 114
  Research Natural Area, 138
Estero de San Antonio Wetland, 60
Estrella River, 44, 46
Etiwanda Wash University of California Natural
  Reserve System, 166
Eugene O'Neill National Historic Site, 92
Eureka, County Seat, 2
Eureka Dunes National Natural Landmark, 140
Eureka Slough Wildlife Area, 106
Eureka Valley Dunes
  Area of Critical Environmental
    Concern, 154, 156
  Outstanding Natural Area, 148
Ewing Oak California Nature Conservancy
  Preserve, 174
Experimental Ecological Areas-National Park
  Service, 148
Experimental Forests-Forest Service, 144
Extinct Animals, 81
Fabian Tract, 56
Fagan Slough Ecological Reserve, 122, 124
Fairfield, County Seat, 2
Fairfield Osborn California Nature Conservancy
  Preserve, 172, 174
Fall River, 46
  Catch and Release Stream, 120
False Klamath Rock Offshore Land, 112
Farallon Islands, 64, 162
  Area of Special Biological Significance, 158
  Biospherere Reserve, 128
  Marine Life Refuge, 114
  National Wilderness, 88
  National Wildlife Refuge, 132
  Offshore Land, 112
Farnsworth Bank
  Ecological Reserve, 122, 124
  Marine Life Refuge, 114
Faults, 36
Fay Slough Wildlife Area, 106
Feather Falls, 50
  Special Interest Area, 146
  Trail, 86
Feather River, 44, 46
  Hatchery, 116
  Wild and Scenic, 48
Federal Land Policy and Management
  Act, 150, 152, 154
Fern Canyon Research Natural Area, 134
Fillmore Hatchery, 116
Fiscalini Ranch Important Habitat Area, 78
Fish and Game Commission, 12, 112
Fish and Game, Department of, 8, 12, 112, 116,
  118, 120, 122, 126, 176
Fish Hatcheries-Department of Fish and Game, 116
Fish Hatcheries-U.S. Fish and Wildlife Service, 132
Fish Refuges, 112, 114
Fish Slough
  Area of Critical Environmental Concern, 154
  Ecological Reserve, 122, 124
  Important Habitat Area, 78
  National Natural Landmark, 140
  University of California Natural Reserve
    System, 166
  Wetland, 62
Fish Springs Hatchery, 116
Fishing Research Stations, 132
Fitzhugh Creek Riparian Habitat Demonstration
  Area, 152
Flying M Ranch California Nature Conservancy
  Preserve, 172, 174
Folsom
  Dam, 54
  Lake, 52
Ford Dry Lake OHV Area, 198
Fordyce Jeep Trail OHV Area, 198
Forest and Range Renewable Resource Act, 144
Forest of Nisene Marks Redwood Park, 104
Forest Service, 8, 88, 194
Foresthill OHV Area, 198
Forestry and Fire Protection,
  Department of, 8, 14, 82
Forests, 80
Forests-Department of Forestry and
  Fire Protection, 82
Forests, National, 80
Fort Bill, 84
Fort Piute Outstanding Natural Area, 148
Fort Point National Historic Site, 92
Fort Ross Underwater Park, 96
Fort Sage OHV Area, 198
Fort Soda-Mojave Chub Area of Critical
  Environmental Concern, 156
Fossil Falls Area of Critical Environmental
  Concern, 154, 156
Foster City Crescent Shell Bar, 58
Fourth Street Overlook Sanctuary, 170
Frank Raines OHV Area, 198
Franks Tract, 56
Fremont Peak Habitat Site, 74
Fremont Valley Ecological Reserve, 122, 124
Fremont Wier Wildlife Area, 106
Frenzel Creek Research Natural Area, 134
Freshwater Lagoon Wetland, 62
Fresno County, 2
  County Seat, 2
  Williamson Act County, 188
Fresno River, 44, 46
Friant Dam, 54
Friends Of The River, 165
Frog Meadow OHV Area, 198
Gabrielino Trail, 86
Gallinas Creek, 58
Game Refuges, 112
Garcia River, 44, 46
  Wetland, 60
Garden, Mary, 7
Garfield Research Natural Area, 138
Garlock Fault, 38
Garner Valley Important Habitat Area, 78
General Law Counties, 2
General Sherman tree, 104
Gerstle Cove Area of Special Biological
  Significance, 158
  Marine Life Refuge, 114
Giant Kelp Forests, 68
Giant Sequoia, 104
Glacier Divide Glacier, 32
Glaciers and Glaciated Areas, 32
Glamis-Gecko OHV Area, 198
Glass Mountain, 28
  Glass Flow Special Interest Area, 146
  Volcanic Area, 40
Glen Helen OHV Area, 198
Glen Hot Springs, 42
Glenn County, 2
  Williamson Act County, 188
Goddard, Mount - see Mount Goddard
Gold Lake OHV Area, 198
Golden Gate National Recreation Area, 58, 194
  Biosphere Reserve, 128
  National Park, 92
Golden Trout, State Symbol, 7
Golden Trout National Wilderness, 88, 90
Goleta Slough
  Ecological Reserve, 122, 124
  Wetland, 60
Goose Creek Wild and Scenic River, 48
Goose Lake, 52
  Area of Critical Environmental Concern, 154
  Important Habitat Area, 78
Goosenest Volcanic Area, 40
Gorman Post Road Important Habitat Area, 78
Grand Island, 56
Granite Chief National Wilderness, 88, 90
Granite Creek Research Natural Area, 138
Granite Mountains University of California Natural
  Reserve System,166
Grass Lake Wildlife Area, 106
Grasslands National Wildlife Refuge, 132
Grasslands, Vernal Pools, and Meadows, 80
Grasslands Water District Wetland, 62
Gray Davis-Dye Creek California Nature
  Conservancy Preserve, 172, 174
Gray Lodge
  Wildlands Program Area, 110
  Wildlife Area, 106
Gray Whale, State Symbol, 7
Great Basin Desert, 26, 70
Great Falls Basin
  Area of Critical Environmental Concern, 154
  Research Natural Area, 136

Great (Central) Valley, 26
Greenwater Canyon Area of Critical Environmental Concern, 154, 156
Grimshaw Lake Area of Critical Environmental Concern, 156
Grizzly Bay, 58
Grizzly Bear, State Symbol, 7
Grizzly Creek Redwoods, 104
Grizzly Falls, 50
Grizzly Island
  Habitat Site, 76
  Wildlands Program Area, 110
  Wildlife Area, 106
Guadalupe River, 46
Gualala River, 44, 46
Guide to Wildlife Habitats of California, 80
Gulf of the Farallones
  Biosphere Reserve, 128
  National Marine Sanctuary, 162
Habitat Losses, 81
Habitat Sites, 74
Habitat Types-California Interagency Wildlife Task Group, 80
Habitat Types-University of California, 80
Hagen Canyon Natural Preserve, 100
Half Dome, 28
Halloran Wash Area of Critical Environmental Concern, 156
Hamilton California Nature Conservancy Preserve, 174
Hamilton, Mount - see Mount Hamilton
Hammond Trail, 178
Hanford, County Seat, 2
Harper Dry Lake Area of Critical Environmental Concern, 156
Harrison Grade Ecological Reserve, 122, 124
Hartman Bar Trail, 86
Harvey Monroe Hall Research Natural Area, 134
Harvey O. Banks Delta Pumping Plant, 56
Hass Slough, 56
Hastings Natural History Reservation University of California Natural Reserve System, 166
Hastings Tract, 56
Hat Creek Catch and Release Stream, 120
Hauser National Wilderness, 88, 90
Havasu
  Lake, 52
  National Wildlife Refuge, 132
Hawley Grade Trail, 86
Hayfork OHV Area, 198
Hayward Bay Regional Shoreline Park, 58
Hayward Fault, 36, 38
Heart Lake Trail, 86
Heater Cave Important Habitat Area, 78
Heather Lake Research Natural Area, 138
Heenan Lake
  Catch and Release Lake, 120
  Wildlife Area, 106
Heisler Park
  Area of Special Biological Significance, 158
  Ecological Reserve, 122, 124
  Marine Life Refuge, 114
Hendy Woods Redwood Park, 104
Henry Cowell Redwoods, 104
Henry W. Coe State Park, 20
  State Wilderness, 102
Herbaceous Dominated Habitats, 80
Hermosa Beach Artificial Reef, 126, 127
Hermosa Beach Trail, 178
Heron Rookery Natural Preserve, 100
Hibberd California Nature Conservancy Preserve, 172, 174
Hickman Vernal Pools Important Habitat Area, 78
Hidden Palms Ecological Reserve, 122, 124
High Desert Trail, 86
High Rock Canyon (Nevada) Area of Critical Environmental Concern, 154, 156
High Sierra Primitive Area, 150
Hill Slough Wildlife Area, 106
Hilton Creek Fault, 36
Historic Parks-Department of Parks and Recreation, 94
Hite Cove OHV Area, 198
Holland Tract, 56
Hollister, County Seat, 2
Hollister Hills OHV Area, 198
Honey Lake, 52
  Wildlife Area, 106
Honker Bay, 58
Hoover, Herbert, 112
Hoover National Wilderness, 88, 90
Hope Valley
  OHV Area, 198
  Wildlife Area, 106
Hopkins Marine Life Refuge, 114
  Area of Special Biological Significance, 158
Hopper Mountain National Wildlife Refuge, 132
Horse Trail Ridge Trail, 86
Horseshoe Ranch Wildlife Area, 106
Hot Creek
  Catch and Release Stream, 120
  Hatchery, 116
  Hot Springs, 42
  Volcanic Area, 40
Hot Springs, 42
Hotchkiss Tract, 56
Hotlum Glacier, 32
Huasna River, 46
Huber Dunes OHV Area, 198
Humboldt Bay
  National Wildlife Refuge, 132
  Wetland, 60
Humboldt County, 2
  Association of Governments, 4
  Williamson Act County, 188
Humboldt Lagoons
  Important Habitat Area, 78
  Redwood Park, 104
Humboldt Redwoods, 104
Humphreys, Mount - see Mount Humphreys
Hungry Valley
  Oak Woodland Natural Preserve, 100
  OHV Area, 198
Hunter-Leggett Habitat Site, 76
Hunter Rocks Offshore Land, 112
Huntington Beach Artificial Reef, 126, 127
Huntington Beach Tire Artificial Reef, 126, 127
Huron Cycle Park OHV Area, 198
Ides Cove Loop Trail, 86
Illilouette Waterfall, 50
Immigrant Primitive Area, 150
Imperial
  County, 2
  Fault, 38
  National Wildlife Refuge, 132
  Reservoir, 52
  Wildlife Area, 106
Imperial Dunes
  Outstanding Natural Area, 148
  Research Natural Area, 136
Imperial Sand Hills National Natural Landmark, 140
Imperial Valley Warmwater Hatchery, 116
Imperial-Wister Unit Wildlands Program Area, 110
Important Habitat Areas-Fish and Wildlife Service, 78
Inaja Nature Trail, 86
Independence, County Seat, 2
Indian Grinding Rock Historic Park, 94
Indian Pass Area of Critical Environmental Concern, 156
Indian Springs Outstanding Natural Area, 148
Indian Tom Wildlife Area, 106
Indian Valley
  Dam, 54
  Wildlife Area, 106
Indian Valley Brodia Area of Critical Environmental Concern, 154, 156
Indiana Summit Research Natural Area, 134
Inglewood Fault, 38
In-ko-pah Mountains Area of Critical Environmental Concern, 156
Inks Creek Ranch California Nature Conservancy Preserve, 172, 174
Interior Wetland Losses, 81
International Biological Programme, 128
International Historic Site, 92
International Union for Conservation of Nature and Natural Resources, 128
International Wildlife, 177
Inyo
  County, 2
  National Forest, 84
Inyo Craters Volcanic Area, 40
Inyo-Mono Association of Governmental Entities, 4
Ione Manzanita Area of Critical Environmental Concern, 154, 156
Ione Tertiary Oxisol Soil Area of Critical Environmental Concern, 154, 156
Iron Gate Hatchery, 116
Irvine Coast Marine Life Refuge, 114
  Area of Special Biological Significance, 158
Isabella Dam, 54
  Reservoir, 52
Ishi National Wilderness, 88, 90
Jackass Creek Trail, 86
Jackson, County Seat, 2
Jackson State Forest, 82
  Biosphere Reserve, 128
Jacumba Mountain Outstanding Natural Area, 148

Jaeger California Nature Conservancy Preserve, 174
James San Jacinto Mountains University of California Natural Reserve System, 166
James V. Fitzgerald Marine Reserve
  Area of Special Biological Significance, 158
  Marine Life Refuge, 114
Jawbone-Butterbredt Area of Critical Environmental Concern, 154, 156
Jawbone Canyon
  Habitat Site, 76
  OHV Area, 198
Jedediah Smith Redwoods, 104
  Biosphere Reserve, 128
  World Heritage Site, 20, 130
Jennie Lakes National Wilderness, 88, 90
Jepson Prairie
  California Nature Conservancy Preserve, 172, 174
  University of California Natural Reserve System, 166
Jersey Island, 56
Joan Hamann Dole Sanctuary, 170
John Little State Reserve, 98
John Muir
  National Historic Site, 92
  National Wilderness, 88, 90
Johnson Valley OHV Area, 198
Jones Tract-Upper and Lower, 56
Joseph M. Long Wildlife Sanctuary, 170
Joshua Tree National Monument, 70
  Biosphere Reserve, 128
  National Park, 92
  National Wilderness, 88, 90
  World Heritage Site, potential, 130
Juanita Cook Trail, 178
Jug Handle State Reserve, 98
Julia Pfeiffer Burns Redwood Park, 104
Julia Pfeiffer Burns Underwater Park, 96
  Area of Special Biological Significance, 158
Juniper Flats Area of Critical Environmental Concern, 156
Kaiser National Wilderness, 88, 90
Kaslow Natural Preserve, 100
Kaweah Basin Research Natural Area, 138
Kaweah Oaks California Nature Conservancy Preserve, 172, 174
Kaweah River, 44, 46
  Catch and Release Stream, 120
  Ecological Reserve, 122, 124
Kellogg Creek, 56
Kelso Creek Important Habitat Area, 78
Kelso Dunes Outstanding Natural Area, 148
Kelso Peak Wildlife Area, 106
Kendall-Frost Mission Bay Marsh University of California Natural Reserve System, 166
Kennedy Meadows OHV Area, 198
Keough Hot Ditch Hot Spring, 42
Kern
  Hot Spring, 42
  National Wildlife Refuge, 132
Kern Canyon Fault, 38
Kern County, 2
  Council of Governments, 4
  Williamson Act County, 188
Kern Kaweah Chapter, Sierra Club, 180
Kern Lake California Nature Conservancy Preserve, 172, 174
Kern River, 44, 46
  California Nature Conservancy Preserve, 172, 174
  Catch and Release Stream, 120
  Planting Base Fish Hatchery, 116
  South Fork Important Habitat Area, 78
  Wild and Scenic, 48
Kesterson National Wildlife Refuge, 132
King, Clarence, 32
King Island, 56
King Range National Conservation Area, 150
  Area of Special Biological Significance, 158
Kings Beach OHV Area, 198
Kings Canyon National Park, 88
  Biosphere Reserve, 128
  National Park, 92
  Redwood Park, 104
  World Heritage Site, potential, 130
Kings Cavern Special Interest Area, 146
Kings County, 2
  Regional Planning Agency, 4
  Williamson Act County, 188
Kings River, 44, 46
  Catch and Release Stream, 120
  OHV Area, 198
  Trail, 86
  Wild and Scenic, 48
Kingston Range Area of Critical Environmental Concern, 156
Kinsman Flat Wildlife Area, 106
Kirman Lake Catch and Release Lake, 120
Klamath Mountains, 26, 28, 30
Klamath National Forest, 84
Klamath River, 44, 46
  Wetland, 60
  Wild and Scenic, 48
Knox-Nisbet Acts, 6
Knoxville OHV Area, 198
Koip Peak Glacier, 32
Konocti, Mount - see Mount Konocti
Konwakiton Glacier, 32
Kopta California Naure Conservancy Preserve, 172, 174
Kramer Hills Area of Critical Environmental Concern, 156
Kruse Rhododendron State Reserve, 98
Kuna Peak Glacial Area, 32
La Bellona Estuary Wetland, 60
La Grange OHV Area, 198
La Jenelle Artificial Reef, 126, 127
La Jolla Caves, 34
La Jolla Valley Natural Preserve, 100
Lafayette-Moraga Trail, 178
LAFCO, 6
Laguna Beach Marine Life Refuge, 114
Laguna de Santa Rosa
  Ecological Reserve, 122, 124
  Wetland, 62
Laguna Seca OHV Area, 198
Lake Almanor, 52
  Dam, 54
Lake Arrowhead OHV Area, 198
Lake Berryessa, 52
Lake Cahuilla Area of Critical Environmental Concern, 156
Lake County, 2
  Williamson Act County, 188
Lake County-City Areawide Planning Commission, 4
Lake Davis OHV Area, 198
Lake Earl
  Wetland, 60, 62
  Wildlands Program Area, 110
  Wildlife Area, 106
Lake Matthews Ecological Reserve, 122, 124
Lake McClure, 52
Lake Mead National Recreation Area, 194
Lake Norconian Important Habitat Area, 78
Lake of the Woods Wildlife Area, 106
Lake Oroville, 52
Lake Perris State Recreation Area Underwater Park, 96
Lake Pillsbury
  Habitat Site, 74, 76
  OHV Area, 198
Lake Shasta Caverns, 34
Lake Sonoma Wildlife Area, 106
Lake Tahoe, 52
  Bike-Pedestrian Trail, 86
  Dam, 54
  Rim Trail, 86
Lake Talawa Wetland, 60
Lakeport, County Seat, 2
Lakes, 52
Land Ownership, 8
Landels-Hill Big Creek Reserve
  Biosphere Reserve, 128
  California Nature Conservancy Preserve, 172, 174
  Redwood Park, 104
  University of California Natural Reserve System, 166
Landforms, 26
Langley, Mount - see Mount Langley
Lanphere-Christensen Dunes California Nature Conservancy Preserve, 172, 174
Lark Canyon OHV Area, 198
Larkspur to Corte Madera Bike Path, 178
Las Cruses Hot Springs, 42
Las Posadas State Forest, 82
Las Tunas Grasslands Important Habitat Area, 78
Lassen
  National Forest, 84
  National Wilderness, 88, 90
  Volcanic National Park, 40, 92
Lassen County, 2
  Water Bank County, 192
  Williamson Act County, 188
Lassen, Mount - see Mount Lassen
Last Chance Canyon Area of Critical Environmental Concern, 154, 156
Last Chance (Meadow) Research Natural Area, 134
Latigo Point Area of Special Biological Significance, 158

Latour State Forest, 82
Lava Beds National Monument, 92
National Wilderness, 88, 90
Volcanic Area, 40
Lava Lakes Wildlife Area and Nature Center, 177
Laytonville Habitat Site, 76
League of California Cities, 4
Least Tern Natural Preserve, 100
Leavitt Creek Wild and Scenic River, 48
Lee Vining Canyon Habitat Site, 76
Lehamite Waterfall, 50
Leonard's Hot Spring, 42
Liberty Canyon Natural Preserve, 100
Lighthouse Island Offshore Land, 112
Limantour Bay Wetland, 60
Limantour Spit and Estero, 58
Limestone Salamander
Area of Critical Environmental Concern, 154, 156
Ecological Reserve, 122, 124
Lincoln, Abraham, 20
Lion Rock Offshore Land, 112
Little Butte Ecological Reserve, 122, 124
Little Caliente Hot Spring, 42
Little Lostman Creek Research Natural Area, 138
Little Mill Creek Wild and Scenic River, 48
Little Panoche Reservoir Wildlife Area, 106
Little River, 46
Little Rock OHV Area, 198
Little Sand Springs Area of Critical Environmental Concern, 154
Little Sur River, 46
Little Truckee Summit OHV Area, 198
Little Tule River, 46
Local Agency Formation Commissions, 6
Lokern California Nature Conservancy Preserve, 172, 174
Loma Prieta Chapter, Sierra Club, 180
Long Valley Creek Riparian Habitat Demonstration Area, 152
Loon Lake OHV Area, 198
Los Angeles County, 2
County Seat, 2
Williamson Act County, 188
Los Angeles State and County Arboretum State Reserve, 98
Los Angeles River, 44, 46
Los Banos
Wildlands Program Area, 110
Wildlife Area, 106
Los Osos Oaks State Reserve, 98
Los Padres Chapter, Sierra Club, 180
Los Padres National Forest, 84
Los Penasquitos Lagoon Wetland, 60
Los Penasquitos Marsh Natural Preserve, 100
Los Ranchos Wildlife Area, 106
Lost Lake County Park Sanctuary, 170
Lost River, 46
Loughry Forest, 82
Lousetown Volcanic Area, 40
Lovers Cove Reserve Marine Life Refuge, 114
Lower American Wild and Scenic River, 48
Lower Klamath
Lake, 52
National Wildlife Refuge, 132
Lower San Francisco Bay, 58
Lower Sherman Island Wildlife Area, 106
Lower Tubbs Island, 58
Lyell, Mount - see Mount Lyell
Mace Mill OHV Area, 198
MacKerricher Underwater Park, 96
Macklin Creek Ecological Reserve, 122, 124
Mad River, 44, 46
Hatchery, 116
Mad River Slough Wildlife Area, 106
Madera County, 2
County Seat, 2
Williamson Act County, 188
Magnesia Spring Ecological Reserve, 122, 124
Mailliard Redwoods, 104
State Reserve, 98
Maintop Island, 64
Malakoff Diggins Historic Park, 94
Malibu Artificial Reef, 126, 127
Mammoth Mountain, 40
Mammoth Wash OHV Area, 198
Man and the Biosphere Programme (MAB), 66, 128
Manchesna Mountain National Wilderness, 88, 90
Manchester State Beach Underwater Park, 96
Mandeville Island, 56
Manix Fault, 38
Manzanita Lake Catch and Release Lake, 120
Marble Mountain(s), 28
National Wilderness, 88, 90
Wildlife Area, 106
Marble Mountains Fossil Bed Area of Critical Environmental Concern, 156
Marin County, 2
Williamson Act County, 188
Marin Municipal Water District Biosphere Reserve, 128
Marina del Rey Artificial Reef, 126, 127
Marine Life Refuges and Reserves-Department of Fish and Game, 112, 114
Marine Protection, Research, and Sanctuaries Act, 162
Marine Reserves, 112
Mariposa County, 2
County Seat, 2
Williamson Act County, 188
Mariposa Grove of Big Trees, 20, 104
Markleeville, County Seat, 2
Marsh Creek, 56
Martinez, County Seat, 2
Martinez Regional Shoreline Park, 58
Martis Creek Reservoir
Catch and Release Lake, 120
Wild Trout Lake, 118
Marysville, County Seat, 2
Masonic Caves, 34
Mastings Cut, 56
Matthews, Lake - see Lake Matthews
Mattole River, 44, 46
McArthur-Burney Waterfall, 50
McCain Resource Conservation Area, 150
McCain Valley OHV Area, 198
McCloud Glacier, 32
McCloud Lake Catch and Release Lake, 120
McCloud River, 44, 46
California Nature Conservancy Preserve, 172, 174
Catch and Release Stream, 120
Wild Trout Stream, 118
McClure, Lake - see Lake McClure
McCormack Williamson Tract, 56
McDonald Tract, 56
McGinty Mountain California Nature Conservancy Preserve, 174
McGrath Lake Wetland, 60
McNamee's Cave Important Habitat Area, 78
McVicar Preserve Sanctuary, 170
McWay Cove Waterfall, 50
Meadow Creek Important Habitat Area, 78
Medford Island, 56
Medicine Lake OHV Area, 198
Medicine Lake Glass Flow Special Interest Area, 146
Medicine Lake Highlands Volcanic Area, 40
Mendocino
National Wildlife Refuge, 132
Mendocino County, 2
Council of Governments, 4
County Seat, 2
Williamson Act County, 188
Mendota Wildlife Area, 106
Mercalli Scale, 36
Merced County, 2
Association of Governments, 4
County Seat, 2
Water Bank County, 192
Merced National Wildlife Refuge, 132
Merced River, 44, 46
Catch and Release Stream, 120
Fish Installation, 116
Research Natural Area, 138
Wild and Scenic, 48
Wild Trout Stream, 118
Mercer Caverns, 34
Merrill's Landing Wildlife Area, 106
Merritt Island, 56
Mesquite Hills Area of Critical Environmental Concern, 156
Mesquite Lake Area of Critical Environmental Concern, 156
Methuselah Trail, 86
Miami Creek OHV Area, 198
Mid Hills Outstanding Natural Area, 148
Middle Farallon Island, 64
Middle Palisade, 30
Middle River, 46
Mildred Island, 56
Mill Canyon OHV Areas, 198
Mill Creek Wild and Scenic River, 48
Mill Valley Bike Trail, 178
Milpitas Wash Research Natural Area, 136
Milton Lake Catch and Release Lake, 120

Miner Slough Wildlife Area, 106
Miramar Mounds National Natural Landmark, 140
Mission Bay
Artificial Reef, 126, 127
Wetland, 60
Mitchell Caverns, 34
National Natural Landmark, 140
Natural Preserve, 100
Moaning Cavern, 34
Moccasin Creek Hatchery, 116
Modesto, County Seat, 2
Modified Mercalli Scale, 36
Modoc
National Forest, 84
National Wildlife Refuge, 132
Modoc County, 2
Water Bank County, 192
Modoc Plateau, 26
Volcanic Area, 40
Mojave Desert, 26, 70
Biosphere Reserve, 128
Mojave Desert Camp Wetland, 62
Mojave Fishhook Cactus Area of Critical Environmental Concern, 156
Mojave River, 44, 46
Hatchery, 116
Mokelumne National Wilderness, 88, 90
Mokelumne River, 44, 46
Fish Installation, 116
North and South Forks, 60
Monache Meadows OHV Area, 198
Monarch Butterfly, 74
Monarch National Wilderness, 88, 90
Monitor Pass OHV Area, 198
Mono
County, 116
Lake, 52
Mono Basin National Forest Scenic Area, 150
Mono Lake
Important Habitat Area, 78
Tufa State Reserve, 98
Volcanic Area, 40
Montecito Hot Springs, 42
Monterey Bay, 162
Monterey County, 2
Williamson Act County, 188
Monterey Pennisula Recreation Trail, 178
Montgomery Woods
Redwood Park, 104
State Reserve, 98
Monticello Dam, 54
Mopah Spring(s)
Area of Critical Environmental Concern, 156
Outstanding Natural Area, 148
Moreno Paleontological Area of Critical Environmental Concern, 154, 156
Morgan Summit OHV Area, 198
Moro Cojo Wildlife Area, 106
Morro Bay Wetland, 60
Morro Beach Pismo Clam Preserve Marine Life Refuge, 114
Morro Dunes Ecological Reserve, 122, 124
Morro Rock
Ecological Reserve, 122, 124
Marine Life Refuge, 114
Natural Preserve, 100
Moss Landing, 160
Marine Life Refuge, 114
Wildlife Area, 106
Mother Lode Chapter, Sierra Club, 180
Motte Rimrock University of California Natural Reserve System, 166
Mount Abbot Glacier, 32
Mount Baxter Habitat Site, 76
Mount Conness Glacier, 32
Mount Dana Glacier, 32
Mount Diablo, 28
Fault, 36
National Natural Landmark, 140
Mount Goddard Glacier, 32
Mount Hamilton Habitat Site, 76
Mount Humphreys Glacier, 32
Mount Konocti Volcanic Area, 40
Mount Langley, 30
Mount Lassen, 28
Glaciated Area, 32
Volcanic Area, 40
Mount Lowe Railroad Trail, 178
Mount Lyell, 28
Glacier, 32
Mount Muir, 30
Mount Patterson, 28
Mount Ritter, 28
Mount Russell, 30
Mount Saint Helens, 40
Mount San Antonio, 28
Mount San Jacinto State Wilderness, 102
Mount Shasta, 28, 30
Fish Hatchery, 116
Glacier and Glaciated Area, 32
National Natural Landmark, 140
National Wilderness, 88, 90
Volcanic Area, 40
Mount Sill, 30
Mount Tamalpais, 28
Biosphere Reserve, 128
Redwood Park, 104
Mount Tyndall, 30
Mount Whitney, 28, 30
Fish Hatchery, 116
Mount Williamson, 28, 30
Mount Zion State Forest, 82
Mountain Home State Forest, 82
Mountain Peaks, 28, 30
Mountain Ranges, 28, 30
Mountain View Shoreline Park, 58
Mouth of Cottonwood Creek Wildlife Area, 106
Mud Lake
Research Natural Area, 134
Wildlife Area, 106
Mugu Lagoon
Area of Special Biological Significance, 158
Wetland, 60
Muir, John, 180
Muir, Mount - see Mount Muir
Muir Woods National Monument, 92
Redwood Park, 104
Mule Mountains Area of Critical Environmental Concern, 156
Multiple Use Sustained Yield Act, 84
Nacimiento
Dam, 54
Fault, 38
River, 44, 46
Napa County, 2
County Seat, 2
Williamson Act County, 188
Napa Marshes Wildlife Area, 106
Napa River, 44, 46
Ecological Reserve, 122, 124
National Audubon Society, 170
National Capital Park, 92
National Conservation Areas, 150
National Environmental Policy Act, 84
National Estuarine Reserve Research System-National Oceanic and Atmospheric Administration, 160
National Estuarine Sanctuaries, 112
National Estuarine Sanctuary Program, 160
National Forest Management Act, 84, 134
National Forests-Forest Service, 84
National Grassland, 92
National Historic Park, 92
National Historic Site, 92
National Historic Trails, 86
National Mall, 92
National Marine Sanctuaries-National Oceanic and Atmospheric Administration, 66, 112, 162
National Marine Sanctuary Study Areas, 162
National Natural Landmarks-National Park Service, 140
National Oceanic and Atmospheric Administration, 160, 162
National Park Nomenclature, 92
National Park Service, 8, 66, 88, 90, 92, 140, 194
National Park System, 92
National Recreation Areas, 194
National Recreation Trails, 86
National Registry of Natural History Landmarks, 140
National Registry of Natural Landmarks, 140
National Scenic Trails, 86
National Seashore, 92
National Trails Act, 84
National Trails System, 86, 178
National Wild and Scenic Rivers System, 48
National Wilderness Preservation System, 88, 90
National Wildlife, 177
National Wildlife Federation, 177
National Wildlife Refuge System, 148
National Wildlife Refuges-U.S. Fish and Wildlife Service, 132
Native Gold, State Symbol, 7
Natural Bridges Monarch Butterfly Natural Preserve, 100
Natural Diversity Data Base, 176
Natural Land and Water Reserve System, 166

Natural Preserves-Department of Parks and Recreation, 20, 100
Natural Resources Defense Council, 154
Natural Vegetation, 72
Nature Conservancy - see California Nature Conservancy
Navarro River, 44, 46
Nelson Creek Wild Trout Stream, 118
Nevada City
County Seat, 2
OHV Area, 198
Nevada County, 2
Williamson Act County, 188
Nevada Waterfall, 50
New Bullards Bar Dam, 54
New Exchequer Dam, 54
New Hogan Dam, 54
New Hope Tract, 56
New Melones Dam, 54
New River, 44, 46
New York Mountains Area of Critical Environmental Concern, 156
Newport Beach
Area of Special Biological Significance, 158
Artificial Reef, 126, 127
Marine Life Refuge, 114
Newport Fault, 38
Niagara Ridge OHV Area, 198
Niguel Marine Life Refuge, 114
Nimbus Hatchery, 116
Nipomo Dunes
California Nature Conservancy Preserve, 172, 174
Important Habitat Area, 78
National Natural Landmark, 140
Noble Canyon Trail, 86
Nojoqui Waterfall, 50
Noonday Rock, 64
Nooning Creek Riparian Habitat Demonstration Area, 152
Norconian, Lake - see Lake Norconian
North Bay Aqueduct, 56
North Coast Range, 28
California Nature Conservancy Preserve, 172, 174
North Farallon Island, 64
North Fork National Wilderness, 88, 90
North Mountains Experimental Area Experimental Forest, 144
North Palisade, 30
North Shore Trail, 86
North Tahoe OHV Area, 198
Northern California Chaparral Area of Critical Environmental Concern, 154, 156
Northern California Coast Range Preserve
Area of Critical Environmental Concern, 154
Biosphere Reserve, 128
Research Natural Area, 136
University of California Natural Reserve System, 166
Northern Coast Ranges, 30
Glaciated Area, 32
Noyo River, 44, 46
Egg Collecting Station Fish Hatchery, 116
Oakland, County Seat, 2
Observatory Trail, 86
Obsidian Buttes Volcanic Area, 40
Oceanside Artificial Reef, 127
O'Connor Lakes Ecological Reserve, 122, 124
Ocotillo Wells, 198
Off-Highway Motor Vehicle Recreation Act, 198
Off-Highway Motor Vehicle Recreation Commission, 16, 198
Off-Highway Vehicle Areas, 198
Oh My God Hot Well Hot Spring, 42
Ojai Valley Trail, 178
Olancha Dunes, 198
Old Dad Mountains Wildlife Area, 106
Old Railroad Grade, 178
Old River, 44, 46
Olompali Historic Park, 94
O'Neill Forebay Wildlife Area, 106
Onion Creek Experimental Forest, 144
Open Space Subvention Act, 188
Orange County, 2
Williamson Act County, 188
The Organic Act, 84, 144
Oroville
County Seat, 2
Dam, 54
Wildlife Area, 106
Oroville, Lake - see Lake Oroville
Ortega Trail OHV Area, 198
Orwood Tract, 56
Osborne County Park OHV Area, 198
O'Shaughnessy Dam, 54
Oso Flaco Wetland, 60
Otay River, 46
Outer Continental Shelf, 26
Outstanding Natural Areas-Bureau of Land Management, 148
Owens Lake, 52
Owens River, 44, 46
Catch and Release Stream, 120
Wild Trout Stream, 118
Owens Valley
Fault, 36, 38
Habitat Site, 74, 76
P.L. Boyd - see Philip L. Boyd
Pacific Beach Artificial Reef, 127
Pacific Crest Trail, 86
Pacific Grove Marine Gardens Fish Refuge, 112
Area of Special Biological Significance, 158
Pacific Power and Light Company, 116
Pacific Southwest Forest and Range Station, 134, 144
Packsaddle Cavern Special Interest Area, 146
Paine Wildflower-Semi Tropic Ridge California Nature Conservancy Preserve, 172, 174
Pajaro River, 44, 46
Palawan Artificial Reef, 126, 127
Palen Dry Lake Area of Critical Environmental Concern, 156
Palisades Glacier, 32
Palm Tract, 56
Palo Alto Baylands and Wildlife Refuge, 58
Palos Verdes Peninsula, 68
Panamint Dry Lake OHV Area, 198
Panoche-Coalinga Area of Critical Environmental Concern, 154
Paradise Cove Artificial Reef, 126, 127
Paradise Memorial Trailway, 178
Park Moabi OHV Area, 198
Parker Dam, 54
Parks and Recreation, Department of, 8, 16, 18, 20, 64, 94, 96, 98, 100, 102
Patterson, Mount - see Mount Patterson
Patton's Iron Mountain Divisional Camp Area of Critical Environmental Concern, 156
Paul L. Wattis Sanctuary, 170
Paul M. Dimmick Wayside Campground Redwood Park, 104
Pelican Island, Florida, 132
Pelican Rock Offshore Land, 112
Pendleton Artificial Reef, 126, 127
Peninsula Ranges, 26, 28, 30
Pescadero Marsh
Natural Preserve, 100
San Francisco Bay Access Area, 58
Wetland, 60
Petaluma Marsh Wildlife Area, 106
Petaluma River, 46
Wetland, 60
Peytonia Slough Ecological Reserve, 122, 124
Pfeiffer Big Sur Redwood Park, 104
Philip L. Boyd Deep Canyon Desert Research Center
Biosphere Reserve, 128
University of California Natural Reserve System, 166
Phillip Burton National Wilderness, 88, 90
Phoenix Field Ecological Reserve, 122, 124
Piedra Blanca Trail, 86
Pierson District, 56
Pillsbury, Lake - see Lake Pillsbury
Pilot Knob Area of Critical Environmental Concern, 156
Pine Creek
Important Habitat Area, 78
National Wilderness, 88, 90
Pine Flat Dam, 54
Pine Hill Ecological Reserve, 122, 124
Pinecrest Trail, 86
Pinnacles National Monument, 34, 92
National Wilderness, 88, 90
Pinto Basin Volcanic Area, 40
Piru Creek Catch and Release Stream, 120
Pisgah Volcanic Area, 40
Pismo Dunes
Natural Preserve, 100
OHV Area, 198
Pismo Invertebrate Reserve Marine Life Refuge, 114
Pismo Lake Ecological Reserve, 122, 124
Pismo-Oceano Beach Pismo Clam Preserve Marine Life Refuge, 114

Pit River, 44, 46
Catch and Release Stream, 120
Hatchery, 116
Pit River Canyon Important Habitat Area, 78
Pitas Point Artifical Reef, 126, 127
Pitkin Creek Marsh Wetland, 62
Piute Canyon Trail, 86
Piute Creek
Area of Critical Environmental Concern, 156
Ecological Reserve, 122, 124
Riparian Habitat Demonstration Area, 152
Piute Range Outstanding Natural Area, 148
Pixley National Wildlife Refuge, 132
Pixley Vernal Pools
California Nature Conservancy Preserve, 172, 174
National Natural Landmark, 140
Wetland, 62
Placer County, 2
Williamson Act County, 188
Placerville, County Seat, 2
Plank Road Area of Critical Environmental Concern, 156
Planning & Conservation League, 183
Plaster City OHV Area, 198
Plumas County, 2
Water Bank County, 192
Williamson Act County, 188
Plumas National Forest, 84
Point Año Nuevo Habitat Site, 76
Point Cabrillo Reserve Marine Life Refuge, 114
Point Edith Wildlife Area, 106
Point Fermin Marine Life Refuge, 114
Point Isabel Regional Shoreline Park, 58
Point Lobos
Area of Special Biological Significance, 158
Ecological Reserve, 122, 124
Marine Life Refuge, 20, 114
National Natural Landmark, 142
State Reserve, 98
Underwater Park, 96
Point Loma Reserve Marine Life Refuge, 114
Point Reyes
Habitat Site, 76
National Wilderness, 88
Point Reyes Farallon Islands National Marine Sanctuary Marine Life Refuge, 114
Point Reyes Headland(s) Reserve
Area of Special Biological Significance, 158
Marine Life Refuge, 114
Research Natural Area, 138
Point Reyes National Seashore, 88
Biosphere Reserve, 128
National Park, 92
World Heritage Site, potential, 130
Point Sal Area of Critical Environmental Concern, 154
Poleta OHV Area, 198
Pony Express Trail, 86
Porter-Cologne Water Quality Control Act, 62
Porterville OHV Area, 198
Portola Redwoods, 104
Potter Valley Habitat Site, 74
Pozo Habitat Site, 74
Pozo-LaPanza OHV Area, 198
Prado F. C. Basin Lake, 52
Prairie City OHV Area, 198
Prairie Creek Redwood Park, 104
Biosphere Reserve, 128
World Heritage Site, 20
Prestons Island Offshore Land, 112
Primitive Area, 150
Prince Island(s) Offshore Land, 112
Pronghorn Antelope, 74
Prospect Island, 56
Prosser Hills OHV Area, 198
Protected Natural Areas-National Park Service, 148
Public Use Natural Areas-U.S. Fish and Wildlife Service, 148
Pushawalla Palms Wetland, 62
Putah Creek Riparian Corridor Important Habitat Area, 78
Putah Creek Wildlife Area, 106
Pygmy Forest
California Nature Conservancy Preserve, 172, 174
National Natural Landmark, 140
Redwood Park, 104
University of California Natural Reserve System, 166
Pygmy Forest Ecological Staircase Area of Special Biological Signficance, 158
Quimby Island, 56
Quincy, County Seat, 2
Rails to Trails Conservancy, 178
Rainbow Basin
National Natural Landmark, 140
Outstanding Natural Area, 148
Rainbow Basin-Owl Canyon Area of Critical Environmental Concern, 156
Rainbow Falls, 50
Rancheria Falls Trail, 86
Rancheria Waterfalls, 50
Rancho La Brea National Natural Landmark, 140
Rancho Dos Palmos Important Habitat Area, 78
Ranger Rick, 177
Rasor OHV Area, 198
Reclamation Board, 24
Red Bluff, County Seat, 2
Red Buttes National Wilderness, 88, 90
Red Cliffs Natural Preserve, 100
Red Cones Volcanic Area, 40
Red Hills
Area of Critical Environmental Concern, 154, 156
OHV Area, 198
Red Lake Wildlife Area, 106
Red Mountain
Area of Critical Environmental Concern, 154, 156
Research Natural Area, 136
Red Rock Canyon OHV Area, 198
Redding, County Seat, 2
Redondo Beach Artificial Reef, 126, 127
Redondo-Palos Verdes Artificial Reef, 126, 127
Redwood Chapter, Sierra Club, 180
Redwood City, County Seat, 2
Redwood Experimental Forest Biosphere Reserve, 128
Redwood National Park, 92, 104
Area of Special Biological Significance, 158
Biosphere Reserve, 128
World Heritage Site, 130
Redwood Parks and Preserves, 104
Redwood Shores
Ecological Reserve, 122, 124
Marine Life Refuge, 114
Redwoods State Parks Biosphere Reserve, 128
Reef Ridge Research Natural Area, 136
Regional Councils of Government, 4
Remington Hot Springs, 42
Research Natural Areas-Bureau of Land Management, 136
Research Natural Areas-Forest Service, 134
Research Natural Areas-National Park Service, 138
Research Natural Areas-U.S. Fish and Wildlife Service, 138
Resources Planning Act, 84
Revelation Trail, 86
Reynolds Wayside Campground Redwood Park, 104
Ribbon Waterfall, 50
Rice Valley Dunes, 198
Richardson Bay, 58
Wildlife Sanctuary, 170
Richardson Grove Redwood Park, 104
Richter Magnitude Scale, 36
Rincon Island Artificial Reef, 126, 127
Rindge Tract, 56
Ring Mountain California Nature Conservancy Preserve, 172, 174
Rio Blanco Tract, 56
Rio Honda, 46
Riparian Habitat Demonstration Areas-Bureau of Land Management, 152
Riparian Habitats, 80
Riparian Woodland Losses, 81
Rising River, 46
Ritter, Mount - see Mount Ritter
Ritter Range Glacial Area, 32
Riverfront Park OHV Area, 198
Rivers and Streams, 44, 46
Riverside County, 2
County Seat, 2
Williamson Act County, 188
Roaring River, 46
Waterfall, 50
Robert W. Crown Reserve Marine Life Refuge, 114
Roberts Island, 56
Robyn Hot Spring, 42
Rock Creek Wild and Scenic River, 48
Rodeo Lagoon, 58
Rodman Mountain Petroglyph Trail, 86
Rogue River National Forest, 84
Roosevelt, Franklin D., 66
Roosevelt, Theodore, 132
Rose Spring Area of Critical Environmental Concern, 154
Rough and Ready Island, 56
Round Rock Offshore Land, 112

Rowdy Creek Wild and Scenic River, 48
Rowher Flat OHV Area, 198
Royal Arch Cascade, 50
Rubber Boa Habitat Important Habitat Area, 78
Rubicon River, 44, 46
Wild Trout Stream, 118
Russell, Mount - see Mount Russell
Russian Gulch
Redwood Park, 104
Underwater Park, 96
Russian Peak National Wilderness, 88, 90
Russian River, 44, 46
California Nature Conservancy
Preserve, 172, 174
Ryan Oak Glen University of California Natural
Reserve System, 166
Ryer Island, 56
Saber-Toothed Tiger, State Symbol, 7
Sacramento
Area Council of Governments, 4
Bypass Wildlife Area, 106
Deep Water Ship Channel, 56
National Wildlife Refuge, 132
River, 44, 46
Sacramento County, 2
County Seat, 2
Williamson Act County, 188
Sacramento Mountains University of California
Natural Reserve System, 166
Sacramento-Rio Linda Bikeway, 178
Sacramento River, 56
Catch and Release Stream, 120
Wildlife Area, 106
Sacramento River Oxbow California Nature
Conservancy Preserve, 172, 174
Sacramento-San Joaquin Delta, 56
Saint Francis Dam, 54
Saint George Reef Offshore Land, 112
Saint Johns River, 46
Salinas
County Seat, 2
Wildlife Area, 106
Salinas Lagoon National Wildlife Refuge, 132
Salinas River, 44, 46, 160
State Beach, 58
Saline Valley
Area of Critical Environmental
Concern, 154, 156
Ecological Reserve, 122, 124
Saline Valley Salt Marsh Important Habitat Area, 78
Sally Pekarek Trail, 178
Salmon Creek-Ocean Area Surrounding the Mouth-
Area of Special Biological
Significance, 158
Salmon River, 44, 46
Wild and Scenic, 48
Salmon Summit Trail, 86
Salt Creek (Dumont) Area of Critical Environmental
Concern, 156
Salt Creek Pupfish-Rail Area of Critical
Environmental Concern, 156
Salt Point Underwater Park, 96
Salt River, 46
Salton Sea
Lake, 52
National Wildlife Refuge, 132
Wetlands, 62
Samoa Spit OHV Area, 198
Samuel P. Taylor State Park
Biosphere Reserve, 128
Redwood Park, 104
San Andreas, County Seat, 2
San Andreas Fault, 36, 38
National Natural Landmark, 140
San Andreas Fault Scarp Research
Natural Area, 136
San Antonio
Dam, 54
River, 44, 46
San Antonio, Mount - see Mount San Antonio
San Benito County, 2
Council of Governments, 4
Williamson Act County, 188
San Benito River, 44, 46
San Bernardino County, 2
County Seat, 2
Williamson Act County, 188
San Bernadino Mountains Glaciated Area, 32
San Bernadino National Forest, 84
Biosphere Reserve, 128
San Bruno Mountain Important Habitat Area, 78
San Clemente Island, 66
Area of Special Biological
Significance, 158
Offshore Land, 112
San Diego Bay Wetland, 60
San Diego Chapter, Sierra Club, 180
San Diego County, 2
Association of Governments, 4
Williamson Act County, 188
San Diego-La Jolla Ecological Reserve, 122, 124
Area of Special Biological
Significance, 158
Marine Life Refuge, 114
San Diego Marine Life Refuge, 112, 114
Area of Special Biological
Significance, 158
San Diego River, 44, 46
San Dieguito Lagoon
Ecological Reserve, 122, 124
Wetland, 60
San Dieguito River, 44, 46
San Dimas Experimental Forest, 144
Biosphere Reserve, 128
San Elijo Lagoon
Ecological Reserve, 122, 124
Wetland, 60
San Felipe Creek
Area of Critical Environmental
Concern, 156
Riparian Habitat Demonstration Area, 152
Wetland, 62
San Felipe Creek Area National Natural
Landmark, 140
San Francisco Airport, 58
San Francisco Bay, 58, 60
Conservation and Development
Commission, 58
National Wildlife Refuge, 58, 132
Wetland, 60
San Francisco Bay Area Ridge Trail, 86
San Francisco Bay Chapter, Sierra Club, 180
San Francisco County, 2
County Seat, 2
Williamson Act County, 188
San Gabriel
Fault, 38
National Wilderness, 88, 90
San Gabriel Canyon OHV Area, 198
San Gabriel River, 44, 46
Catch and Release Stream, 120
San Gorgonio
Mountain, 28
National Wilderness, 88, 90
River, 46
San Gorgonio Chapter, Sierra Club, 180
San Jacinto
Fault, 38
National Wilderness, 88, 90
River, 44, 46
Wildlands Program Area, 110
Wildlife Area, 106
San Jacinto, Mount - see Mount San Jacinto
San Joaquin
Desert Research Natural Area, 138
Freshwater Marsh University of California
Natural Reserve System, 166
Hatchery, 116
Marsh Wetland, 62
Natural Area Important Habitat Area, 78
River, 44, 46, 56
San Joaquin County, 2
Council of Governments, 4
Williamson Act County, 188
San Joaquin Experimental Range
Biosphere Reserve, 128
Experimental Forest, 144
San Jose, County Seat, 2
San Leandro Regional Shoreline Park-Doolittle
Pond, 58
San Lorenzo River, 46
Catch and Release Stream, 120
San Luis
Dam, 54
Reservoir, 52
Natural Wildlife Refuge, 132
San Luis Island Important Habitat Area, 78
San Luis Obispo
Area Council of Governments, 4
Wildlife Area, 106
San Luis Obispo County, 2
Artificial Reef, 126, 127
County Seat, 2
Williamson Act County, 188
San Luis Refuge Habitat Site, 76
San Luis Reservoir
Lake, 52
Wildlife Area, 106

San Luis Rey River, 44, 46
San Mateo County, 2
Williamson Act County, 188
San Mateo Canyon National Wilderness, 88, 90
San Mateo Creek Wetland, 60
Natural Preserve, 100
San Mateo, San Francisco Bay Access Area, 58
San Miguel Island, 66, 162
Area of Special Biological Significance, 158
Ecological Reserve, 122, 124
Habitat Site, 76
Marine Life Refuge, 114
Offshore Land, 112
San Nicolas Island, 66
Area of Special Biological Significance, 158
Habitat Site, 76
Offshore Land, 112
San Pablo Bay
Marine Life Refuge, 114
National Wildlife Refuge, 132
San Francisco Bay Feature, 58
Wildlife Area, 106
San Rafael
County Seat, 2
National Wilderness, 88, 90
San Rafael Canal and Bayfront, San Francisco Bay Access Area, 58
San Ramon Valley Iron Horse Trail, 178
San Sebastian Marsh
Area of Critical Environmental Concern, 156
Important Habitat Area, 78
Outstanding Natural Area, 148
Riparian Habitat Demonstration Area, 152
San Simeon, 20
Sand Canyon
Area of Critical Environmental Concern, 154, 156
Riparian Habitat Demonstration Area, 152
Sand Ridge Wildflower Preserve
California Nature Conservancy Preserve, 172, 174
National Natural Landmark, 140
Santa Ana, County Seat, 2
Santa Ana River, 44, 46
Santa Barbara
Area of Special Biological Significance, 158
Artificial Reef, 126, 127
OHV Area, 198
Santa Barbara County, 2
County-Cities Area Planning Council, 4
County Seat, 2
Williamson Act County, 188
Santa Barbara Island, 66, 162
Area of Special Biological Significance, 158
Ecological Reserve, 122, 124
Habitat Site, 76
Marine Life Refuge, 114
Offshore Land, 112
Santa Catalina Island, 66
Area of Special Biological Significance, 158
Offshore Land, 112
Santa Catalina Research Natural Area (Arizona), 134
Santa Clara County, 2
Motorcycle Park OHV Area, 198
Williamson Act County, 188
Santa Clara Estuary Natural Preserve, 100
Santa Clara River, 44, 46
Wetland, 60
Santa Cruz County, 2
County Seat, 2
Williamson Act County, 188
Santa Cruz Island, 66, 162
Area of Special Biological Significance, 158
California Nature Conservancy Preserve, 172, 174
Offshore Land, 112
University of California Natural Reserve System, 166
Santa Cruz Longtoed Salamander Ecological Reserve, 122, 124
Santa Cruz y Aliso Trail, 86
Santa Lucia Chapter, Sierra Club, 180
Santa Lucia National Wilderness, 88, 90
Santa Margarita River, 44, 46
Wetland, 60
Santa Maria River, 46
Santa Monica Artificial Reef, 126, 127
Santa Monica Bay, 162
Artificial Reef, 126, 127
Santa Monica Mountains National Recreation Area, 194
National Park, 92
Santa Rosa
County Seat, 2
National Wilderness, 88, 90
Wildlife Area, 106
Santa Rosa Island, 66, 162
Area of Special Biological Significance, 158
Offshore Land, 112
Santa Rosa Mountain(s)
California Nature Conservancy Preserve, 174
State Wilderness, 102
Santa Rosa Plateau California Nature Conservancy Preserve, 174
Santa Ynez Fault, 38
Santa Ynez River, 44, 46
Wetland, 60
Saratoga Springs Wetland, 62
Sargent Barnhart Tract, 56
Saunders Reef-Kelp Beds Area of Special Biological Significance, 158
Save the San Francisco Bay Association, 58
Savoy Creek Wild and Scenic River, 48
Sawtooth Mountain Glacier, 32
Sawtooth Ridge Glacier, 32
Schonchin Lava Tubes Research Natural Area, 138
Scorpion Rock Important Habitat Area, 78
Scott Mountain OHV Area, 198
Scott River, 44, 46
Wild and Scenic, 48
Scripps Shoreline Underwater University of California Natural Reserve System, 166
Sea Otter(s), 66, 74
Seal Beach National Wildlife Refuge, 132
Seal Slough, San Mateo, 58
Sentinel Falls, 50
Sentinel Meadow Research Natural Area, 134
Sequoia-Kings Canyon National Parks
Biosphere Reserve, 128
National Park, 92
National Wilderness, 88, 90
Redwood Park, 104
Sequoia National Forest, 84
World Heritage Site, potential, 130
Serpentine, State Symbol, 7
Sespe Creek Wild Trout Stream, 118
Sespe Hot Springs, 42
Shadow Mountain OHV Area, 198
Shadow of the Giants Trail, 86
Shag Slough, 56
Sharktooth Hill National Natural Landmark, 140
Shasta
Dam, 54
Historic Park, 94
Lake, 52
National Forest, 84
National Recreation Area, 92, 194
OHV Area, 198
River, 44, 46
Shasta County, 2
Regional Transportation Planning Agency, 4
Williamson Act County, 188
Shasta Mudflow Research Natural Area, 134
Shasta River, 44, 46
Salmon Spawn Area of Critical Environmental Concern, 154, 156
Shastina Volcanic Area, 40
Shaver Lake OHV Area, 198
Sheep Mountain National Wilderness, 88, 90
Sheepy Ridge Wildlife Area, 106
Shepard Canyon Trail, 178
Sherman Island, 56
Shima Tract, 56
Shin Kee Tract, 56
Short Canyon Area of Critical Environmental Concern, 154, 156
Shrub and Chaparral, 80
Shrub Dominated Habitats, 80
Sierra Buttes, 28, 30
The Sierra Club, 180
Sierra County, 2
Planning Organization and Economic Development District, 4
Water Bank County, 192
Williamson Act County, 188
Sierra National Forest, 84

Sierra Nevada, 26, 30
  Fault, 38
  Glaciated Area, 32
Sierra Valley Marsh Important Habitat Area, 78
Sierraville OHV Area, 198
Signal Mountain Research Natural Area, 136
Sill, Mount - see Mount Sill
Silurian Dry Lake OHV Area, 198
Silver Mocassin Trail, 86
Silver Strand
  Bikeway, 178
  Natural Preserve, 100
  Waterfall, 50
Silverado Fisheries Base Fish Hatchery, 116
Silverwood Wildlife Sanctuary, 170
Simon Rodia State Park, 20
Singer Geoglyphs Area of Critical Environmental Concern, 156
Sinkyone Wilderness Redwood Park, 104
Siskiyou
  National Forest, 84
  National Wilderness, 88, 90
Siskiyou County, 2
  Association of Government Entities, 4
  Water Bank County, 192
  Williamson Act County, 188
Sisquoc River, 44, 46
Sisson-Callahan Trail, 86
Sisson Hatchery, 116
Sister Rocks Offshore Land, 112
Six Rivers National Forest, 84
Sled Ridge Motorcycle Trail, 86
Slinkard-Little Antelope Wildlife Area, 106
Smith River, 44, 46
  Wetland, 60
  Wild and Scenic, 48
Smithe Redwoods, 104
  State Reserve, 98
Smithneck Creek Wildlife Area, 106
Sno-Park Sites-Department of Parks and Recreation, 196
Snow Creek Cascade, 50
Snow Mountain
  National Wilderness, 88, 90
  Special Interest Area, 146
Soda Lake
  Area of Critical Environmental Concern, 154, 156
  Important Habitat Area, 78
Soggy Dry Lake Area of Critical Environmental Concern, 156
Solano County, 2
  Williamson Act County, 188
Soledad Fault, 38
Sonoma Barrens, 176
Sonoma Coast State Beach Underwater Park, 96
Sonoma County, 2
  Williamson Act County, 188
Sonora, County Seat, 2
Sonora Desert Biosphere Reserve, 128
Soquel Cove Artificial Reef, 126, 127
Sotcher Lake Catch and Release Lake, 120
South Bay Aqueduct, 56
South Bay Pumping Plant, 56
South Coast Ranges, 28, 30
South East Farallon Island Habitat Site, 76
South Fork of Gallinas Creek, 58
South Kelsey Trail, 86
South Laguna Beach Marine Life Refuge, 114
South San Francisco, 58
South San Francisco Bay Sanctuaries, 170
South Sierra National Wilderness, 88, 90
South Warner National Wilderness, 88, 90
South Yuba Independence Trail, 86
Southeast Farallon Island, 64
Southern California Association of Governments, 4
Southern East Mesa Flat-Tailed Horned Lizard Area of Critical Environmental Concern, 156
Southern San Benito Habitat Site, 74
Spangler Hills OHV Area, 198
Spannus Gulch Wildlife Area, 106
Special Interest Areas-Forest Service, 146
Spencer Meadow Trail, 86
Spenceville Wildlife Area, 106
Spindrift Point California Nature Conservancy Preserve, 172, 174
Split Mountain, 30
Springville Clarkia Ecological Reserve, 122, 124
Squaw Spring Area of Critical Environmental Concern, 154, 156
Staircase Cascade, 50
Standish-Hickey Redwood Park, 104
Stanislaus
  Area Association of Governments, 4
  Experimental Forest, 144
  National Forest, 84
Stanislaus County, 2
  Williamson Act County, 188
Stanislaus River, 44, 46
  Catch and Release Stream, 120
  Wild Trout Stream, 118
Stanislaus-Tuolumne Experimental Forest Biosphere Reserve, 128
Starr Ranch Audubon Sanctuary, 170
State Beach, Park, Recreational, and Historical Facilities Bond Act, 60
State Beaches, 20
State Coastal Conservancy, 164
State Colors, 7
State Ecological Reserves, 112
State Flag, 7
State Forest Program, 82
State Forests-Department of Forestry and Fire Protection, 82
State Historic Monument, 20
State Historic Parks, 20
State Historic Units, 20
State Indian Museum Historic Park, 94
State Lands Commission, 8
State Park System, 16, 20, 104
State Parks Commission, 20
State Recreation Areas, 20
State Reserves-Department of Parks and Recreation, 20, 98
State Symbols, 7
State Vehicular Recreation Area and Trails System, 20, 86
State Water Resources Control Board, 22, 52
State Wilderness Areas-Department of Parks and Recreation, 20, 102
Staten Island, 56
Steam Well Area of Critical Environmental Concern, 154, 156
Steamboat Slough, 56
Stebbins Cold Canyon University of California Natural Reserve System, 166
Steinberger Slough, 58
Stewart Tract, 56
Stockton, County Seat, 2
Stoddard Valley, 198
Stone Lagoon Wetland, 60, 62
Stony Creek, Middle Fork
  Catch and Release Stream, 120
  Special Interest Area, 146
Struve Pond California Nature Conservancy Preserve, 172, 174
Stumpy Meadows OHV Area, 198
Subway Cave, 34
Sugarloaf Trail, 86
Suisun Bay, 58
Suisun Marsh Wetland, 60
Summit Trail, 86
Sunset Bay Wetland, 60
Superstition Hill Fault, 38
Surprise Canyon Area of Critical Environmental Concern, 154, 156
Surprise Valley Wildlife Area, 106
Susan River, 44, 46
Susanville, County Seat, 2
Sutil Island Offshore Land, 112
Sutter Buttes, 28
Sutter
  Bypass Wildlife Area, 106
  Island, 56
  National Wildlife Refuge, 132
Sutter County, 2
  Water Bank County, 192
Swain Mountain Experimental Forest, 144
Sweetwater River, 44, 46
Sykes Hot Spring, 42
Table Bluff Ecological Reserve, 122, 124
Table Mountain Area of Critical Environmental Concern, 156
Tahoe, Lake - see Lake Tahoe
Tahoe National Forest, 84
Talawa, Lake - see Lake Talawa
Tamalpais, Mount - see Mount Tamalpais
Tamarack Ridge OHV Area, 198
Teakettle Creek Experimental Forest, 144
Tehama County, 2
  Williamson Act County, 188
Tehema Wildlife Area, 106
Tehipite Chapter, Sierra Club, 180
Ten Mile River, 46
  Wetland, 60
Tenaya (Pyweak) Cascade, 50
Terminous Tract, 56
Theodore J. Hoover Natural Preserve, 100

Thompson Peak Glacier, 32
Thousand Lakes National Wilderness, 88, 90
Thunderbolt Peak, 30
Tiburon Lineal Park, 178
Tidal Invertebrate Act, 112
Tijuana Estuary
National Natural Landmark, 140
Natural Preserve, 100
Wetland, 60
Tijuana River National Estuarine Reserve Research System, 160
Tijuana Slough National Wildlife Refuge, 132
Tilden Botanic Garden, 176
Timberland Production Zones, 190
Todd, William, 7
Toiyabe Chapter, Sierra Club, 180
Toiyabe National Forest, 84
Tokopah Cascade, 50
Tomales Bay
Biosphere Reserve, 128
Ecological Reserve, 122, 124
Marine Life Refuge, 114
Sanctuary, 170
Wetland, 60
Topanga Artificial Reef, 126, 127
Torrey Pines
Artificial Reef, 126, 127
National Natural Landmark, 140
State Reserve, 98
Trails, 86
Transverse Ranges, 26, 28, 30
Traveler's Home Trail, 86
Travertine Hot Springs Area of Critical Environmental Concern, 154, 156
Tree Dominated Habitats, 80
Trestle Wetlands Natural Preserve, 100
Tri-County Area Planning Council, 4
Trinidad Head-Kelp Beds Area of Special Biological Significance, 158
Trinity
Dam, 54
National Forest, 84
National Park, 92
National Recreation Area, 194
Trinity Alps
Glaciated Area, 32
National Wilderness, 88, 90
Trinity County, 2
Williamson Act County, 188
Trinity River, 44, 46
Hatchery, 116
Wild and Scenic, 48
Trona Pinnacles
Area of Critical Environmental Concern, 154, 156
National Natural Landmark, 140
Outstanding Natural Area, 148
Trout and Steelhead Conservation and Management Planning Act, 118, 120
Truckee OHV Area, 198
Truckee River, 44, 46
Bike Trail, 178
Catch and Release Stream, 120
Wild Trout Stream, 118
Trust for Public Land, 182
The Tub Hot Spring, 42
Tulare County, 2
Council of Governments, 4
Williamson Act County, 188
Tule Elk, 74
State Reserve, 98
Tule-Klamath Basin Wetland, 62
Tule Lake National Wildlife Refuge, 132
Tule Lake Sump, 52
Tule River, 44, 46
Catch and Release Stream, 120
OHV Area, 198
Tuolumne
Experimental Forest, 144
River, 44, 46
Wild and Scenic River, 48
Tuolumne County, 2
Williamson Act County, 188
Tupman Habitat Site, 76
Turtle Mountains
Natural Area National Natural Landmark, 140
Volcanic Area, 40
Twin Lakes Catch and Release Lake, 120
Twitchell Island, 56
Tyler Island, 56
Tyndall, Mount - see Mount Tyndall
Ubehebe Craters Volcanic Area, 40
Udell Gorge Natural Preserve, 100
Ukiah, County Seat, 2
Underwater Parks-Department of Parks and Recreation, 20, 96
UNESCO, 128, 130
Union Island, 56
University of California Natural Reserve System, 64, 74, 166
Upper Johnson Valley Yucca Rings Area of Critical Environmental Concern, 156
Upper Newport Bay
Ecological Reserve, 122, 124
Marine Life Refuge, 114
Wetland, 60
Wildlands Program Area, 110
Upper Santa Clara River Important Habitat Area, 78
U.S. Army Corps of Engineers, 116
U.S. Fish and Wildlife Service, 8, 78, 88, 132, 148
Valentine Eastern Sierra University of California Natural Reserve System, 166
Valley Grassland Losses, 81
Valley Vernal Pools Important Habitat Area, 78
Van Arsdale Fisheries Station Fish Hatchery, 116
Van Damme Redwood Park, 104
Van Damme Underwater Park, 96
Van Duzen River, 44, 46
Wild and Scenic, 48
Van Sickle Island, 56
Veale Tract, 56
Venice Island, 56
Ventana Chapter, Sierra Club, 180
Ventana National Wilderness, 88, 90
Ventura
Artificial Reef, 126, 127
River, 44, 46
Ventura County, 2
County Seat, 2
Williamson Act County, 188
Vernal Pool Losses, 81
Vernal Waterfall, 50
Victoria Island, 56
Vina Plains California Nature Conservancy Preserve, 172, 174
Vine Hill Preserve, 176
Visalia, County Seat, 2
Volcan Mountains Important Habitat Area, 78
Volcanic Areas, 40
Volta Wildlife Area, 106
Walker River, 44, 46
Warm Springs Dam, 54
Warm Springs Hatchery, 116
Warm Sulfur Springs Area of Critical Environmental Concern, 154
Wassama Round House Historic Park, 94
Water Bank Counties, 192
Water Resources, Department of, 24, 52, 116
Waterfalls, 50
Waterfowl Management Areas, 108
Waterfowl Production Areas, 132
Waterwheel Waterfall, 50
Watsonville Slough
Important Habitat Area, 78
Wetland, 60
Watsonville Wildlife Area, 106
Waucoba Mountain, 28
Weaverville, County Seat, 2
Webb Tract, 56
Week's Law, 84
Weir Canyon Important Habitat Area, 78
West Anacapa Island Research Natural Area, 138
West Fork Trail, 86
West Marin Island Important Habitat Area, 78
West Mesa Area of Critical Environmental Concern, 156
West Mojave Desert Ecological Reserve, 122, 124
West Waddell Creek State Wilderness, 102
West Walker River, 44, 46
Wild and Scenic, 48
West Waterfall, 50
Western Rand Mountains Area of Critical Environmental Concern, 154, 156
Wetlands-Coastal, 60
Wetlands-Interior, 62
Whaleback Field Volcanic Area, 40
Whaler Island Offshore Land, 112
Wheeler Crest OHV Area, 198
Whipple Mountains Area of Critical Environmental Concern, 156
Whiskeytown National Recreation Area, 194
National Park, 92

White Mountain, 28, 30
Fault, 36
Research Natural Area, 134
Scientific Special Interest Area, 146
White Mountain City Area of Critical
Environmental Concern, 154, 156
White Mountains Glaciated Area, 32
White River, 44, 46
White Rock Offshore Land, 112
White Slough Wildlife Area, 106
White Wolf Fault, 38
Whitewater Canyon Area of Critical Environmental
Concern, 156
Whitewater River, 44, 46
Whitewater River Marsh Important Habitat Area, 78
Whitney Creek Research Natural Area, 138
Whitney Glacier, 32
Whitney, Mount - see Mount Whitney
Whitney Portal Trail, 86
Wild and Scenic Rivers, 48
Wild and Scenic Rivers Act, 48, 84
Wild Trout Waters-Department of
Fish and Game, 118, 120
Wildcat Waterfall, 50
Wilder Beach Natural Preserve, 100
Wilder Ranch Redwood Park, 104
Wilderness Act, 84
Wilderness Areas-Department of Parks and
Recreation, 20, 102
Wildlands Program Areas-Department of Fish and
Game, 110
Wildlife Areas-Department of Fish and Game, 106
Wildlife Conservation Act, 106
Wildlife Management Areas-Department of
Fish and Game, 106
Wildlife Rehabilitation Centers, 184, 186
Wildlife Research Centers, 132
Wildomar OHV Area, 198
Williams Audubon Sanctuary, 170
Williams Wildlife California Nature Conservancy
Preserve, 172, 174
Williamson Act, 188, 190
Williamson Act Counties, 188
Williamson, Mount - see Mount Williamson
Willow Creek-Lurline National
Wildlife Refuge, 132
Willows, County Seat, 2
Wilson Rock, Offshore Land, 112
Winding Stair Cave National
Natural Landmark, 140
Woodbridge Ecological Reserve, 122, 124
Woodland, County Seat, 2
Woodlands, 80
Woodson Bridge Natural Preserve, 100
Woodward Island, 56
Wooley Creek Wild and Scenic River, 48
World Heritage Sites-UNESCO, 130
Wright-Elmwood Tract, 56
Yaudanchi Ecological Reserve, 122, 124
Yellow Creek
Catch and Release Stream, 120
Wild Trout Stream, 118
Yolla Bolley-Middle Eel National
Wilderness, 88, 90
Yolo County, 2
Water Bank County, 192
Williamson Act County, 188
Yosemite Falls, 50
Yosemite National Park, 20, 88, 90, 92
National Wilderness, 88, 90
Redwood Park, 104
World Heritage Site, 130
Younger Lagoon University of California Natural
Reserve System, 166
Yreka, County Seat, 2
Yuba City, County Seat, 2
Yuba County, 2
Yuba River, 44, 46
Catch and Release Stream, 120
Yuha Basin Area of Critical Environmental
Concern, 156
Yurok Redwood
Experimental Forest, 144
Research Natural Area, 134
Zamboni Hot Spring, 42
Z'Berg-Warren-Keene-Collier Forest Taxation
Reform Act, 190